5th International Conference on Materials and Applications for Sensors and Transducers (IC-MAST2015)

IOP Conference Series: Materials Science and Engineering
Volume 108

Mykonos, Greece
27 – 30 September 2015

Editors:

Evangelos Hristoforou
Dimitrios S. Vlachos

ISBN: 978-1-5108-2227-6
ISSN: 1757-8981

Printed from e-media with permission by:

Curran Associates, Inc.
57 Morehouse Lane
Red Hook, NY 12571

Some format issues inherent in the e-media version may also appear in this print version.

Printed by Curran Associates, Inc. (2016)

For permission requests, please contact the Institute of Physics at the address below.

Institute of Physics
Dirac House, Temple Back
Bristol BS1 6BE UK

Phone: 44 1 17 929 7481
Fax: 44 1 17 920 0979

techtracking@iop.org

Additional copies of this publication are available from:

Curran Associates, Inc.
57 Morehouse Lane
Red Hook, NY 12571 USA
Phone: 845-758-0400
Fax: 845-758-2633
Email: curran@proceedings.com
Web: www.proceedings.com

Table of contents

Volume 108

5th International Conference on Materials and Applications for Sensors and Transducers (IC-MAST2015)
27–30 September 2015, Mykonos, Greece

Accepted papers received: 22 December 2015
Published online: 18 March 2016

Preface

Papers

012050
OPEN ACCESS
Attractive mechanical properties of a lightweight highly sensitive bi layer thermistor: polycarbonate/organic molecular conductor

E Laukhina, V Lebedev, C Rovira, V Laukhin and J Veciana   pg. 299

IOP Conf. Series: Materials Science and Engineering **108** (2016) 011001 doi:10.1088/1757-899X/108/1/011001

Editors:

Evangelos Hristoforou, National Technical University of Athens, Greece.
Dimitrios S.Vlachos, University of Peloponnese, Greece.

Co-Editors:

Ioanna Giouroudi, Vienna University of Technology, Austria.
Sohini Kar-Narayan, University of Cambridge, United Kingdom.
Stelios Potirakis, Piraeus University of Applied Sciences (TEI of Piraeus), Greece.

IOP Conf. Series: Materials Science and Engineering **108** (2016) 011001 doi:10.1088/1757-899X/108/1/011001

Preface

The 5th International Conference on Materials and Applications for Sensors and Transducers, Mykonos island, Greece, hosted about 110 oral and poster papers and more than 90 participants. IC-MAST, as an international annual conference which tries to meet the needs for various types of sensors, particularly those which may be manufactured by low cost methods (i.e. hybrid sensors, smart specialization devices, particular applications not necessarily requiring integrated micro-nano technologies), covering all types of materials and physical effects, appears to be a necessity. IC-MAST has been established as a high quality international conference by:

I. Gathering together multinational researchers from all over the world, working in different materials for sensors and transducers and technical applications of sensors, but also in some cases in the management of the data coming from sensors and transducers. The careful selection of the conference place (like Aegean Sea, Budapest, Prague, Bilbao, Mykonos etc) allows for enjoying the local hospitality and sightseeing

II. Emphasizing in hybrid sensors and smart specialization devices produced by inexpensive methods, without excluding of course micro-nano technology, from all kinds of solid state, liquid and gaseous materials, as well as in particular transducer applications (design and development, as well as use of sensing data)

III. Innovatively implementing the Virtual Paper Concept, allowing for large impact of research works presented in the conference by authors who either have no time or no funding support for visiting a conference; this year more than 12 virtual papers are presented in the 5th IC MAST, following a standardized procedure via the our robust and reliable Conference Site (www.icmast.net)

IV. Allowing for lengthy technical and managerial discussions in terms of sensor, material and instrumentation development; furthermore, the different research groups gathered together are offered the particular advantage of arranging and concluding research proposals and projects, otherwise not having a visible possibility of such realization

The 5th IC-MAST organizing committee is proud that the Conference Keynote Speaker was Prof George Hadjipanayis, University of Delaware.

We are also proud for the invited speakers of the conference:

- Stergios Logothetidis, Aristotle University of Thessaloniki, Greece
- Dimitris Tsoukalas, National Technical University of Athens, Greece
- Susana Cardoso de Freitas, INESC Microsistemas e Nanotecnologias

- Yuris Dzenis, University of Nebraska–Lincoln, USA

The IC-MAST 2015 organizers believe that the target of the Conference has been successfully met by enhancing knowledge in sensors by all participants, accelerating the achievement of results and optimizing the under design products, in a quite friendly way! Therefore, participants made an appointment for the next year in Athens, Greece, where the 6th International MAST Conference will be realized!

The Conference Chairmen

Evangelos Hristoforou
National Technical University of Athens, Greece

Dimitrios Vlachos
University of Peloponnese

IOP Conf. Series: Materials Science and Engineering **108** (2016) 011001 doi:10.1088/1757-899X/108/1/011001

International Advisory Committee

- **Shehab Ahmed,** Texas A&M University at Qatar, Doha, Qatar
- **Faramarz Hossein-Babaei,** K.N. Toosi University of Technology, Iran
- **José Manuel Barandiarán,** BCMAN, Basque Center for Materials, Applications and Nanostructures

International Scientific Committee

- **Mihai Popescu,** Editor-in-Chief, Journal of Optoelectronics and Advanced Materials
- **Spyridon Diplas,** Sintef Oslo, Norway
- **Jon Gutiérrez,** Universidad del País Vasco UPV/EHU, Spain
- **C.N.R. Rao,** Jawaharlal Nehru Centre for Advanced Scientific Research, India
- **Gascon, J.,** Catalysis Engineering / Delft University of Technology
- **Athouël, Laurence, Dr.,** IMN-CNRS, University of Nantes, France
- **Peter Svec,** Slovak Academy of Sciences
- **Horia Chiriac,** National Institute for R&D in Technical Physics, Iasi, Romania
- **Dimitris Kouzoudis,** University of Patras, Greece
- **Saci Messaadi,** Université de Batna, Algeria
- **Pavel Ripka,** Czech Technical University, Czech Republic
- **Christophe Dolabdjian,** University of CAEN, France
- **Raffaele Battaglia,** Novaetech Srl
- **Ioannis Raptis,** NCSR "Demokritos", Athens, Greece
- **Roberto Germano,** Promete, Naples, Italy
- **Papakonstantinou, P.,** University of Ulster, School of Engineering, Engineering Research Institute
- **Morent, Rino,** Applied Physics - Ghent University
- **Anthopoulos, T.,** Imperial College London
- **Ajay Sood,** Indian Institute of Science, India
- **Valiyaveettil, S.,** National University of Singapore
- **Carlos, L.D.,** Department of Physics and CICECO, University of Aveiro, Portugal
- **Qifeng Zhang,** University of Washington
- **Jiujun Zhang, Dr., Prof.,** National Reserach Council of Canada, Vancouver, BC, V6T 1W5 Canada
- **Yong Zhang,** Department of Bioengineering, National University of Singapore
- **Francesco Stellacci,** Ecole Polytechnique Fédérale de Lausanne, Switzerland
- **Wang, E.,** Changchun Institute of Applied Chemistry,Chinese Academy of Sciences

- **Dr. Zhigadlo, N.D.,** Laboratory for Solid State Physics, ETH Zurich, Switzerland
- **Dediu, Valentin Alek,** ISMN-CNR, via Gobetti 101, 40129 Bologna, Italy
- **Velonia, K.,** Department of Materials Science and Technology, University of Crete
- **Dr. Vladimir Sivakov**, Institute of Photonic Technology, Albert Einstein Str. 9, 07745 Jena, Germany
- **Woo Lee,** Korea Research Institute of Standards and Science
- **Mei, Yongfeng,** Fudan University, Shanghai China
- **Chunyi Zhi,** City University of Hong Kong, Hong Kong
- **Horowitz, Gilles,** CNRS - Ecole Polytechnique, France

Peer review statement

All papers published in this volume of *IOP Conference Series: Materials Science and Engineering* have been peer reviewed through processes administered by the proceedings Editors. Reviews were conducted by expert referees to the professional and scientific standards expected of a proceedings journal published by IOP Publishing.

IOP Conf. Series: Materials Science and Engineering **108** (2016) 012001 doi:10.1088/1757-899X/108/1/012001

Comparison of different feature reduction methods in the improvement of gas diagnosis of a temperature modulated resistive gas sensor

S M Hosseini-Golgoo[1,3] and N Ebrahimpour[2]

[1] Department of Electrical Engineering, University of Guilan, Rasht 41635-3756, Iran

[2] Guilan Science and Research Branch, Islamic Azad University

E-mail: smhosseini@guilan.ac.ir

Abstract. The present study aims to analyze dynamic responses of a temperature modulated resistive gas sensor with the emphasis on the comparison of different feature reduction methods. For this purpose, four selected feature reduction methods consist of Linear Discriminant Analysis (LDA), Principal Component Analysis (PCA), Generalized-LDA (GDA) and Kernel-PCA (KPCA) are applied and compared. The sensor selected for the experiment is a tin oxide based sensor, FIS commercial type. A staircase voltage with the step length of 40 s and voltage range of 1-5 V constitutes the input of the sensor. Sensor system was modeled by ARMAX linear model. The effects of induced gases were recorded as parameter vectors in the data obtained by the model. After applying the methods of feature reductions, the performance of gas separation was compared. It was found out that LDA and GDA yielded the best data classification.

1. Introduction

In the last few decades, scientists in the machine olfaction field have developed a tool called electronic nose which its name was taken from arranging a number of sensors in an array along with a sophisticated pattern recognition engine [1]. Today, development of an electronic nose, similar to human nose is highly important. Already commercial systems from several companies are targeting applications, present and potential, that range from quality assurance of food and drugs to medical diagnosis, environmental monitoring, safety and security, and military use.

If the sensors are able to detect aromas in an environment contaminated with different gases, one of the most important goals of the development of a machine olfaction systems is to increase the quality of gas detection by a certain sensor or a set of sensors. The question whether it's better to use more sensors or to develop a more sophisticated sensor has brought debates among researches who study on the machine olfaction. Although it's possible to increase the quality of analysis by adding more sensors or additional properties, it's not easy to answer the above question [3-6]. On the other hand, it's necessary to use combined sensors for the pattern of an electronic nose. A combined sensor possesses more selectivity and the power of analysis than a single sensor. Additional sensors are useful for decreasing detection error and increasing the sensitivity of array. However, it increases dimensions of pre-processing, the complexity of calculations and also causes a high level of noise [7]. Attempts

[3] Corresponding author. Tel.: +98 13 336 90387; Fax: +98 13 336 90447. E-mail addresses: smhosseini@guilan.ac.ir, smhosseini1@yahoo.com (S.M. Hosseini-Golgoo).

have been made to create selectivity in a single sensor [8]. One of the techniques that's commonly used for creating the capability of selectivity in single sensors is the modulation of sensor temperature [9-10]. Various studies have shown that gas sensors with modulated working temperature yield more discriminant features. This method allows one to feed sensor's heater with variable voltages instead of using sensors at fixed temperature, which is produced by applying fixed voltage on the sensor's heater. This leads to a change in the planned temperature of sensors during the measurement stage [11]. The stage of feature identification includes a vector of inputs which is produced by different methods of linear and non-linear feature reduction either with or without supervision for the next stage (i.e. gas classification). Babaei and Hosseini-Golgoo applied temperature modulation to the micro-heater of an RGS (as input) and recording temporal responses of a sensor (as output), and modelled the system by a linear system identification technique. Then, they used linear discriminant analysis (LDA) for decreasing dimensions of feature vector of the model [10]. In this work, output model could classify four target gases: Methanol, Ethanol, 2-Propanol and 1-Butanol. In another work by Babaei and Amini, they applied temperature modulation in the form of different voltage waves and used principal component analysis (PCA) method to reduce dimensions of feature vector. By doing so, they managed to reduce the dimensions of feature vectors to three dimensions and identify four Butanol isomers [12].

The present article is organized as follows. In the first section, the stages of sensor response recording in the presence of 7 target gases are discussed. Different stages of study on recorded response and the identification of a feature vector for the classification of target gases are considered in sec. 3. Then, feature extraction and the four methods of dimension reduction – LDA , PCA , GDA and KPCA – are explained and compared for the classification of gases. Lastly, results from above mentioned methods are analyzed and compared.

2. Experimental studies

Seven target gases including methanol, ethanol,1-propanol, 2-propanol, 1-butanol, acetone and hydrogen were imposed on the sensor in 11 concentration levels from 250 to 2000 ppm. A commercial sensor from FIS family (SP3-AQ2) based on tin oxide is selected for the present study. Stages used for the classification of data from a single sensor are the same used in electronic nose. Resistive gas sensors are the most common gas sensors which their basic performance principle is based on conductivity changes of the sensitive layer in the presence of target gas. Sensor input is a staircase temperature modulation as shown in figure 1. The modulation is imposed on sensors microheater. Then, the changes which occur in sensor's resistance are recorded by the circuit shown in the figure 1. If the applied voltage on the sensor's micro-heater, V_s, and the value of the series conductance, G_o are constant, then the changes of sensors conductance, G_s, will precisely calculated by measuring output voltage, y(t),as:

$$G_S(t) = G_O\left(\frac{V_S}{y(t)} - 1\right) \quad (1)$$

Using the above expression, the changes of responses in different gas concentrations were computed by recording y(t). For instance, figure 2 shows the recorded response of FIS sensor in the presence of 11 concentration levels of methanol. The staircase heater voltage of the sensor in the gas atmosphere is shown in figure 2, as well. Sensors responses are converted into digital signals by an analog to digital convertor and are transferred into a computer. These responses are in transient state. Sensor's recorded responses to 11 concentration levels of seven target gases are, then, sent to pre-processing unit in order to be processed, identified and separated.

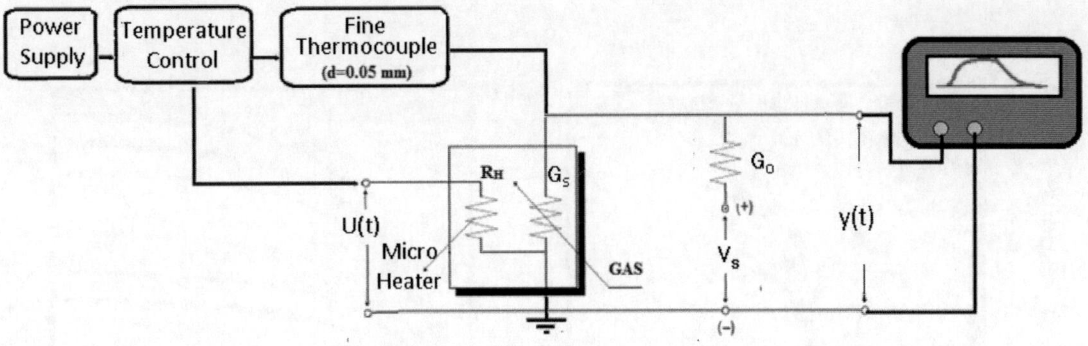

Figure 1. Measurement system paradigm in a controlled atmosphere chamber for measuring sensor's response to a target gas.

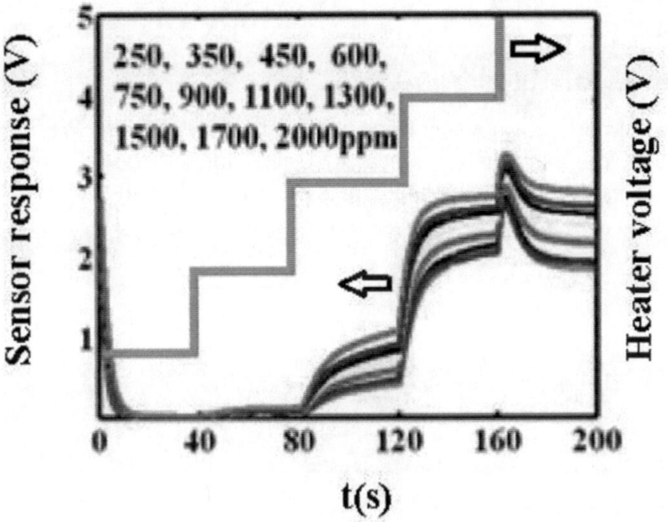

Figure 2. Temporal response of a temperature modulated FIS gas sensor subjected to 11 concentration levels of methanol.

After recording sensors responses by the circuit shown in the figure 1, the responses are filtered and normalized in order to prevent effects such as environmental changes like temperature changes and moisture in atmosphere as well as the effects of high frequency. The noise from high frequency is reduced for all gases by a digital moving low pass filter with 28 sample window length. Then filtered responses, y_n, were normalized as the following expression:

$$y_n(t) = \frac{y(t) - y_{min}}{y_{max} - y_{min}} \quad (2)$$

in which y_{max} and y_{min} are maximum and minimum values of sensor response, respectively. In order to enter sensor's non-linear responses into a linear modelling system, we used 5 segment of sensor responses and 5 corresponding input heater voltage as 5 systems. Figure 3(a) and (b) show total normalized sensor response and its second segment for methanol, respectively. As can be seen in this figure, sensor responses are limited between 0 and 1 after normalization and the length of each output step will be equal to its input modulation, i.e., 40s.

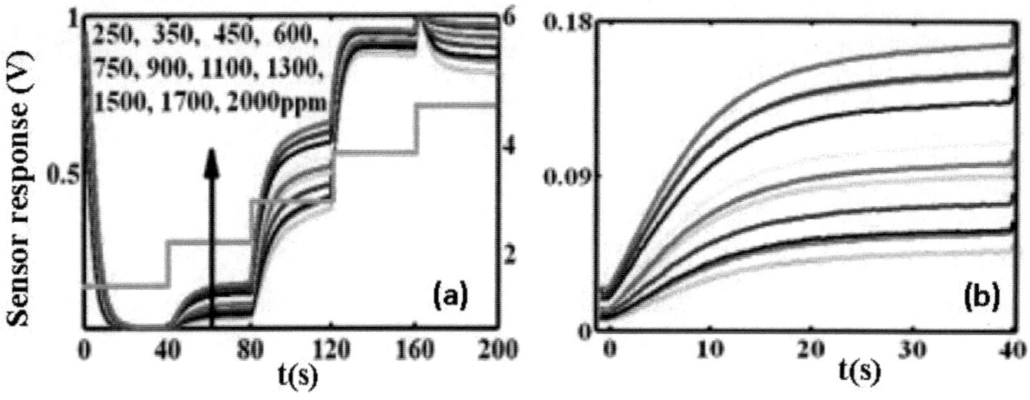

Figure3: Filtering and normalizing sensor's output response to methanol gas (a) all segments (b) second segment.

3. Modelling and feature extraction

A single input- single output (SISO) system was selected for modelling. We used n-order, n_k delay dynamic linear ARMAX (n,n,n,n_k) in our research. In this model, the relation between input, u(t), output, y(t), and noise transfer function e(t) are expressed in the following:

$$y(t) = \frac{B(q^{-1})}{A(q^{-1})} u(t - n_k) + \frac{C(q^{-1})}{A(q^{-1})} e(t) \tag{3}$$

in which

$$\begin{aligned} A(q^{-1}) &= 1 + a_1 q^{-1} + a_2 q^{-2} + \ldots + a_{na} q^{-na} \\ B(q^{-1}) &= b_1 + b_2 q^{-1} + b_3 q^{-2} + \ldots + b_{nb} q^{-nb+1} \\ C(q^{-1}) &= 1 + c_1 q^{-1} + c_2 q^{-2} + \ldots + c_{nc} q^{-nc} \end{aligned} \tag{4}$$

where coefficients of a_i, b_i, and c_i are arranged in a feature vector

$$\theta = (a_1 \ldots a_{na} b_1 \ldots b_{nb} c_1 \ldots c_{nc})^T \tag{5}$$

In order to compare measured output and the output of a trained network, *fitness* criterion is computed for each segment [9]. Based on the results of fitness computations, the smallest and best model order for each five defined systems is selected. The values for *n* are one or two. According to the relation (4), for n=2, the number of features is 6 and for n=1, the number of features is 3. In this way, segment- I, II and V are modelled by 2nd order model, while segment-III and IV are modelled by 1st order model. Thus, we'll have 24 features in total. Parameter vector from each recorded response is used as a feature vector for encoding operations in feature reduction methods like PCA, which is the representative of linear unsupervised training methods and LDA, which is the representative of linear supervised training methods. Utilization of kernel functions let us use linear feature extraction methods such as LDA and PCA for non-linear systems. Kernel PCA and generalized discriminate analysis (GDA) methods are two examples of feature extraction methods which have been generalized from LDA and PCA methods by kernel function. In the present study, we used a kernel function as the following expression:

$$\phi(x) = [\, x_1, \; x_2, \; \ldots, \; x_n, \; x_1x_1, \; x_1x_2, \; \ldots, \; x_1x_n, \; x_2x_2, \; \ldots, \; x_2x_n, \; \ldots, \; x_nx_n]^T \qquad [6]$$

Thus, dimension of feature vector, first increases and then reduce in a new feature space (3-dimensions). As a matter of fact, the system must be able to separate new concentrations (test data), as well. Therefore, three randomly selected test data are also introduced into the system. Training and test data after LDA, PCA, GDA and KPCA methods are given in the figure 8(a)-(d), respectively. In these figure, test data shown by filled markers, could join their class in LDA and GDA methods. It should be noted that LDA linear method could perform better than non-linear GDA method in terms of data separation and dispersion. This means that LDA is a more suitable method for separating classes of both training and test data.

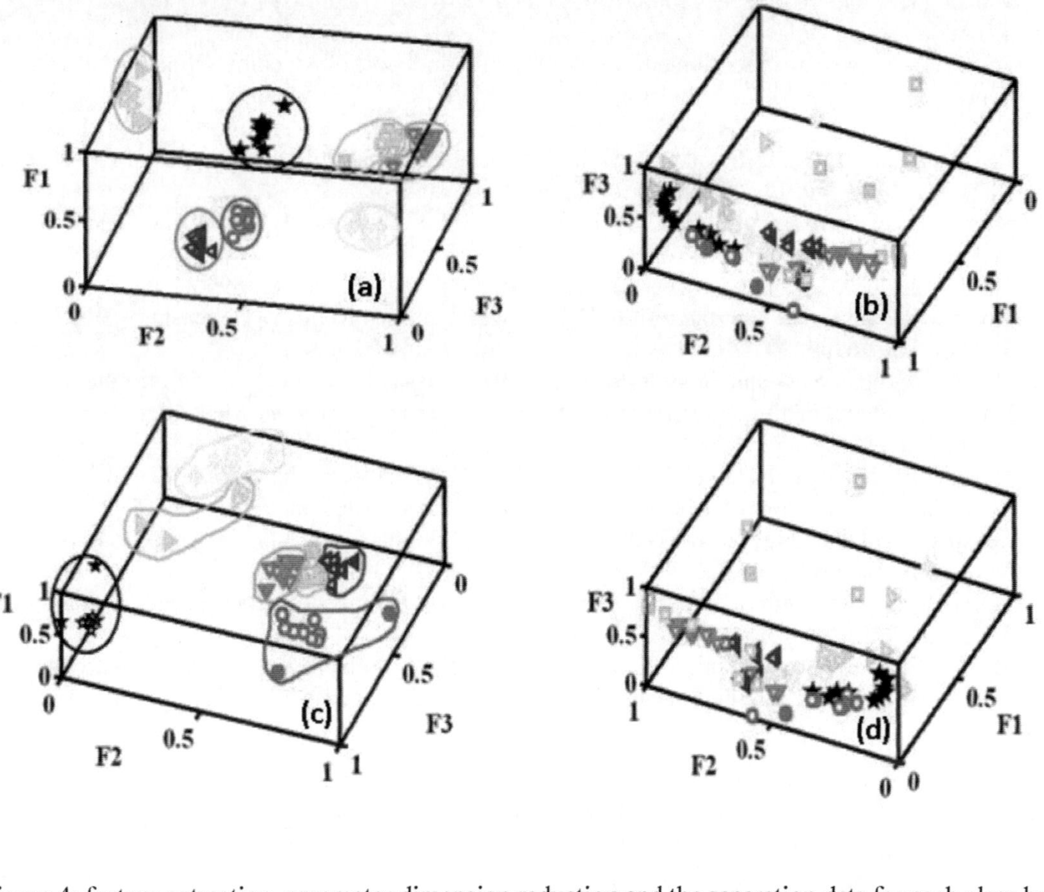

Figure 4: feature extraction, parameter dimension reduction and the separation data for each class by encodings (a) LDA, (b) PCA,(c) GDA, (d) KPCA after selecting two different concentrations of 450 ppm and 1700 ppm.

4. Conclusion

In this paper, staircase temperature modulations with 5 voltage steps and plateau length of 40s were imposed on sensor's micro heaters as the inputs of the sensor. Then sensor's output responses were recorded. Then, each input and output increment was imposed on a model called ARMAX, a feature vector with 24 dimensions was obtained. Using four selected feature reduction methods, 24 dimensions were reduced to 3 dimensions. We tried to compare linear and non-linear feature reduction methods for extracting features from responses of a resistive sensor. Also we evaluated the separation of gases in test and training concentrations. Although linear LDA methods performed better than non-linear GDA methods, results suggested that LDA could separate both test and training data better than other methods.

5. References

[1] Schiffman S S and Pearce T C "Introduction to olfaction: perception, anatomy, physiology, and molecular biology" in Pearce T C, Schiffman S S Nagle H T and Gardner J W 2003 "Handbook of machine olfaction: electronic nose technology" WILEY-VCH Verlag, Weinheim, pp.55, 105-106, 133-143 and 329-330.

[2] Nagle H T schiffman S S and Gutierrez-Osuna R "The how and why of electronic noses" September 1998 IEEE Spectrum Vol. 38 No. 9 pp. 22-34.

[3] Wilson D M and Gaffod S D 2002 IEEE Sens. J 2, 169. (b) Zubritsky E 2000 Anal. Chem. 72, 421A.

[4] Pearce T C and Sanchez-Montanes M 2003 "In Handbook of Machine Olfaction" Pierce T C Schiffman, S Nagle H T, and Gardner J W, Eds.; Wiley VCH: Weinheim, New York, pp. 347-375.

[5] Li J, Hermann B, Hobbs S and Severin E 2000 In 7th International Symposium on Olfaction and Electronic Nose (ISOEN), Brighton, U.K., pp. 81.

[6] Stetter J R and Penrose W R, 2002 Sensors Update, Vol. 10; Baltes H, Korvink J, Fedder G, Eds.; Wiley VCH: Weinheim, New York, pp. 189.

[7] Hierlemann A 2008 "Higher-Order Chemical Sensing" vol 570, pp:565-599.

[8] Hossein-Babaei F and Orvatinia M 2004 "A novel approach to hydrogen sensing", IEEE Sensor J., Vol. 4, No. 6, pp. 802-806.

[9] Hosseini-Golgoo S M and Hossein-Babaei F 2011 "Assessing the diagnostic information in the response patterns of the temperature modulated tin oxide gas sensor" Meas. Sci. Technol., vol. 22, no.3, 03520.

[10] Hossein-Babaei F and Hosseini-Golgoo S M 2008 "Analizing the response of the thermally modoulated gas sensor using a linear system identification technique for gas diagnosis" IEEE Sensors J., vol 8, no. 11, pp. 1837-1847.

[11] Vergara A, Llobet E, Brezmes J, Vilanova X, Ivanov P, Grácia I, Cané C and Correig X, 2005 "Optimized temperature modulation of microhotplate gas sensors through pseudorandom binary sequences" IEEE Sensors J., vol. 5, no. 6, pp. 1369–1378.

[12] Hossein-Babaei F and Amini A 2013 "Obtaining Highly Selective Responses from a Bulk Tin Oxide Gas Sensor" Key Engineering Materials, Vol. 543, pp. 239-242.

IOP Conf. Series: Materials Science and Engineering **108** (2016) 012002 doi:10.1088/1757-899X/108/1/012002

Formaldehyde gas sensor based on nanostructured nickel oxide and the microstructure effects on its response

D. Lahem[1], F.R. Lontio[2], A. Delcorte[2], L. Bilteryst[3], M. Debliquy[4]

[1] Department Material Science, Materia Nova ASBL Mons, Belgium

[2] Institute of Condensed Matter and Nanosciences, Université Catholique de Louvain, UCL,Louvain-La-Neuve, Belgium

[3] Certech, R&D Institue, Seneffe, Belgium

[4] Service Science des Matériaux, Faculté Polytechnique de Mons, UMONS, Mons, Belgium

Driss.Lahem@materianova.be

Abstract. NiO nanostructures can be used as a promising material for semiconductor gas sensor to detect formaldehyde at low concentrations (< 1 ppm). Here, the effect of the morphology of the synthesized NiO nanostructures on gas sensing properties is studied and discussed. NiO nanostructures have been synthesized by thermal decomposition of precursors obtained by two different chemical precipitation methods and a sol-gel technique. Thick films of the synthesized NiO nanostructures were deposited by spray coating on alumina substrates fitted with gold interdigitated electrodes and a platinum heater. The gas sensing properties of those NiO films were studied for low concentrations of formaldehyde gas at different working temperatures. A clear difference in response characteristics was observed between the samples prepared by different synthesis routes.

1. Introduction

In Europe, formaldehyde is now considered by the authorities as one of the priority pollutants in indoor air because of its carcinogenic character and because of the many sources in our closed environment: paper products, stiffeners, wrinkle resisters and water repellents, insulation, combustion devices, pressed-wood products, etc. The WHO guideline for indoor air formaldehyde concentration is 0.08 ppm (0.1 mg/m^3) [1].

It seems therefore necessary to measure the concentration of formaldehyde in closed environments because this data is essential for preventive actions (choice of the materials/furniture treatment, ventilation or indoor air purification). There exist many chemical and biochemical sensors (indicator tubes, electrochemical cells…) that are the subject of numerous works [2]. However, these inexpensive and miniature sensors lack in selectivity and possess relatively high detection limits which are not compatible for "indoor" detection for which the useful range is typically 10 ppb to 1 ppm. NiO was already pointed as a sensitive material for formaldehyde at low concentrations [3, 4]. Because of its applications in various fields (smart windows, supercapacitors, catalysis), a lot of synthesis routes for NiO nanopowders or films were proposed [5].

We present in this paper the characteristics of NiO layers synthesized by three different methods and compare the performances towards formaldehyde detection.

2. Experimental

2.1. Synthesis of the Nickel Oxide powders

NiO powders were synthesized by three different methods, sol-gel from nickel chloride and two precipitation methods from nickel malonate and nickel sulfate precursors.

2.1.1. By Sol-gel method [6]. $NiCl_2.6H_2O$ (1.5 g, 0.0063 mol) was dissolved in 70 ml of absolute ethanol at room temperature, leading to a clear green coloured solution. In another beaker, NaOH (0.5g, 0.0125 mol) was dissolved in 100 ml absolute ethanol and the NaOH solution was added to the nickel chloride solution dropwise. The mixture was stirred at room temperature for 2 h. During this time, the reaction mixture was found to form light green coloured gel. After 2 h, the gel was filtered, washed thoroughly with distilled water and then finally with ethanol. The precipitate was air dried, yielding a light green coloured powder. The TGA data showed that this green powder decomposes at ~294 °C.

2.1.2. By precipitation from nickel sulfate [6]. 0.083 mol of urea was added to 5.2 mmol of nickel sulphate dissolved in 80 ml of distilled water. The mixture was heated to 80 °C for 4 h. The green precipitate formed was filtered, washed with distilled water several times to remove the unreacted nickel salt and urea and finally with absolute ethanol, and dried at room temperature. The TGA data showed that the green powder decomposes at ~369 °C.

2.1.3. By precipitation from nickel malonate [7]. The NiO here was obtained via two experimental steps: the synthesis of the precursor by precipitation in aqueous solution containing nickel ions using the malonate ligand as precipitating agent and its subsequent decomposition. Using nickel malonate precursor, lithium malonate was first prepared in aqueous solution by adding lithium hydroxide (10 mmol) to malonic acid (5 mmol) and the pH adjusted to 7. After 5 min, an aqueous solution containing nickel chloride was poured drop by drop into the previously prepared solution of lithium malonate with continuous stirring. The resulting green solution was then stirred for 3h at 90 °C under reflux. The obtained precipitate was filtered, washed successively with distilled water, isopropanol and acetone to ensure the total removal of impurities and dried in an oven at 70-80 °C to obtain a light green powder. The as-prepared precursor powder was calcined in a ceramic crucible at 500 °C in a muffle oven (10 °C/min) for 1 h under air flow. The TGA data showed that green powder decomposes at ~343 °C.

2.2. Sensor preparation

For measuring gas sensing, synthetized NiO powders were subjected to calcination at 400 °C for 1 h and each coated on alumina substrates fitted with interdigitated gold electrodes and platinum heating element (Fig. 1) by simple spray-coating of the corresponding ink. The ink was prepared by mixing the metal oxide powder with distilled water and the resulting solution (charged with 10 % of metal oxide) was placed for 24 h in a TURBULA Shaker-Mixer to separate all the agglomerate particles. The thickness of the sensitive layers is ~8-10 μm (checked with an optical profiler). Finally the NiO coating was annealed at 400 °C for 1h in air before their exposure to the gases in a Teflon chamber.

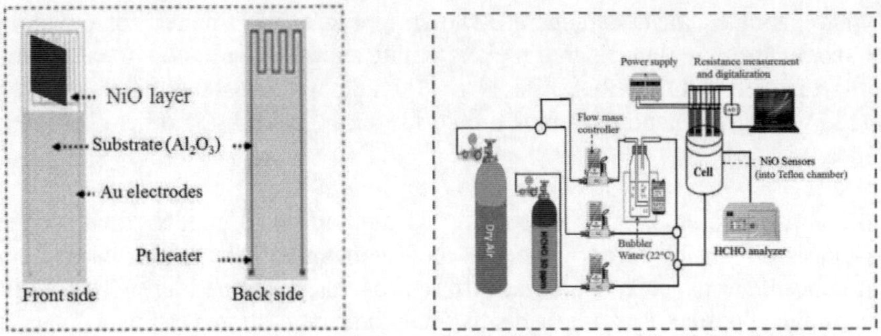

Fig.1. Substrate used for sensing measurement and setup used for formaldehyde sensing measurements

3. Results and discussion

In order to obtain the insight information about the surface morphology and particle size of the samples, FE-SEM analysis was performed. Fig.2 shows the FE-SEM images of three synthetized NiO.

It can be observed that the film obtained from sol-gel method had a uniform distribution of small size particles approximatively 15-20 nm on the surface (Fig.2A). In the sample obtained with precipitation from nickel malonate (Fig.2C), the film is composed of particles with various sizes between 20 and 100 nm. Whereas in the case of the sulfate precursor (Fig.2B), the particles possess an agglomerated hexagonal sheet-like morphology.

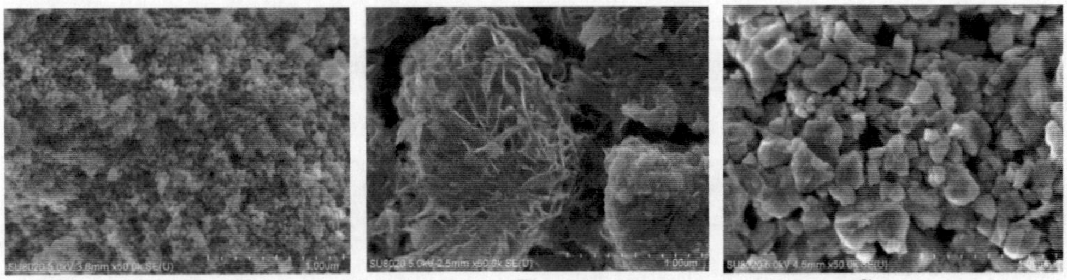

A B C

Fig.2. FE-SEM image showing nanostructures NiO obtained from precursor prepared by (A) sol-gel from nickel chloride, (B) precipitation from nickel sulfate and (C) precipitation from nickel malonate

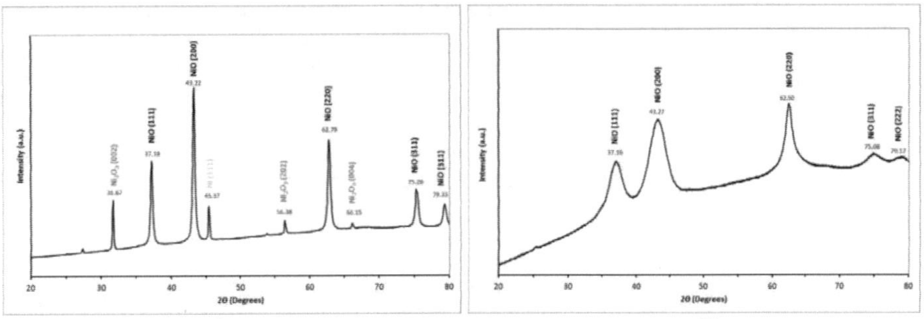

A B

Fig.3. XRD patterns of NiO nanoparticles synthetized by sol-gel method from nickel chloride (A) and by precipitation from nickel sulfate (B).

The XRD patterns of calcined samples are shown in Fig.3. XRD pattern of powder prepared by precipitations from nickel malonate and nickel sulfate (Fig.3B) revealed the formation of face-centered-cubic structured NiO (JCPDS file N°. 78-0428). Whereas the XRD pattern of calcined sample prepared by sol-gel method from nickel chloride shows in addition other peaks (Fig.3A) which were identified as Ni_2O_3 [8].

In order to compare the effect of synthetized NiO method on formaldehyde response of sensors, electrical resistances of sensing layers were recorded with formaldehyde diluted in humid clean air (RH=50 %) at concentrations under 1 ppm at different operating temperatures (between 150 and 280 °C). The setup for the exposure tests of sensors to formaldehyde is shown in Fig. 1.

Since NiO is a p-type semiconductor, the electric resistance of NiO sensor increases when formaldehyde gas is introduced to testing chamber. The sensitivity, S, described in this paper was defined, according to the following formula:

$$S = \frac{\text{Rgas} - \text{Rair}}{\text{Rair}} X\ 100$$

Where, R_{gas} and R_{air} are the resistance of sensor after exposure to formaldehyde and clean air, respectively.

The Fig. 4 and Fig. 5 show the dynamic response of different NiO-based sensors operated at 280 and 220 °C. It can be seen that at higher temperature, the NiO sample prepared by precipitation from malonate shows the better sensitivity to formaldehyde. At lower temperature, the sample fabricated by sol-gel method from nickel chloride shows a better response. Whereas the formaldehyde response of NiO sample prepared by precipitation from nickel sulfate is very slow in the studied temperature range. It can be observed that for all temperatures the response and recovery times of the sensors based on the chloride route are shorter.

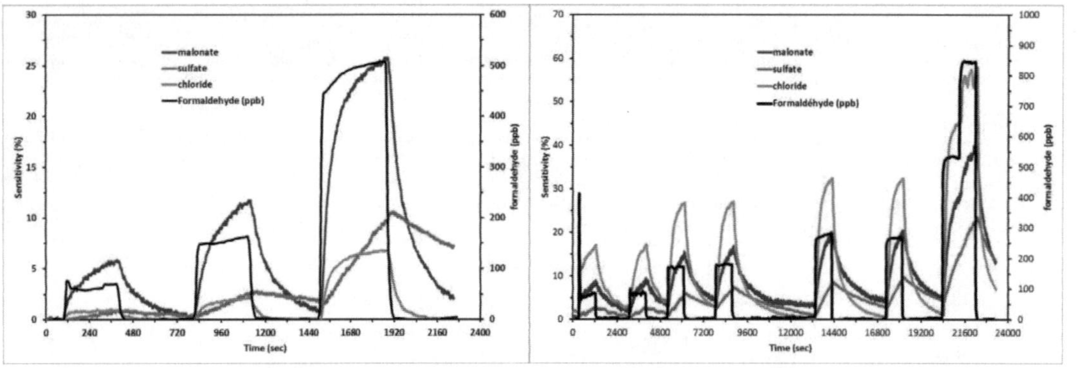

(A) (B)

Fig.4. Dynamic response of different type of NiO sensors at different operating temperatures (A) 280 °C and (B) 220 °C.

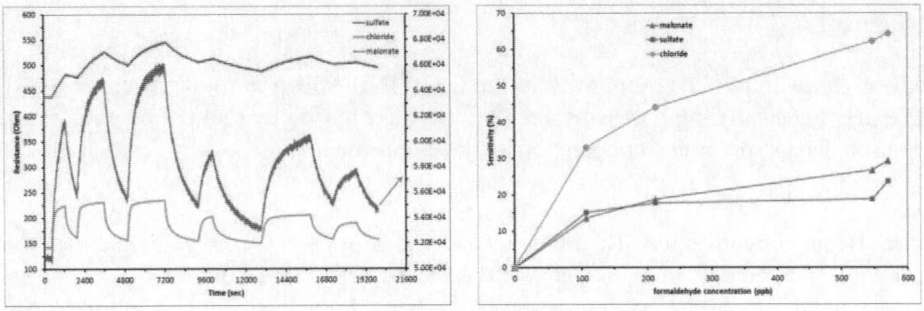

Fig. 5. Dynamic response of different NiO sensor exposed to 532, 547, 571, 197, 212 and 107 ppb of formaldehyde at 200 °C.

The formaldehyde sensing characteristics of these NiO-films are strongly influenced by the surface microstructure and the synthesis method. The more favourable microstructure corresponds to that of NiO sample obtained by sol-gel method from nickel chloride. A possible reason is the morphology of the layers showing very fine particles in a porous stacking. XPS analyses are underway to complete understanding.

Fig. 6 shows the variation of sensitivities against formaldehyde in NiO sensor based on this sample, which was measured at different operating temperatures. The sensor exhibits a maximum sensitivity at a temperature of 200 °C.

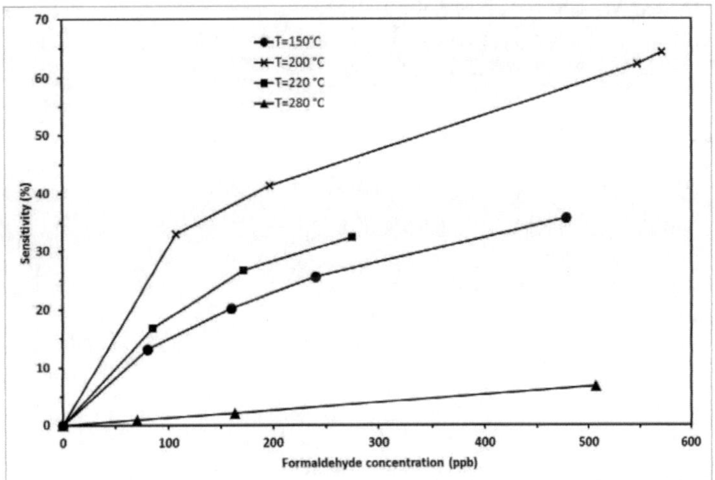

Fig.6. Variation of sensitivity of NiO sensor prepared by precipitation from nickel chloride with formaldehyde concentration at different operating temperature.

4. Conclusion

This study has demonstrated that chemical sensors based on nanostructures NiO are able to detect very low concentration of formaldehyde (<1 ppm) with good performance at working temperature around 200 °C. The effect of the synthesis routes was clearly put in evidence. These results are promising for development of simple and inexpensive devises allowing the analysis of formaldehyde in parts per billion (ppb) in indoor air.

Acknowledgment

This work was carried out in the framework of the CAPTINDOOR project of Materia Nova, UMONS and CERTECH, financially supported by the Walloon Region of Belgium. The authors also thank UCL for the scholarship program "cooperation au développement".

References

[1] World Health Organization, Regional Office for Europe, "WHO guidelines for indoor air quality: selected pollutants", Geneva 2010, ISBN: 9789289002134.

[2] P.-R. Chung, C.-T. Tzeng, M.-T. Ke and C.-Y. Lee, "Formaldehyde Gas Sensors: A Review", *Sensors* **13** (2013) 4468-4484.

[3] C.Y. Lee, C.M. Chiang; Y.H. Wang, R.H. Ma, "A self-heating gas sensor with integrated NiO thin-film for formaldehyde detection", *Sens. Actuators B Chem.* **122** (2007) 503–510.

[4] U. Cindemir, Z. Topalian, L. Österlund, C. G. Granqvist and G. A. Niklasson, "Porous Nickel Oxide Film Sensor for Formaldehyde", *J. Phys.: Conf. Ser.* **559** (2014) 1-5.

[5] Y. Wu, Y. He, T. Wu, T. Chen, W. Weng, H. Wan, "Influence of some parameters on the synthesis of nanosized NiO material by modified sol–gel method", *Materials Letters* **61** (2007) 3174–3178.

[6] P. Jeevanandam & V.R.R. Pulimi, " Synthesis of nanocrystalline NiO by sol-gel and homogeneous precipitaion methods", Indian Journal Chemistry **51A** (2012) 586-590.

[7] F.R. Lontio, J.L. Ngolui, A. Delcorte, M. Debliquy, D. Lahem, V. Dupont, "Synthesis, characterization and gas sensing properties of $Ni_{1-x}Zn_xO$ nanoparticles and $Ni_{1-x}Zn_xO/ZnO$ nanocomposites", International Conference on Advanced Functional Materials (AFM 2015), New York, USA, June 29th to July 3rd 2015.

[8] N. Srivastava and P.C. Srivastava, "Realizing NiO nanocrystals from a simple chemical method", *Bull.Mater.Sci.* 33 (2010) 653-656.

IOP Conf. Series: Materials Science and Engineering **108** (2016) 012003 doi:10.1088/1757-899X/108/1/012003

Nanostructured SnO_2 thick films for gas sensor application: analysis of structural and electronic properties

Goran Miskovic[1], Obrad S Aleksic[2], Maria V Nikolic[2], Johann Nicolics[1], Goran Radosavljevic[1], Zorka Z Vasiljevic[3], Miloljub D Lukovic[2] , Walter Smetana[1]

[1] Department of Applied Electronic Materials, Institute of Sensor and Actuator Systems, Technische Universität Wien, Vienna, Austria
[2]Institute for Multidisciplinary Research, University of Belgrade, 11000 Belgrade, Serbia.
[3]Institute of Technical Sciences, Serbian Academy of Science and Arts, 11000 Belgrade, Serbia

goran.miskovic@tuwien.ac.at

Abstract. This research is focused on structural and electrical characterisation of tin oxide (SnO_2) applied as a thick film and investigation of its properties as gas sensitive material. Micron sized SnO_2 powder was milled in an agate mill for six hours to fabricate SnO_2 nanopowder, which was afterwards sieved by 325 mesh sieve and characterized by XRD and SEM. This powder was used as functional part in the production of thick film tin oxide paste containing a resin vehicle with 4 wt. % nanosize glass frits acting as permanent binder. The glass frits where additionally milled for twelve hours in the agate mills to nanosized powder and sieved by a 325 mesh sieve as well. The achieved thick film paste was screen printed on alumina and fired at 850°C peak temperature for 10 minutes in air. After the sintering process, thick film samples where characterized by X-ray powder diffraction (XRD) and scanning electron microscopy (SEM). The reflectivity was measured on the same samples by UV-VIS spectrophotometer: the band gap was determined from the slope of reflectance. After that a matrix of different interdigitated electrode structure of PdAg paste was printed and sintered using the mentioned sintering conditions. The tin oxide thick film was printed over the interdigitated electrodes as a top layer and sintered again under the same conditions. The total electrical resistance was measured as a function of the electrode spacing and temperature. A negative temperature coefficient (NTC) was identified and measured in the range from room temperature (27°C) to 180°C in a climate chamber. Finally the samples were placed into a gas reactor with NOx and CO gas and the resistance was measured in the same temperature range (27°C-200°C).

1. Introduction

Solid state gas sensors are one of the most used and the most present type of gas sensors for commercial applications in auto industry, airplane industry, heavy industry, environmental monitoring etc. In particular gas sensors based on semiconducting sensitive materials, such as metal oxides (MOX), have attracted the attention of the scientific community for the last few decades [1]. However, not all materials are suitable for all kinds of gas detection and the right choice of a suitable and effective solid state sensor depends on analysis and correlation between various metal oxide parameters, such as electro-physical (band gap, conductivity etc.) and electronic, structural and other properties.

SnO_2-based gas sensors have been the subject of much research [2] due to the advantage of a relatively low operating temperature, long term stability and low cost. Thin and thick films of SnO_2 have been deposited using a variety of techniques [2-4]. Thick film gas sensors have a certain advantage such as low cost, simple construction and high sensitivity [5]. One of the factors affecting gas sensitivity is the microstructure [6, 7]. Nanostructured materials for gas sensor applications have been the subject of much research as the large surface to volume ratio of these materials enhances gas response characteristics thus having a great influence on the sensor performance.

In this work we have used milling as a simple, low cost method to reduce the grain size of the starting SnO_2 powder and also glass frit used in thick film paste. Structural and electronic properties of SnO_2 thick films were analyzed in view of application as inexpensive gas sensors.

2. Experimental

SnO_2 powder (Sigma Aldrich, 99.9% purity, 325 mesh) was milled in an agate mill for six hours in order to reduce grain size and obtain nanopowder. The milled powder was sieved through a 325 mesh sieve. XRD analysis was performed on a Philips PW1050 diffractometer with CuKα radiation step f 0.05s and hold time of 10 s, 2θ 10-90°. A SEM analysis was performed on a TESCAN Electron Microscope VEGA TS 5130MM device.

Thick film paste was prepared by mixing the milled powder with a resin vehicle and 4 wt.% nanosized glass frits acting as a permanent binder. The glass frit was additionally milled for twelve hours in an agate mill to reduce grain size and then sieved through a 325 mesh sieve. The prepared SnO_2 thick film paste was screen printed on an alumina substrate and sintered in a hybrid conveyor furnace at 850°C for 10 minutes in air. Thick film samples were characterized by XRD and SEM analysis.

Diffuse reflectance spectra of thick film samples were measured on an UV/Vis Shimadzu UV-2600 with an ISR2600-Plus integrating sphere attachment in the measuring range 220-1000 nm.

Interdigitated test matrices were made by screen printing an interdigitated electrode structure of PdAg paste with an electrode spacing of 0.25 μm and 0.35 μm. The tin oxide thick film was screen printed as a top layer and sintered in a hybrid conveyor furnace at 850°C for 10 minutes in air. The electrical resistance was measured as a function of the electrode spacing value and the input voltage. The DC resistance versus temperature was measured in a climate chamber in the temperature range of 27°C to 200°C enabling the determination of the material constant B characteristic for negative temperature coefficient (NTC) materials.

3. Results and Discussion

XRD patterns of the milled SnO_2 powder and thick film are shown in Fig. 1.

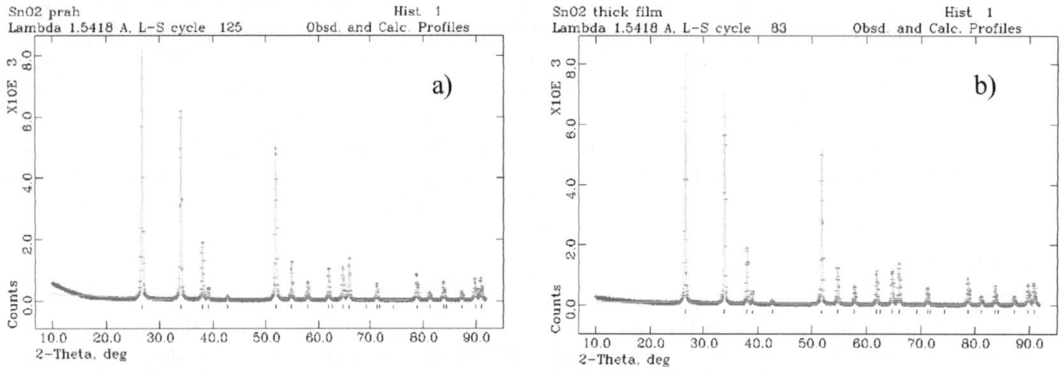

Figure 1 XRD patterns of milled SnO_2 powder (a) and thick films (b)

Structural refinement was carried out by the Rietveld method using the GSAS [8] package with the EXPGUI graphical user interface [9]. SnO_2 (cassiterite) with a tetragonal structure (space group - $P4_2/mnm$) was determined. Starting values for tetragonal SnO_2 were taken from Bolzan et al [10]. The

IOP Conf. Series: Materials Science and Engineering **108** (2016) 012003 doi:10.1088/1757-899X/108/1/012003

lattice parameters determined were a=4.7364(3), c=3.1853(3), crystallite size 59.6 nm for milled SnO_2 powder and a=4.7419(2), c=3.1894(2), crystallite size 67.3 nm for SnO_2 thick film samples.

SEM micrographs of SnO_2 powder and thick film are shown in Fig 2 and Fig. 3, respectively, confirming the small grain size of both. The thick film is porous with a small relatively uniform grain size.

Figure 2. SEM micrograph of milled SnO_2 powder.

Figure 3. SEM micrograph of SnO_2 thick film sample.

The measured diffuse reflectance specter of a tin oxide thick film sample is shown in Fig. 4. It was converted to the Kubelka-Munk function that is proportional to the absorption coefficient using the UV-Probe software. The optical band gap was estimated using a Tauc plot relating the Kubelka-Munk function to the density of optically absorbing energy transitions using the relationship of Davis and Mott:

$$h\vartheta \cdot F(R_\infty) = A(h\vartheta - E_g)^m,$$

where *hv* is the absorbed photon energy, A is a constant related to the density of electronic states above and below the band gap, E_g is the optical band gap where m=1/2 is direct allowed transition and m=2 is indirect allowed transition. In case of SnO_2 the allowed transition is direct [11]. The band gap of the tin oxide thick film was determined as 3.95 eV (as shown in the inset in Fig. 4) that is in accordance with literature data where the optical band gap of SnO_2 varies in the range between 3.4 and 4.6 eV [5].

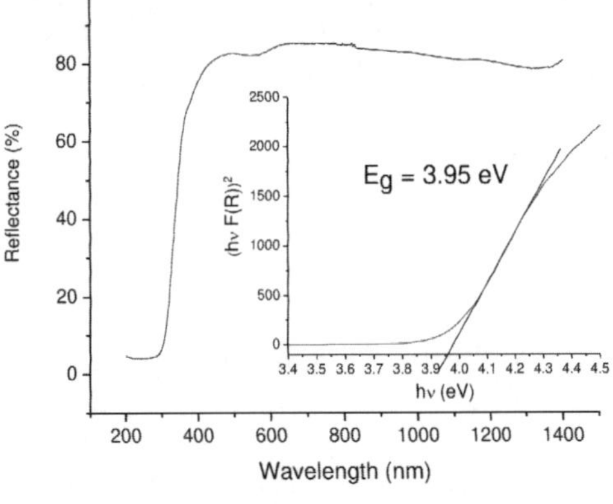

Figure 4. Diffuse reflectance spectra of SnO_2 thick film sample (inset: Tauc plot for the allowed direct transition).

The measured resistance in a climate chamber in the temperature range 27-180°C is shown in Fig. 5.

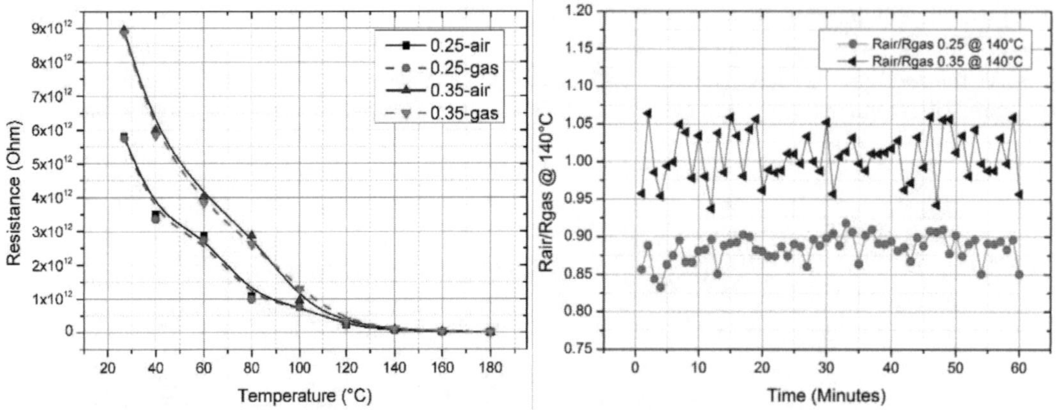

Figure 5. Resistance of SnO_2 thick film sample exposed to air and to 5% of CO measured in a climate chamber.

As Fig. 5 shows, the measured values are quite high. Up to 120°C the measured values were unstable so 100 measurements were made and the average value has been taken for further analysis. From 120°C the measured values are stable and 10 measurements were taken for averaging. Clearly samples with bigger spacing between electrodes (0.35mm) have higher values of resistance, but both types of samples exhibit the same temperature dependence. In both atmospheres a decrease in the resistance with increase in temperature can be noted. In case of air, the B-value (material constant) can be determined from the slope of the ln$R$-1000/T (in Kelvin) characteristic over the specified temperature range [12]. $B_{27/180}$ for both analysed electrode spacing values (0.25 and 0.35 mm) was determined as 5931 K and 8081 K, respectively. After exposing samples to 5% of CO gas, measurements were performed for 1 hour. As an example, the ratio Rair/Rgas at 140°C for both samples is given in Fig. 5. As it can be seen, there is a lack of reaction towards CO gas and similar values were obtained for other temperatures as well as for 5% of NO gas.

4. Conclusion

In this paper a complete analysis and characterisation of nanostructured SnO_2 thick film was made. Micron sized SnO_2 powder was milled in an agate mill to fabricate SnO_2 nanopowder, which was afterwards characterized by XRD and SEM. This powder was used as functional part in the production of thick film tin oxide paste. The achieved thick film paste was screen printed on alumina and fired at 850°C peak temperature for 10 minutes in air. After the sintering process, thick film samples where characterized by X-ray powder diffraction (XRD) and SnO_2 (cassiterite) with a tetragonal structure (space group - $P4_2/mnm$) was determined. Scanning electron microscopy (SEM) confirmed the small and relatively uniform grain size. The reflectivity was measured on the same samples by UV-VIS spectrophotometer: the band gap of SnO_2 was determined from the slope of reflectance as 3.95 eV. As for the response toward CO and NO, the samples exhibited a lack of sensitivity, which is explained with the low temperature needed for electro-chemical reactions. Plans for future work is to add catalytic dopants (Pt, Pd) and to perform measurements at higher temperatures.

5. Acknowledgement

The authors would like to express their gratitude to Dr. M. Mitric for XRD measurements. This work was performed as part of project III45007 financed by the Serbian Ministry for Education, Science and Technological Development and as a part of the FP7 SENSEIVER project (Low-cost and energy-efficient LTCC sensor/IR-UWB transceiver solutions for sustainable healthy environment, project reference 289481).

IOP Conf. Series: Materials Science and Engineering **108** (2016) 012003 doi:10.1088/1757-899X/108/1/012003

References

[1] Das S, Jayaraman V 2014 *Prog.Mater. Sci.* **66** 112
[2] Ansari S G, Gosavi S W, Gangal S A, Karekar R N and Aiyer R C 1997 *J. Mater. Sci.: Mater. Electron.* **8** 23
[3] Lee S W, Tsai P P and Chen H 2000 *Sensor. Actuat. B* **67** 122
[4] Gardeshzadeh A R, Raissi B, Marzbanrad E and Mohebbi H 2009 *J. Mater. Sci.: Mater. Electron.* **20** 127
[5] Khun K K, Mahajan A and Bedi R K 2009 *J. Appl. Phys.* **106** 1245009
[6] Ansari S G, Boroojerdian P, Sainkar S R, Karekar R N, Aiyer R C and Kulkarni S K 1997 *Thin Solid Films* **295** 271
[7] Zhang G and Liu M 2000 *Sensor. Actuat. B* **69** 144
[8] Larson A C and Von Dreele V B 2004 General Structure Analysis System (GSAS) los Alamos National Laboratory report LAYR 86-748
[9] Toby B H 2001 *J. Appl. Cryst.* **34** 210
[10] Bolzan A A, Fong C, Kennedy B J and Howard C J 1997 *Acta Crystal.* **53** 373
[11] Akgul F A, Gumus C, Er A O, Farha A H, Akgul G and Ufuktepe Y 2013 *J. Alloys Compd.* **579** 50
[12] Ouyang P, Zhang H, Zhang Y, Wang J and Li Z 2015 *J. Mater. Sci.: Mater. Electron.* **26** 6163

Copper oxide thin films for ethanol sensing

M Lamri Zeggar[1], F Bourfaa[1], A Adjimi[1], M S Aida[1] and N Attaf[1]

[1]Laboratory of thin films and interfaces, Faculty of science, University Constantine 1, Algeria

meryemlamri@gmail.com

Abstract. The present is a study of a new active layer for ethanol (C2H5OH) vapour sensing devices based on copper oxide (CuO). CuO films were prepared by spray ultrasonic pyrolysis at a substrate temperature of 350 °C. Films microstructure was examined by X-ray diffraction and atomic force microscopy. Vapour-sensing testing was conducted using static vapour-sensing system, at different operating temperatures in the range of 100°C to 175°C for the vapour concentration of 300 ppm. The results show a high response of 45% at relatively low operating temperatures of 150°C towards ethanol vapour.

1. Introduction

Gas sensors based on semiconducting metal oxides are widely studies for solving applications spanning from the detection of toxic, dangerous and pollutant gas. They are having several advantages such as easy and low cost production, controlled size and simple measure electronics [1, 2].

However the performance of such gas sensors is considerably influenced by the proprieties of sensing materials such as the morphology and structure. Among the used sensing materials, copper oxide have been the subject of interest for the detection of many different gas such as H_2 [3], NO_3, CO and C_5H_2OH [4].CuO is generally p-type semiconductor with bad gap in the rage of 1, 21 to 1, 51 eV [5, 6]. Several researches in the field of gas sensing have been reported, they are based on the association of metal oxide with CuO, and forming of the p-n heterojuction such as CuO/ZnO, CuO/Cu_2O and CuO/SnO_2. These structures have achieved a high sensitivity and selectivity [7–9].CuO films have been deposited by various conventional deposition methods; with include electrodeposition [10]; dip coating [11]; chemical vapor deposition [12] and reactive sputtering [13].Ultrasonic spray pyrolysis technique has been used extensively because it is very attractive method to deposit adherence and stochiometric thin films. The important features of this technique are the low cost and the simplicity of equipments.

In this work, we have synthesized CuO films by ultrasonic spray pyrolysis technique and investigate their ethanol vapor sensing performance.

2. Experimental

2.1. Preparation of CuO thin film

Copper oxide thin film has been prepared on glass substrates by ultrasonic spray pyrolysis. The precursor solution was prepared by dissolving 0.05 M copper chloride ($CuCl_2.2H_2O$) in distilled water. Then, the precursor solution sprayed in fine droplets of 40 μm in diameter, by an ultrasonic generator on heated glass substrate. Film was formed by pyrolytic reaction. During deposition, the substrate temperature is kept at 350°C. The deposition time was fixed 10 min. Films structural properties were determined by XRD using Philips X'Pert system with Cu K_{α} radiation (λ_{Cu} = 0.154056 nm). The reflections were taken at room temperature and at 2θ value ranged from 20° to 80°. The film surface morphology was characterized by using atomic force microscopy (AFM) was carried out in air at ambient condition (300 K) using Nanosurf easy scan 2.

2.2. Sensing system for measurement of vapour response

The vapour-sensing studies were carried out using a static gas chamber to ethanol vapor detection in air ambient. The CuO thin film deposited at 350°C was used as the sensing elements. Two gold interdigitated electrodes were evaporated on the top surface of the sensing layer. A thermocouple is mounted to measure the temperature. The output of the thermocouple is connected to a temperature indicator. Gas concentration inside the system is achieved by injecting a known volume of test vapor.

The electrical response of the sensing layer was investigated under two atmospheres: synthetic air and a mixture of air + vapour ethanol (300 ppm) by registering the variations in resistance with Keithley 617 programmable electrometer as a function of time. The response of the sensor was calculated using the expression given in the literature [14]:

$$Sensitivity = \left| R_v - R_{air} \right| / R_{air} \, x \, 100 \quad (1)$$

Where R_{air} is the reference resistance in the presence of air and R_v is the measured resistance in the presence of ethanol.

3. Results and discussion

3.2 Structural characterization of CuO thin films

The XRD study was primarily carried to determinate the structure and orientation of deposited thin films CuO. Figure 1 shows the XRD diffraction pattern of the sensitive CuO thin film. The presence of intense peaks in the XRD patterns of the film indicates that the CuO film is polycrystalline in nature. Two most prominent peaks can be clearly seen at 2θ value 35.5° and 38.7° corresponding to atomic planes (002) and (111), respectively of CuO phase. No peak corresponding to Cu2O phase of copper oxide has appeared in the XRD pattern thereby indicating the formation of pure CuO films.

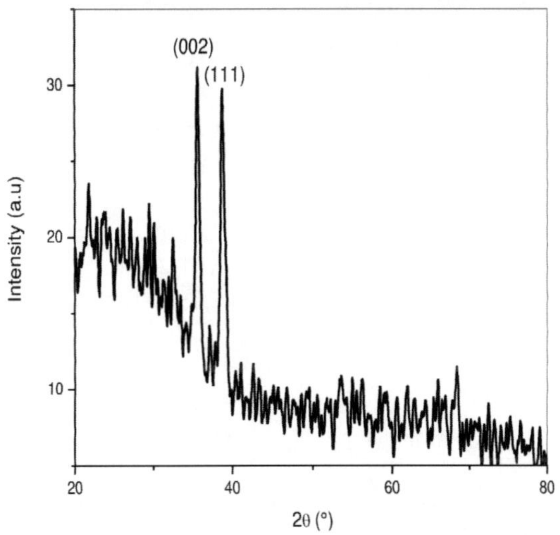

Figure 1. XRD pattern of the sensitive layer CuO thin film.

Atomic force microscopy (AFM) has been proved to be a unique method to analyze surface morphology of a film. Figure 2(a and b) shows typical two dimensional (2-D) and three dimensional (3-D) AFM image of CuO film deposited on glass substrate. The (3-D) image indicate that the surface evolution of the film shows hills and valley like structures, which are uniformly distributed over the entire substrate surface. The 2-D images of the deposits show compact and granular morphology with presence of pores in the surface this is a good argument to increase the sensitivity of the gas sensor. It is recognized that porous materials have been extensively investigated due to high surface area and accordable pore size in the fields of catalysis and gas sensing.

Figure 2. (a) The AFM 2-D image , (b) the AFM 3-D image of the deposited CuO film.

3.3 Ethanol sensing characteristics of CuO thin films

The response of semiconducteur based on sensor is greatly influenced by its operating temperature. In fact, the adsorption of gases is directly related to the temperature at the surface of the sensing layer.
In order to determine the operation temperature for the sensor based on CuO thin film to vapor ethanol, it was exposed to 300 ppm of ethanol at different operation temperature between 100 and 175 °C and the resultant responses are showing in figure 3. A typical operation temperature correspond to maximum response is widely reported in literature. In our case, the evolution of the response of the layer shows that this maximum is located around 150°C.

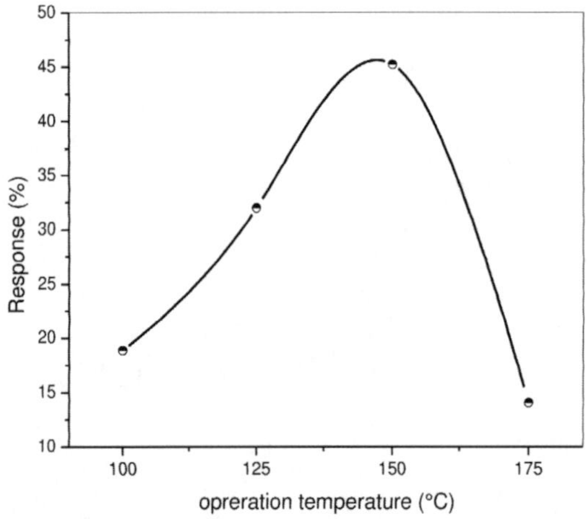

Figure 3. Sensor response curves of the CuO based sensor towards ethanol (300 ppm) at different operating temperatures

Figure 4 shows the resistance transient of CuO layer recorded at the optimum temperature of 150 °C. The decrease in resistance when vapor of ethanol is injected reveals that Rv is lower than Rair. The response is found to be 45% for 300 ppm of ethanol concentration at operation temperature. The CuO based sensor fabricated by Mitesh.al [15] shows a relatively sensor response of 8,3% towards 700 ppm ethanol vapor at operation temperature of 400 °C. It is observed that our thin film CuO based sensor has the lowest operating temperature (150 ◦C) as well as the highest response (45%).The higher response of sensor in this case may be attributed to the large amount of porosity in the material as suggested by Jimenez et al. [16]. The response and recovery time are respectively the duration by which the sensor response reaches to almost 90% of the saturation value and the time to recover its initial value when the vapor is evacuated. The sensing layer exhibits response and recovery times of 2,98 min and 2,11 min respectively. The long response and recovery time observed is probably the result of a complex mechanism of adsorption and desorption of molecules when the layer is exposed to vapor of ethanol. The kinetics of each of this mechanism is strongly influenced and controlled by the morphology and deposition conditions of the film. Modification of these conditions to obtain more porosity and decrease of the layer thickness may be a suitable method to improve the response time.

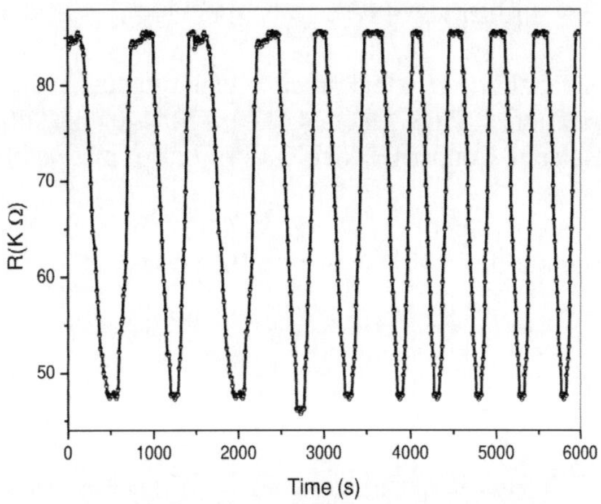

Figure 4. Resistances change vs. time during 300 ppm of ethanol sensing at the operation temperature.

4. Conclusion

A CuO thin film was tested for ethanol vapour sensing. This film was deposited by spray ultrasonic pyrolysis on heated glass substrate. Two gold interdigitated electrodes were used for gas-sensing measurement. The optimal response was obtained at a temperature of 150°C with a value of 45%. These results obtained with a simplified test device are very promising due to the low operating temperature and the high sensitivity of the material. These characteristics are promising for industrial applications, especially in gas and vapour related chemical sensing.

References

[1] Tomchenko, A.A.; Harmer, G.P.; Marquis, B.T.; Allen, J.W. Semiconducting metal oxide sensor array for the selective detection of combustion gases. *Sens. Actuat. B* 2003, *93*, 126-134.

[2] Kanan, S.M.; El-Kadri, O.M.; Abu-Yousef, I.A.; Kanan, M.C. Semiconducting metal oxide based sensors for selective gas pollutant detection. *Sensors* 2009, *9*, 8158-8196.

[3] S.A. Patil, L.A. Patil, D.R. Patil, G.H. Jain, M.S. Wagh, CuOmodified tin titanate thick film resistors as H2gas sensors, Sens. Actuators B: Chem. 123 (2007) 233–239.

[4] P. Samarasekara, N.T.R.N. Kumara, N.U.S. Yapa, J. Phys.: Condens. Matter 18 (2006) 2417

[5] Marabelli F, Parraviciny GB, Drioli FS. Phys Rev B 1995;52:1433.

[6] Ghijsen J, Tjeng LH, Elp JV, Eskes H, Westerink J, Sawatzky GA, Czyzyk MT. Phy Rev B 1988;38:11322.
[7] S.J. Appleyard, Simple photovoltaic cells for exploring solar energy concepts, Phys. Educ. 41 (5) (2006) 409–419.
[8] V.R. Katti, A.K. Debnath, K.P. Muthe, Manmeet Kaur, A.K. Dua, S.C. Gadkari, S.K. Gupta, V.C. Sahni, Mechanism of drifts in H2S sensing properties of SnO2/CuO composite thin film sensors prepared by thermal evaporation, Sens. Actuators B 96 (2003) 245–252.
[9] S. Aygun, D. Cann, Hydrogen sensitivity of doped CuO/ZnO heterocontact sensors, Sens. Actuators B 106 (2005) 837–842.
[10] E.R. Kari, K.S. Brown, Choi, Electrochemical synthesis and characterization of transparent nanocrystalline Cu2O films and their conversion to CuO films, Chem. Commun. (2006) 3311 3313.
[11] N. Serin, T. Serin, S. Horzum, Y. Celik, Annealing effects on the properties of copper oxide thin films prepared by chemical deposition, Semicond. Sci. Technol. 20 (5) (2005) 398.
[12] T. Maruyama, Copper oxide thin films prepared from copper dipivaloylmethanate and oxygen by chemical vapor deposition, Jpn. J. Appl. Phys. 37(1998) 4099–4102.
[13] V.F. Drobny, D.L. Pulfrey, Properties of reactively-sputtered copper oxide thin films, Thin Soild Films 61 (1979) 89–98.
[14] S. Kar, B.N. Pal, S. Chaudhuri, D. Chakravorty, One-dimensional ZnO nanostructure array: synthesis and characterization. J. Phys. Chem. C 110, 4605–4611 (2006).
[15] Mitesh Parmar and K.Rajanna Copper (II) oxide thin film for methanol and ethanol sensing vol. 4, December 2011 710-725.
[16] I. Jimenez, M.A. Centeno, R. Scotti, F. Morazzoni, J. Arbiol, A. Cornet, J.R. Morante, NH3 interaction with chromium-doped WO3 nanocrystalline powders for gas sensing applications, J. Mater. Chem. 14 (2004) 2412–2420.

A portable air-quality station based on thick film gas sensors for real time detection of traces of atmospheric pollutants

A Fioravanti[1,2], A Bonanno[1], S Gherardi[3], M C Carotta[3] and A N Skouloudis[4]

[1] CNR-IMAMOTER Ferrara, Via Canal Bianco 28, FE, Italy
[2] Dipartimento di Chimica, Università di Parma, Parco Area delle Scienze 17A, PR, Italy
[3] Consorzio Futuro in Ricerca, Via Saragat 1, FE, Italy
[4] Institute for Environment and Sustainability, JRC, European Commission, VA, Italy

E-Mail: a.fioravanti@imamoter.cnr.it

Abstract. Different functional materials, single or mixed nano-crystalline semiconductor oxides, were synthesized via appropriated wet-chemistry routes. The powders were used to fabricate metal oxide (MOX) thick film gas sensors. Portable monitoring stations based on the aforementioned sensors were prepared, including electronics for acquisition, processing and wireless transmission of the data. Results of long term trials in field, carried out locating few units closely to as many conventional fixed-site monitoring stations, have been reported. The comparison was performed between the temporal evolution of the conductivity changes of the sensors with the pollutants' concentrations, as measured by the analytical instruments.

1. Introduction

Several EU directives [1,2] limit anthropogenic emissions of air pollutants. Despite such regulations, air pollution continues to damage human health and the environment with a significant proportion of the European urban population is still exposed to concentrations above limit values.

Atmospheric pollution monitoring is today performed by means of traditional analytical techniques in few fixed-site stations in urban areas. The equipment used is cumbersome and expensive, also requiring frequent maintenance and calibration. Thereby, low cost, portable and versatile equipment is an issue. The technology of thick film gas sensors is an optimal candidate to be implemented in such a device. However, the sensors have to measure traces of gases in the atmosphere with sensitivity and selectivity comparable to those of the analytical instruments. This achievement implies a careful coordination of all different processes involved, such as functional material preparation and its morphological and structural characterization, deposition of sensing layers, fabrication of optimal substrates, and electronics.

Aim of this work is to demonstrate that it is possible to determine with good precision and accuracy the pollutants' concentration in the atmosphere using specific sensors, if the aforementioned characteristics are optimized. To detect the various pollutants, thick film gas sensors, based on specific functional materials, have been prepared and investigated with relation to the gases to detect, taking into account for each material the main interfering gases.

2. Experimental

2.1. Preparation and characterization of nano-ceramic materials
Wet chemical methods for powder synthesis have been investigated to prepare functional materials in form of nano-crystalline powders to obtain a high surface to volume ratio needed to amplify the gas response [3]. Simultaneous thermogravimetry and differential thermal analysis was carried out to study the conversion of precursors to oxides. Scanning or transmission electron microscopy and X-ray diffraction were used to analyze the morphology and the crystalline structure.

Passing to the specific materials, a $Ti_xSn_{1-x}O_2$ solid solution (hereinafter named TS) has been adopted to detect CO, due to the maximization of the response at a specific Ti molar ratio [4]. To detect BTX (benzene, toluene, xylene) the material is a solid solution of titanium, tantalum and vanadium (hereinafter named TTV), as described in [5]. For nitrogen oxides detection, the sensor has been prepared using iron lanthanum oxide [6]. To detect ozone, a sensor based on tungsten trioxide (hereinafter named W) has been used. The synthesis of this material and its microstructural characterization are described in [7].

2.2. Fabrication of thick film gas sensors
Through screen printing technique, the sensing film, the interdigitated contacts and the heater element are deposited on the same substrate. To print the sensing layers, a viscous paste must be prepared by adding to the functional material an organic vehicle and a small amount of a bonding agent to promote the film adhesion to the alumina substrate. After the printing, the thermal processes of drying (150-200 °C) and firing (650-850 °C) give to the film the desired electrical and micro-structural properties.

2.3. Laboratory electrical characterizations
The sensors were placed in a sealed test chamber, equipped with temperature and humidity sensors, maintained at a fixed ambient temperature of 25 °C. The flow-through technique was used to perform conductance measurements maintaining a flow rate of 0.5 l/min using synthetic air as carrier gas in dry or wet ($0 < RH < 60\%$) conditions. Dynamical responses of sensing films were obtained in presence of mixture of different gases by varying the operating temperature from 350 to 650 °C. The sensor response is defined as ratio between the conductance in presence of the target gas and the conductance in air.

2.4. In field electrical characterizations
Arrays of thick film gas sensors, assembled in small remotely controlled units, were used for on-site tests. They consist in small boxes to lodge the sensors, the electronic circuitry of each sensor and a main electronic control unit. The data are recorded in a data logger and transferred via GSM to an input data server. The sensor responses were studied in comparison with the pollutant concentrations as measured by the analytical instruments of fixed-site monitoring stations. In particular, IR and UV spectrophotometers for carbon monoxide and ozone, chemiluminescence analyzer for NO_x and gas chromatograph for BTX.

3. Results

3.1. Laboratory electrical behavior
For all sensors, conductance measurements as a function of operating temperature, in presence of a gas concentrations suitable for environmental monitoring, were performed to determine the best working temperature for each sensor. The measurements to determine the dependence of the conductance on the

gas concentration verified for all reducing gases the power law: $G_{gas} = G_{air} + \alpha C^{\beta}$, where G_{gas} is the conductance in presence of the test gas, G_{air} the conductance in presence of air, α is a constant depending on the measurement units, C is the gas concentration and β is a positive number. The result agrees with the theory of power laws for semiconductor gas sensors developed by Yamazoe and Shimanoe [8].

Electrical stability and repeatability are crucial features for sensors' performance to operate in long term trials. Some features of screen-printing technology, such as the high firing temperature or the opportunity to fabricate in one batch a lot of sensors, are crucial for making them the most reliable among the various kinds of MOX sensors. The experiment of the figure 1 highlights the good repeatability of a TTV sensor, alternately subjected to the two concentrations of 100 and 800 ppb of benzene for 24 hours.

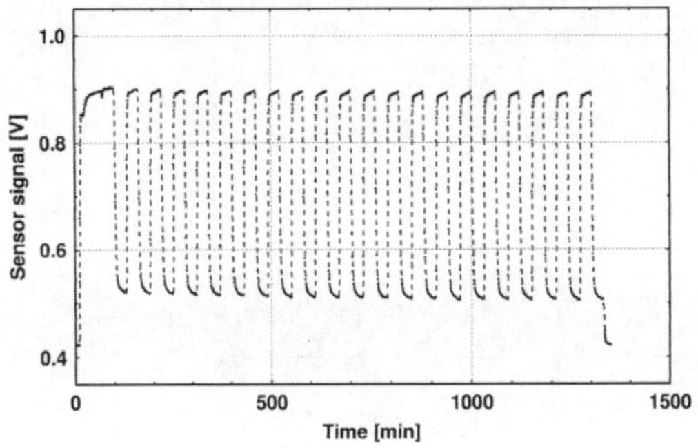

Figure 1. Alternate response of a TTV sensor to 100 and 800 ppb of benzene. The duration of a whole response is 1 h, the duty cycle is 50%.

It is well known that the water vapor partial pressure is the main interfering gas for MOX sensors. This is an important problem especially for carbon monoxide, when it is detected through a tin oxide sensor, as often the literature reports [9]. However, the problem can be approached in two ways: i) to develop functional materials with poor water vapor adsorbability, as in the case of TS sensors; ii) to compensate the signal due to overestimation caused by humidity as described in [10].

3.2. In field experiments

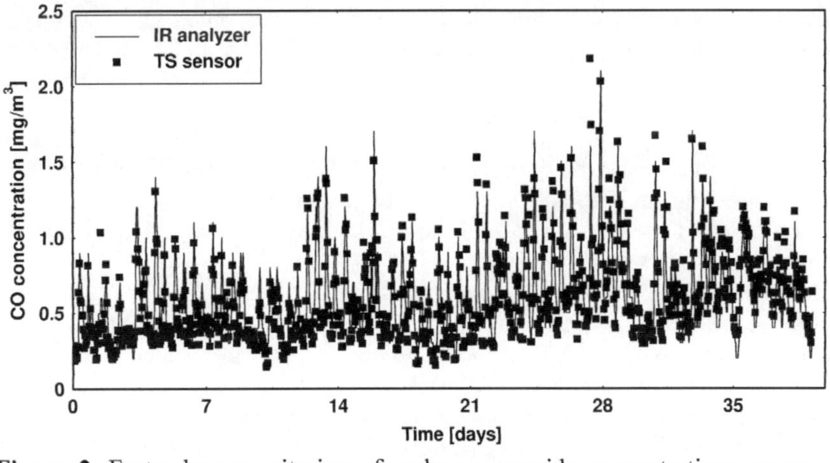

Figure 2. Forty days monitoring of carbon monoxide concentrations measured through a TS sensor and an IR analyzer.

In the figure 2, a comparison between the CO concentrations evaluated through a TS sensor and the measurements of an IR analyzer is reported. Due to the low influence of humidity on TS sensors, it wasn't necessary to compensate the sensor signal, although the low carbon monoxide concentration detected during the monitoring. An experiment devoted to test the reliability of sensors is reported in the figure 3. Two identical monitoring units were located beside a fixed-site monitoring station. Here, the behavior of the two W sensors toward ozone is reported during a monitoring two weeks long. It can be observed that the two sensors exhibited the same trend as well as the same behavior of the conventional analyzer.

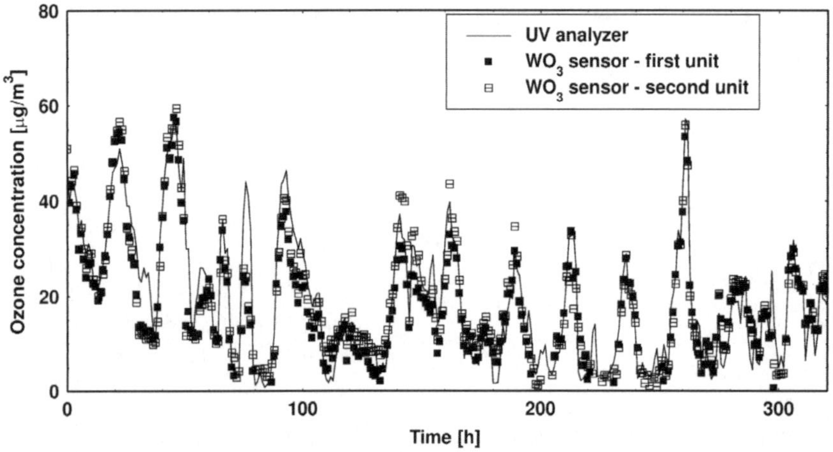

Figure 3. Comparison between the measurements of ozone carried out through two −nominally identical− W sensors and an UV analyzer.

4. Concluding remarks

Arrays of nanostructured thick film gas sensors based on different semiconducting oxides have been successfully used to detect atmospheric pollutants with the same trend observed by the conventional analytical techniques. The proposed instrumentation, due to its portability and low cost, can offer the possibility of enhancing the capacity to protect the population from air pollution and assess health risks.

5. References

[1] Regulations of road traffic emissions: Directives 2009/126/EC, 2012/33/EU, 2009/30/EC
[2] Regulation of industrial emissions: Directives 2010/75/EU, 2001/80/EC, 2004/42/EC
[3] Rothschild A and Komem Y 2004 *J. Appl. Phys.* **95** 6374-6380
[4] Carotta M C, Fioravanti A, Gherardi S, Ghiotti G, Malagù C, Morandi S and Sacerdoti M 2014 *Sensors and Actuators B*, **194** 195– 205
[5] Carotta M C *et al.* 2005 *Sensors and Actuators B* **108** 89-96
[6] Martinelli G, Carotta M C, Ghiotti G and Traversa E 1999 *MRS Bulletin* **24** (6) 30-36
[7] Carotta M C, Cervi A, Fioravanti A, Gherardi S, Giberti A, Vendemiati B, Vincenzi D and Sacerdoti M 2011 *Thin Solid Films* **520** 939-946
[8] Yamazoe N and Shimanoe K 2008 *Sensors and Actuators B* **128** 566–573
[9] Fine G F, Cavanagh L M, Afonja A and Binions R 2010 *Sensors* **10** 5469-5502
[10] Carotta M C, Benetti M, Ferrari E, Giberti A, Malagù C, Nagliati M, Vendemiati B and Martinelli G 2007 *Sensors and Actuators B* **126** 672–677

IOP Conf. Series: Materials Science and Engineering **108** (2016) 012006 doi:10.1088/1757-899X/108/1/012006

Sensing element for detection of polar organic vapours on the base of polyaniline-composite – Effect of substrate surface area

Robert OLEJNIK[1], Dipak Gorakh BABAR[1], Petr SLOBODIAN[1], Jiri MATYAS[1]

[1] Centre of Polymer Systems, University Institute, Tomas Bata University, Trida T. Bati 5678, 760 01 Zlin, Czech Republic

Corresponding author's e-mail address: slobodian@ft.utb.cz

Abstract. Conductive polymer polyaniline (PANI) was synthesized by oxidative polymerization of aniline hydrochloride as a source of aniline and ammonium persulfate as an oxidation agent. The polymerization process is relatively easy and cheap. The reaction was carried out in presence of polymer substrate, in our case polyethylene terephthalate (PET) as a representative of smooth surface substrate and polyvinylidenfluoride (PVDF) nanofibers membrane as a representative of porous substrate. Both these substrates were covered by polyaniline (PANI) and used as a sensing element for organic vapors detection. The detection was made by measuring and the record of the change of resistivity during adsorption and desorption of saturated vapors. The result shows that sensitivity decreases with increasing polarity of chosen solvent in order N,N-Dimethylformamide (DMF), N,N-Dimethylacetamide (DMAc) and Dimethyl sulfoxide (DMSO). The PANI base sensing element on PVDF substrate improves sensitivity, selectivity and it also has good reversibility and repeatability.

1. Introduction

Conducting polymers were first discovered in 1976 by MacDiarmid, Shirakawa and Heeger [1]. Conducting polymers can be sorted into many categories according to their functional groups e.g. poly-acetylene, poly-pyrrole, poly-aniline. Polyaniline (PANI) is a member of conductive polymer family which is widely used. Polyaniline can be synthetized in three different types of oxidation states: leucoemeraldine, pernigraniline and emeraldine. Leucoemeraldine and pernigraniline show the insulating properties and emeraldine base, too. Emeraldine salt is a conductive polyaniline polymer. The basic site (imine and amine group) in the polymer backbone of the emeraldine base can be protonated with strong acid to produce emeraldine salt leading to the increase of conductivity due to the effect of charge transfer in the polymer backbone. Polyaniline in Emeraldine salt form has a unique behavior, including a wide range of electrical [2], electrochemical and optical properties [3], as well as good stability [4].

The changes of physicochemical properties of PANI occur in response to various external stimuli are used in various application e.g. Organic electrodes [5], actuators and sensors [6, 7]. Different membranes has been used to coat PANI and used sensors. We are using PET foil and PVDF membrane as substrate for PANI. Nickel-deposited PET membrane entwined with microbial pili-like poly(vinyl alcohol-co-ethylene) (PVA-co-PE) nanofibers are used for sensing alcohol [8]. Functionalization of track-etched PET membrane, selectively used for hydrogen purification [9]. PVDF membrane also used for sensing purpose. Gold coated PVDF sandwich array are used for detection of ultratrace 2,4-dinitrotoluene (DNT) [10]. Assembly of multi-walled carbon nanotube/gold nanoparticle (CNT/AuNP) nanocomposites on PVDF membrane was used for Surface-enhanced Raman spectroscopy (SERS) [11]. Qiangxian Huang and co-workers used PVDF for Dynamic Micro Force Sensing Probe [12]. Functionalized track-etched PVDF membrane by poly-4-vinyl pyridine (P4VP) used for monitoring trace mercury in water [13]. With poly(aminophenylboronic acid), PVDF membrane used for sensing glucose [14].

The aim of this contribution is synthesis polyaniline emeraldine salt using aniline hydrochloride and ammonium persulfate. Two different kind of substrates are used. One is smooth PET foil and another is porous PVDF nanofiber membrane. Using two different substrates (PET foil and PVDF nanofibers membrane) for in situ polymerization. From the experimental data it is clearly indicates the PVDF membrane is more sensitive as compared to PET foil.

2. Experimental

The aniline hydrochloride (1.29 g, 10 mmol) was dissolved in 50 ml of distillated water. Another 50ml of distillated water was added to ammonium persulfate (APS) (1.56 g, 6.7 mmol). The solutions mentioned above were prepared separately. The solutions were transferred to the 150 ml beaker and mixed together to reach the polymerization mixture. The substrate was inserted in the reaction mixture. Then the solution was stirred at 500 rpm. The reaction was carried out at room temperature. The reaction mixture was left for 24 hours without disturbing. After that the membrane was removed, washed with distilled water for three times and dried in air room temperature. FTIR spectra of the PANI coated and uncoated PET and PVDF membrane were collected on the NICOLET 6700 FTIR Spectrophotometer using attenuated total reflectance ATR-FTIR. Scanning electron microscope Nova NanoSEM 450 (FEI) with Schottky field emission electron source operated at acceleration voltage ranging from 200 V to 30 kV was used for surface morphology. Further, the membrane was cut in small pieces and response was measured by the two point's method technique using multimeter Sefram 7338.

3. Results and discussion

Conductive polymer polyaniline (PANI) was synthetized by oxidative polymerization of aniline hydrochloride with ammonium persulfate (APS) to reach emeraldine salt. The synthesis was made on two different substrates PET foil and PVDF membrane which is made by electrospinning technology. Figure 1 shows the IR spectroscopy of coated and uncoated PET and PVDF membrane. In IR spectra of PANI coated PET foil (Figure 1a), peak at 1585 cm^{-1} is due to the quinoid ring and at 1498 cm^{-1} due to the stretching of the benzoic rings. The peak at 1304 cm^{-1} is due to the C-N stretching. Peak at 810 cm^{-1} is due to the N-H out of plane bending vibrations [15-16]. From these peaks it is clearly indicates the presence PANI on PET foil. The same peaks are also observed with almost the same values (1591 cm^{-1}, 1498 cm^{-1}, 1304 cm^{-1}, 811 cm^{-1}) in the PANI coated PVDF membrane (Figure 1b).

Figure 2 shows the morphology of the pure and PANI coated PET and PVDF membrane. PET foil clearly shows the planer surface which is representer of non-porous substrate (Figure 2a). When coated with PANI, PET foil shows the presence of nano particles having average diameter 165.5nm. In contrast Pure PVDF membrane shows the randomly oriented and porous nanowoven structure which is typical for electrospinning process. The average diameter of nano fibers is 138.24 nm ± 30.41. The average diameter of fibers present in the PANI coated PVDF membrane get increased, and goes upto 266.84 nm ± 67.77 which clearly indicates the increment by 128.60 nm after coating by PANI. It also shows the some spherical particles of PANI.

These two membranes were used as a sensing element for polar solvent, namely: N,N-Dimethylformamide (DMF), N,N-Dimethylacetamide (DMAc) and Dimethyl sulfoxide (DMSO). These three solvents were chosen with respect to their polarity. The polarity index increases in this order and sensitivity decreases (Table1). The measurement was made on the base of the recording of the change of the resistivity of sensing elements during adsorption and desorption cycles. Figure 3 showing four adsorption and desorption recycles for each solvent.

IOP Conf. Series: Materials Science and Engineering **108** (2016) 012006 doi:10.1088/1757-899X/108/1/012006

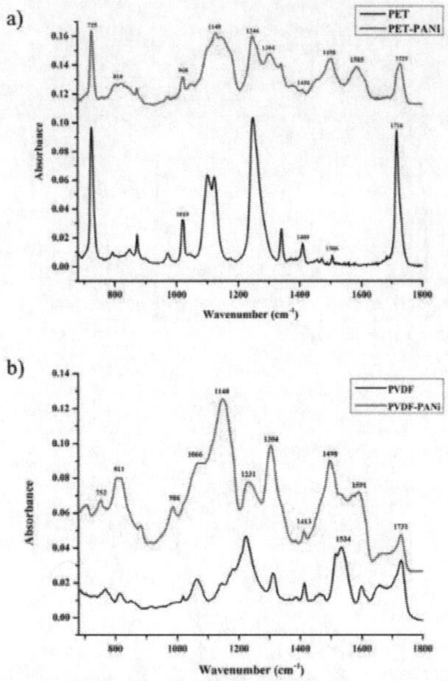

Figure 1. IR spectra of a) pure and PANI coated PET foil; b) pure and PANI coated PVDF membrane.

The percentage of sensing for PVDF membrane is more than that of PET foil in every analyte (Table 1). The nanofiber structure in PVDF membrane is more jugged and porous. The porous structure improves the sensitivity during adsorption cycles when the molecules of analyte are adsorbed. Desorption cycles indicate good reversibility of this process. Both adsorption and desorption are on physical base, so no chemisorption occurs (Figure 3). As compared to PET foil as a substrate there is no porosity and so adsorption and desorption process is lower compared to a PVDF membrane.

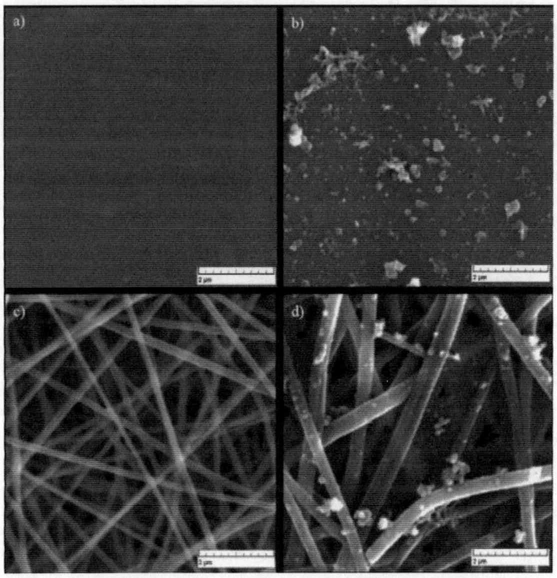

Figure 2. a) SEM analysis of polymer (PVDF) base substrate made by electrospinning method. b) SEM analysis of polyaniline spherical particles attached on nanofibers surface during in situ polymerization. c) SEM analysis of polyaniline spherical particles on a PET foil.

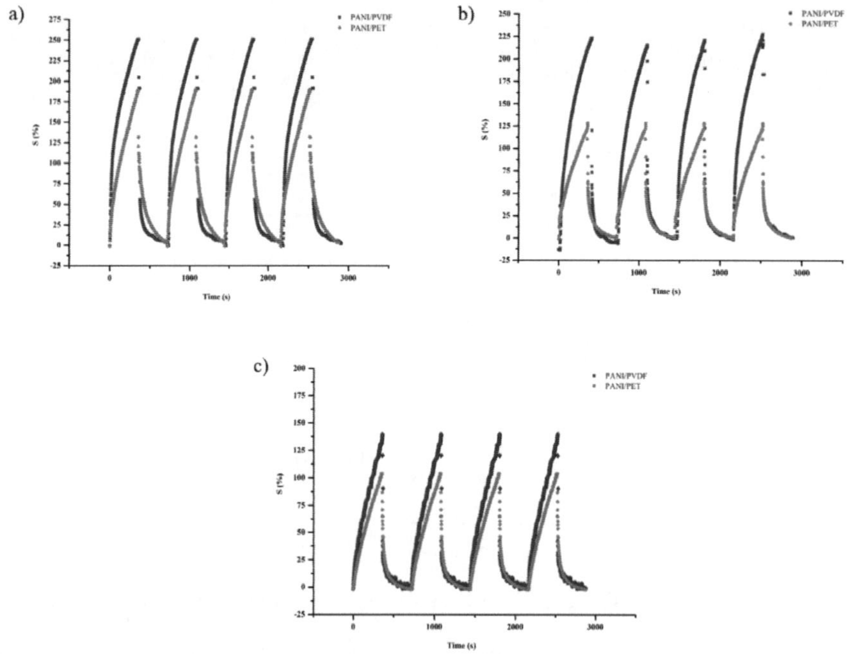

Figure 3. Change of sensitivity S in % during four adsorption/desorption cycles for tree different polar solvents a) N,N-Dimethylformamide (DMF),b) N,N-Dimethylacetamide (DMAc) and c) Dimethyl sulfoxide (DMSO).

Solvent	Vapore pressure at 20 °C [kPa]	Volume in air [%]	Polarity index	PANI on PET foil S [%]	PANI on PVDF membrane S [%]
DMF	0.36	0.36	6.4	185	250
DMAC	0.20	0.20	6.5	124	214
DMSO	0.05	0.05	7.2	104	140

Table.1 Physical properties of chosen solvents and sensitivity response for PANI coated PET and PVDF membrane.

4. Conclusion

Electrically well conductive polymer polyaniline was synthetized by oxidative polymerization. The polymerization process was carried out in the presence of two different substrates, planer PET foil and porous and nano fibrous PVDF membrane. Sensing property of these two membranes were measured against three different solvents and it shows that PANI coated PVDF membrane show good response as compared to the PANI coated PET foil. The surface area plays an important role in improving sensitivity and selectivity of the prepared sensor.

Acknowledgements

This project was supported by the Ministry of Education, Youth and Sports of the Czech Republic - Program NPU I (LO1504).

References

[1] Barros R.A., Azevedo W.M. and Aguiar F.M., Mater. Charact. 2003;50:131-134.

[2] Dong H., Prasad S., Nyame, V. and Jones W., Chem. Mater. 2004;16:371-373.

[3] Wei X.L., Wang, Y.Z., Long S.M., Bobeczko C. and Epsteib A.J., J. Am. Chem. Soc. 1996;118:2545 -2555.

[4] Roy B.C., Gupta, M.D., Bhowmik L. and Ray J.K., Synth. Met., 1999;100: 233-236.

[5] Oyama N., Tatsuma T. and Sotomura T., Nature 1995;373:598-600.

[6] Herod T.E. and Schlenoff J.B., Chemistry of Materials 1993;5:951-955.

[7] Virji S., Huang J., Richard B. and Bruce Weiller H., Nano Letters 2004:4:491-496.

[8] Liu Q., Zhou Z., Xia M., Tao Y., Liua K. and Wang D., RSC Adv. 2014:4:40788-40793.

[9] Awasthi K., Choudhury S., Komber H., Simon F., Formanek P., Sharma A., Stamm M., International journal of hydrogen energy 2014:39:9356-9365.

[10] Wang J., Jin W., Zhang X., Hu C., Luo Q., Lin Y. and Hu S., Anal. Chem. 2014;86:8383−8390.

[11] Zhang K., Ji J., Fang X., Yan L. and Liu B., Analyst 2015;140:134-139.

[12] Huang Q., Ni K., Shi N., Hou M., Wang X., Sensors & Transducers Journal 2010;114:122-131.

[13] Bessbousse H., Zran N., Fauléau J., Godin B., Lemée V., Wade T., doi:10.1016/j.radphyschem.2015.03.011

[14] Manesha K.M., Santhosha P., Gopalana A., Leea K., Analytical Biochemistry 2007;360:189–195.

[15] Menegazzo N., Kahn M., Berghauser R., Waldhauser W. and Mizaikoff B., Analyst 2011;136:1831-1839.

[16] Chauhan N.P.S., Ameta R., Ameta R., Ameta S.C., Indian Journal of Chemical Technology 2011;18:118-122.

Metal composite as backing for ultrasonic transducers dedicated to non-destructive measurements in hostile

R Boubenia[1,2] **,E Rosenkrantz**[1,2] **, F Despetis**[3,4] **, P Combette**[1,2] **and J-Y Ferrandis**[1,2]
[1] Univ. Montpellier, IES, UMR 5214, Montpellier, France
[2] CNRS, IES, UMR 5214, Montpellier, France
[3] Univ. Montpellier, L2C, UMR 5221, Montpellier, France
[4] CNRS, L2C, UMR 5221, Montpellier, France

E-mail: redha.boubenia@ies.univ-montp2.fr

Abstract. Our team is specialized in ultrasonic measurements in hostile environment especially under high temperatures. There is a need for acoustic transducers capable of continuous measurement at temperatures up to 700°C. To improve the performances of acoustic sensors we focus our works on the realisation and characterisation of transducer backings able to operate under very high temperature. Commercially, they are produced by the incorporation of tungsten powder in a plastic matrix, which limits the working temperature. The realisation of ultrasonic transducers for non-destructive measures at high temperatures requires adequate materials, manufacturing and assembly processes. To produce the backings, composites were made using very ductile metals such as tin and tungsten. These composites are manufactured by uniaxial hot pressing. First, we studied the influence of temperature and pressure on the densification of tin pellets. Then, several specimens made of tin/W were made and characterised by measuring the specific weight, speed and attenuation of sound. The acoustic measures were realised by ultrasonic spectroscopy. This test-bench was designed and tested on control samples of PMMA and on standard backings (epoxy / tungsten).

1. Introduction

Currently it does not exist acoustic transducers capable of operating continually at temperatures up to 700°C. However, the need is real in many industrial sectors (Nuclear, Oil, ...). The Institute of Electronics and Systems (IES) of Montpellier has been working for many years, especially with the Commissariat of Atomic Energy (CEA) for the development of acoustic instrumentation in hostile environment (High temperature, neutron flux).
The realisation of ultrasonic transducers devoted to the non-destructive measurement in hostile environments requires finding adequate materials, manufacturing and assembly processes. The ultrasonic transducers are composed of 3 elements: an active element such as a piezoelectric element, a front layer to inject a maximum of acoustic power into the investigated material and finally, a backing. The backing permits to control the bandwidth, the temporal resolution and the sensitivity of the transducer. To increase the bandwidth, the acoustic impedance Z (which is the product of the speed of sound to the density) of the backing must be equal to those of the active element and the backing must be strongly attenuating.
The backings are particulate composite materials made of a matrix with a low acoustic impedance in which spherical scatters with high acoustic impedance are dispersed. The quantity and diameter of scatters control the impedance, attenuation and velocity of the backing. Standard backing mostly used in industry are made of polymer and inclusions of tungsten, because it is simple to fabricate. These mixtures allow acoustic impedances ranging from 0.1 MRayls to 100 MRayls depending on the volume fraction of tungsten introduced [1-2].
However, whichever the polymer used, none of them is able to work at temperatures up to 700°C. In order to increase the work temperature, we propose to use metallic composites. To obtain a suitable

IOP Conf. Series: Materials Science and Engineering **108** (2016) 012007 doi:10.1088/1757-899X/108/1/012007

protocol of manufacture, we start our study on tin-tungsten composite because tin is a very ductile metal allowing working at moderate temperatures and pressures. The melting temperature is the only limiting factor for the sensor. In addition, the protocol implemented using tin may be applied to other ductile metals having higher melting temperatures (zinc, copper ...). In this study all specimens were made by temperature uniaxial pressing.
The study conducted by Sayers [2] on such metallic composites is only theoretical. In the reference [3], the author does not give the attenuation of the composites made nor results on the densification. Thus, the mechanical strength of composite are proposed. It was then decided to start our work by studying the densification of tin pellets as a function of temperature and pressure. Then we give the explanation of the manufacturing process and the characterisation method of Tin-Tungsten backing. Three backings were made and characterised. In this paper we present for each specimen the speed of sound, the attenuation and impedance as a function of the volume fraction of inclusion.

2. Realisation of tin pellets

We manufactured tin pellets with a white tin powder of 99.5% purity, mean diameter of about 149 microns and a density of 7,29g / cm^3.To determine the pressure and temperature leading to samples with good mechanical strength and relative density, it was decided to vary some parameters to have a density close to pure tin. The parameters are the pressing load, the pressing speed and the pressing temperature to observe their effects on the final properties (relative density and mechanical strength) of the pellets.
The pellets and the backing were made by simple uniaxial hot pressing processing, by a single moving piston [Fig 1]. We choose this kind of process for its simplicity, to reduce the costs of manufacturing and to increase density.
The samples were loaded in compression using 1196 mechanical testing machine with 250kN maximum load capacity. It is controlled by a software allowing to choose the maximal pressure and the pressing speed.
Heating is provided by a heating element wrapped around the matrix. To minimize heat losses, an insulating glass wool is wrapped around the heating element. The temperature is controlled by a PID. A thermocouple, inserted into a hole filled with thermal grease drilled into the matrix measures the reference temperature [Fig 1].
The desired amount of powder is weighed and then introduced into the matrix. To determine the amount of powder m_{Sn} to be weighed, we arbitrarily fix a final height h_{Sn} which is used to calculate the final volume of the pellet knowing the diameter D_{Sn} of the matrix.

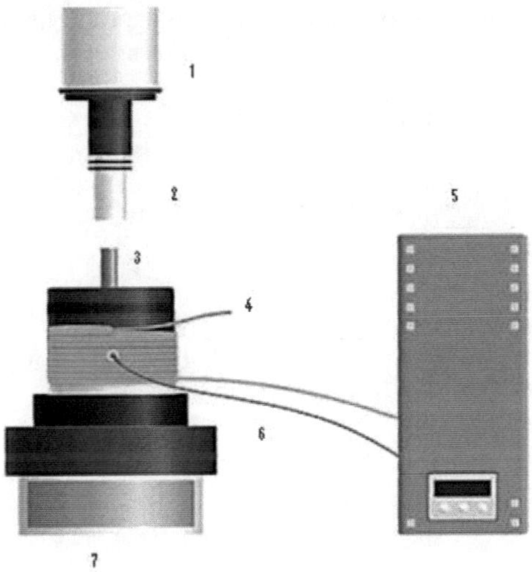

Figure 1: Uniaxial press (1 uniaxial press, 2 press piston, 3 matrix piston, 4 heating element, 5 PID regulator, 6 thermocouple, 7 support platform)

Two various tests are carried out:

- Two cold tests at various pressing speeds (0.1 mm/min, 1mm/min),
- Three tests at constant speed and different temperatures (25°C, 50°C, 100°C).

2.1. Influence of the pressing speed

When pressing, there are frictions on the rim of the pellet which depend on the pressing speed. The pressing speed is then an important parameter to study. Two samples of tin powder were pressed under the same temperature and maximum pressure (1600 Bars) but with different pressing speeds (0,1 and 1 mm/minute).

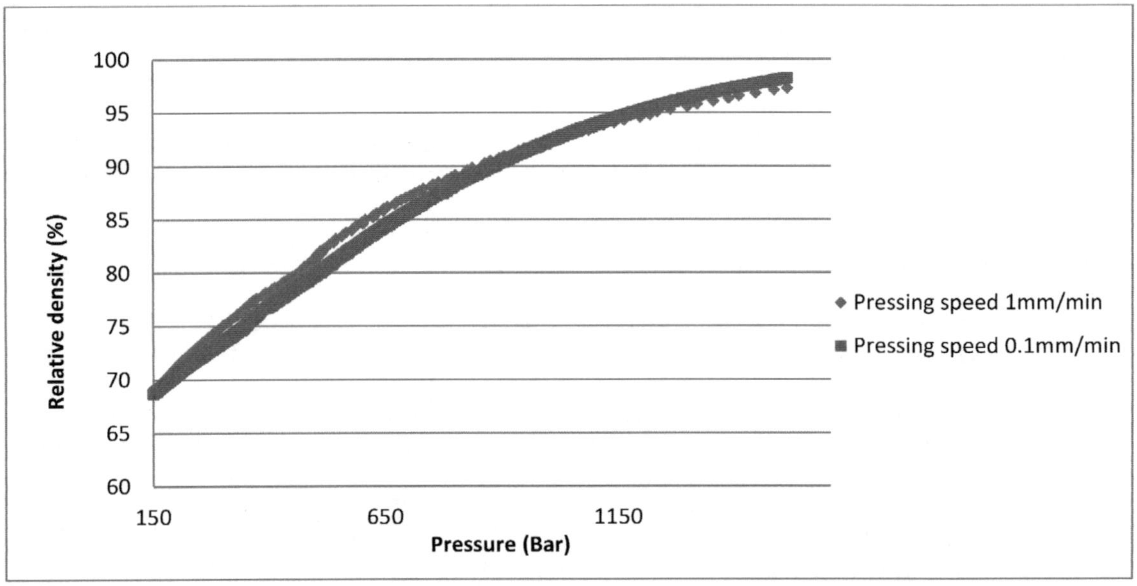

Figure 2: Influence of the pressing speed of the relative density at ambient temperature

It is observed in figure 2 that for both pressing speeds, the changes in relative density as a function of the pressing force is the same. It seems that the pressing speed has not influenced the densification of the tin samples. However, the sample pressed at 1mm/min had a bad mechanical wear, the sample breaks during demolding because of the important radial frictions and therefore cannot be characterise. It was then decided to press at 0.1mm/min.

We can now proceed to the second study which is the influence of temperature on the densification.

2.2. Influence of Temperature

A few works show the dependence of the densification used to make metallic samples [4-5]. However the papers we cite use an isostatic press. We need to check the influence of the temperature in our conditions (uniaxial press). In this part, we will highlight the importance of the temperature on the realisation of dense pellet. Three tin pellets were realised at 25°C, 50°C and 100°C.

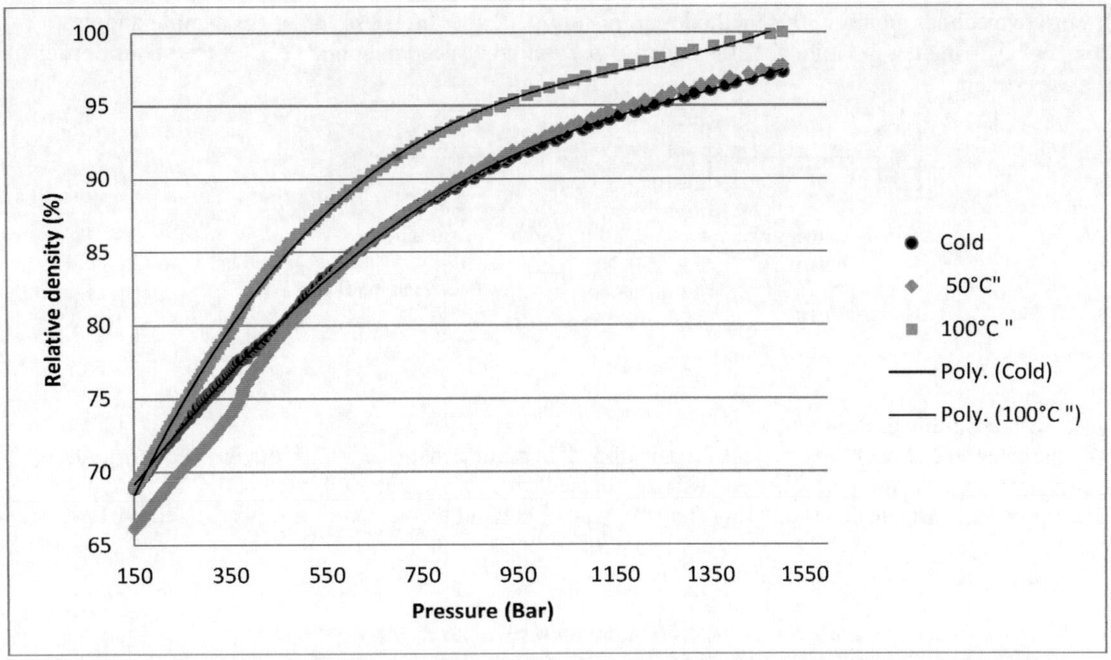

Figure 3: Evolution of relative density versus the pressure for a pressing speed equal to 0,1 mm/minute

The relative density as a function of the pressure is shown in figure 3. We observe a significant difference between an ambient pressing temperature and a 100°C pressing temperature. The relative density grows faster and its value is higher at 100°C and reaches 99% of relative density. We do not find significant difference between the ambient temperature curve and the 50°C curve. However, it would be preferable to press at ambient temperature. At this temperature there is no delaminating. Indeed, the fact that tin is more ductile at high temperature increases the radial friction and consequently the differences of compaction increase the phenomena of delaminating.
Finally we retained the following protocol. The sample is:

1. shaped with a 1600 bars maximum pressure at room temperature,
2. reversed and pressed again (double action),
3. annealed at 100 ° C.

This tin study allowed choosing the pressure, temperature and the pressing speed for the realisation of tin/ tungsten backing.

3. Realisation and characterisation of transducer's backings

3.1. Realisation of tin/tungsten backings

The backings with tungsten particles (size between 140 to 200 micron diameters) are realised according to the following protocol:

1. Weigh the quantity of tin and tungsten desired,
2. Mixture of tin and tungsten,
3. Introduction of the mixture into the matrix, then pressed dynamically until 1600 bars, with holding pressure,
4. Annealing at 100°C.

3.2. Estimation of the phase velocity and attenuation

The method used to characterise the backings is the ultrasonic spectroscopy using the insertion/substitution method [6]. It consists of two measurements [Fig 4], a first reference measurement is performed through water, then a second measurement through the material being studied.

If only two echoes are used, this method requires knowing the thickness "d" of the sample and its density " ρ_S", the temperature "T", the density "ρ_w" and the speed of sound "c_w" of the water during the experiment.

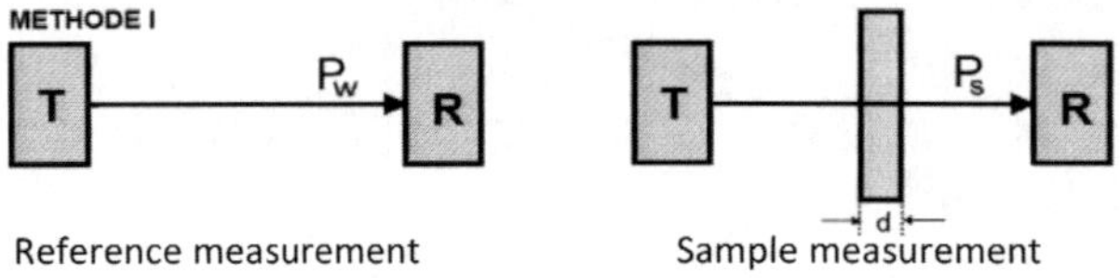

Figure 4: Principle of the ultrasonic spectroscopy immersion method

4. Results and discussion

To characterise our backings, we used a pair of 5MHz plane sensor central frequency, covering the frequency range between 4MHz and 6MHz. We used a pulse generator Olympus 5052 PR and a Tektronix oscilloscope. The sampling frequency is 1GHz and the acquisition is performed on 1000 points.
The characteristics of the three backings are:

Tableau 1: theoretical and experimental densities for three backings with different volume fraction of tungsten

	Theoretical densities(g/cm^3)	Experimental densities (g/cm^3)
Backing1 (5%W)	7.91	7.76
Backing2 (15%W)	9.11	8.98
Backing3 (30%W)	10.91	10.77

For each sample, there is a reference measurement and a measure through the sample. The phase velocity and the attenuation are respectively deduced from the phase and the amplitude.

The figure 5, 6 below compare the experimental and theoretical speed of sound and attenuation depending on the frequency for the three samples (5% W, 15%W and 30%W).

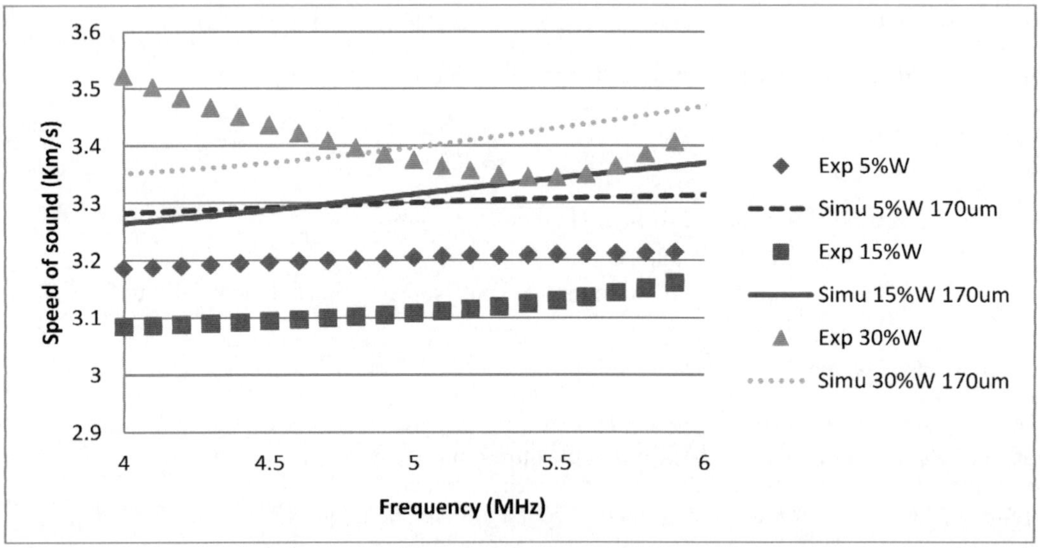

Figure 5: Speed of sound of 3 different fractions (5%, 15%, and 30%) of tungsten in backing as a function of ultrasound frequency at 26°C, for a transducer with a 5MHz center frequency

It is found that the higher the volume fraction of tungsten, the higher the phase velocity. However, the phase velocity for the backing having a volume fraction of 30% tungsten is above the speed of the other two with a lower volume fraction, it is in accord with the theory [1-7]. The average relative differences between the theoretical and the experimental measures for each backing are:

Tableau 2: Average relative difference for the speed of sound

	Average relative difference for the speed of sound[%]
Backing1 (5%W)	2.94
Backing2 (15%W)	4.20
Backing3 (30%W)	2.22

Although the theoretical curve does not seem correctly fit the experimental data the relative difference is quite low. The relative difference is more important for the 15% backing it is probably due to a quantity of porosity more important than the two other.

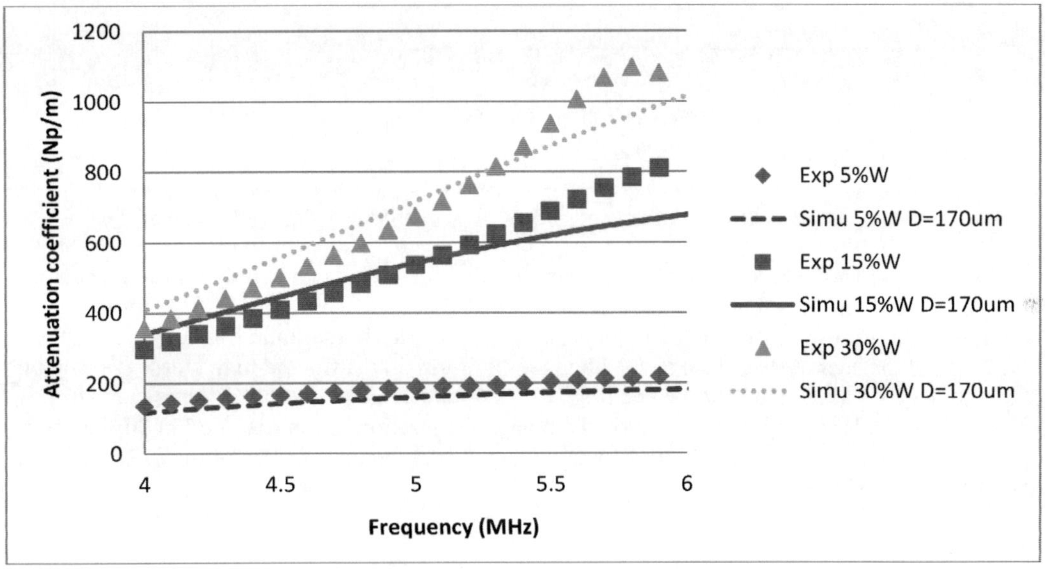

Figure 5: Attenuation of sound in 3 different fraction of tungsten in backing (5%, 15%, and 30%) as function of ultrasound frequency at 26°C, for a transducer with 5MHz center frequency

Concerning the attenuation coefficient shown in figure 5, it is found that the higher the volume fraction of tungsten, the higher the attenuation. The average relative differences between the theoretical and the experimental measures are:

Tableau 3: Average relative difference for the attenuation

	Average relative difference for the attenuation (%)
Backing1 (5%W)	17.63
Backing2 (15%W)	13.57
Backing3 (30%W)	8.57

The differences between the simulation and the experiment are more important for the attenuation than for speed of sound. These differences can be due to several parameters:

• The presence of porosity which reduce the value of speed of sound in the matrix
• The experimental diameter of the diffuser is between 140 and 200 μm, whereas the simulation was made with a single diameter 170 μm (mean of 140-200 μm). Indeed, during the simulation we observed that the values of the attenuation coefficients change with the theoretical values of the particles diameter [Fig 6].

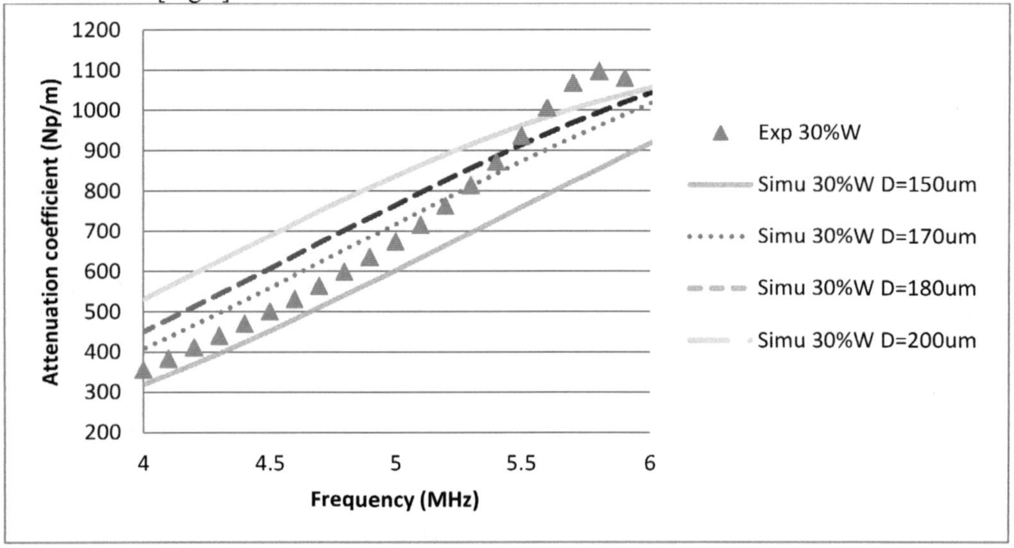

Figure 6: Attenuation of sound in 4 different backings with different diameters of tungsten particle (150 μm, 170 μm, 180 μm and 200 μm) as function of ultrasound frequency at 26°C, for a transducer with a 5MHz center frequency

• Simulation does not take into account the attenuation in the tin matrix which reduces the attenuation coefficient of the backing. It is found an important average relative difference of 17.63% for the attenuation in the backing1 (5%W). Indeed, the backing1 has the lowest volume fraction of tungsten. The attenuation of the backing1 is close to the attenuation of the tin matrix and in this case we cannot neglect the tin matrix attenuation. For the backing3 (30%W) the average relative difference for the attenuation is 8.57%. This result is justified by the high concentration of tungsten and in this case we can neglect the attenuation in the tin matrix because attenuation is a consequence of the scattering on a spherical obstacle (tungsten).

5. Conclusion

We studied the influence of the pressing speed and of the temperature on the manufacture of pure tin pellets which allowed us to deduce a protocol. It was shown that although the pressing speed does not seem to change the densification, it is important to slowly press to have good mechanical strength. The metal is more ductile at 100°C and significant densification rates are achieved quicker. Yet, it is better to press at room temperature and to anneal after pressing at 100 ° C. This significantly reduces the delamination.
The realisation of backings by uniaxial pressing with powder allows increasing the particle size and the volume fraction of tungsten because there is no sedimentation. We realised several samples and characterised them by ultrasonic spectroscopy to determine their velocity, attenuation. We concluded on the effectiveness of the method of manufacture. Even if they are differences between simulation and experiment results, we have acceptable relative differences.

6. Bibliography

[1] M.G. Grewe and T. R. Gururaja 1989 *IEEE Ultrasonics symposium* **Vol 173** Acoustic properties of particle/polymer composites for transducer backing applications
[2] C.M .Sayers and C.E .TAIT 1984 *Ultrasonics*.**Vol 22** .p57-60 Ultrasonic properties of transducer backings

[3] S. Rokhlin, S.Golan, and Y.Gefen 1981 *J. Acous. Soc. Am* **Vol 69** No 5 Acoustic properties of tungsten-tin composites
[4] F.B Swinkels, D.S. Wilkinson. E.Arzt and M.F.Ashby 17 April 1983, *Acta metal.* **Vol 31**, No 11 pp 1829-1840 Mechanisms of hot-isostatic pressing
[5] M.M. Caroll November 86-77 *Metallurgical transaction A,* **Vol 17 A** An Empirical model for hot isostatic pressing of metal powders
[6] Ping He 3, may 1999 IEEE transactions on ultrasonics, ferroelectrics, and frequency control, **Vol. 46**, no3. Experimental Verification of Models for Determining Dispersion from Attenuation
[7] D. G. Aggelis, S. V. Tsinopoulos,and D. Polyzos 2004 *J. Acous. Soc. Am.* **Vol 116** P3443-3452 An iterative effective medium approximation (IEMA) for wave dispersion and attenuation predictions in particulate composites, suspensions and emulsions

7. Annex

Note $p_w(t)$ and $p_s(t)$the respective signals received by the transducer received in the absence of material (shot in water) and in the presence of material (shot in the material) [Figure 4].

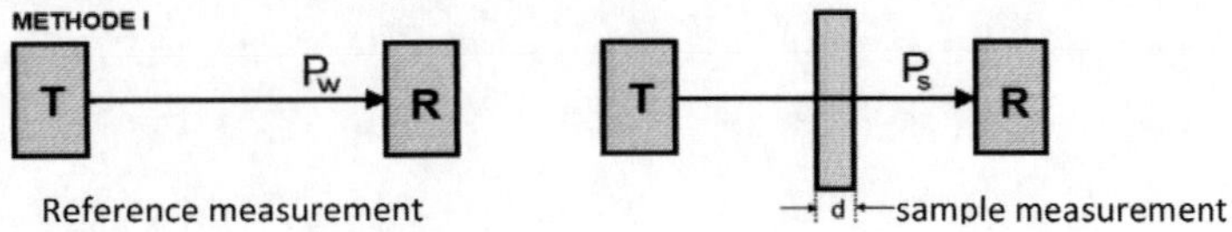

Figure 1: Principle of ultrasonic spectroscopic immersion method

It is found between these two signals a decrease in amplitude and a phase shift [Figure 5]. The phase shift is due to the fact that the material at a different speed of the water $\square = \frac{\omega}{V_s(f)} \times d$

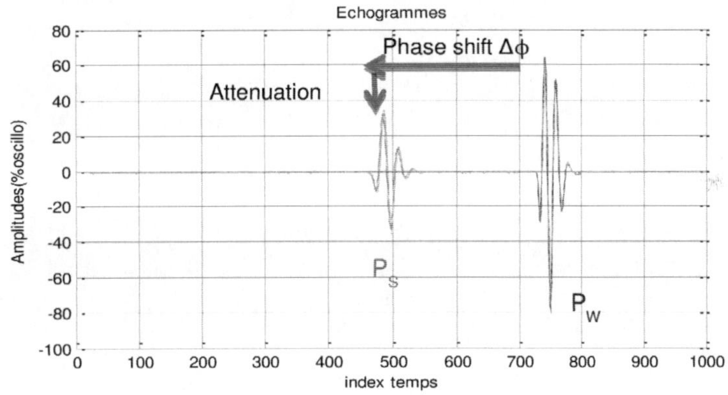

Figure 2: Reference echo in blue and echo passed through the sample in red. The velocity of the wave in the material is greater than the speed of the wave in the water

The decrease in amplitude is related to two phenomena: the attenuation characterized α and losses inserts characterized by the energy transmission coefficient T which is:

$$T = \frac{4(Z_w Z_s)}{(Z_w + Z_s)^2} \quad (1)$$

Where $Z_w = \rho_w\, V_w\ et\ Z_s = \rho_s\, V_s$

Experimentally we deduce the velocity and attenuation from the acoustic transfer function.

$$H\ \omega\ = \frac{P_s(\omega)}{P_w(\omega)} = T\ exp\ -\ \alpha\ \omega\ - \alpha_w\ d\ \ exp\ -j\omega d\ \frac{1}{V_s} - \frac{1}{V_w} \quad (2)$$

Where $P_w\ \omega\ \ et\ P_s\ \omega$ are Fourier transform respectively of $p_w\ t\ $ et $p_s(t)$

From the phase $\Delta\emptyset$ of H (ω) is deduced the speed and amplitude | H (ω) | the attenuation coefficient [7]. The velocity is given by the following relationship:

$$V_s\ \omega\ = \frac{V_w}{1 - \frac{\Delta\emptyset V_w}{\omega d}} \quad en\ \ m/s \qquad (3)$$

Where $\Delta\emptyset = \emptyset_s - \emptyset_w$. Thus it is freed from the phase shifts introduced by the cables and the impulse responses of the transducers.
From the amplitude spectrum we have:

$$\alpha_s\ \omega\ - \alpha_w = \frac{1}{d}\mathrm{Ln}\frac{P_s}{P_w} = \frac{1}{d} *\ \ln\ T\ - \ln\ H\ \omega \qquad (4)$$

The attenuation of the water α_w being negligible compared with that of the material α_s.

Assessment of Material Properties of Gallium Orthophosphate Piezoelectric Elements for Development of Phased Array Probes for Continuous Operation at 580°C

Mario Kostan [1, a], Abbas Mohimi [1, 2, b], Channa Nageswaran [2], Vassilios Kappatos [1], Liang Cheng [1], Tat-Hean Gan [1, 2], Luiz Wrobel [1], Cem Selcuk [1]

[1] Brunel University London, Kingston Lane, Uxbridge, Middlesex, UB8 3PH, UK
[2] TWI Ltd, Granta Park, Great Abington, Cambridge, CB21 6AL UK

[a] mario.kostan@brunel.ac.uk, [b] abbas.mohimi@brunel.ac.uk, bic@brunel.ac.uk

Abstract: In this paper, the thickness extension mode gallium orthophosphate single crystal elements were characterised using the impedance analyser. Impedance characteristics of piezoelectric elements were investigated at temperatures from 25°C up to 580°C at first and then at a constant temperature of 580°C for a period of 25 days. The resonant and anti-resonant frequencies extracted from the impedance characteristics, capacitance (measured at 1 kHz), density and dimensions of the gallium orthophosphate elements were used to calculate electromechanical, piezoelectric and elastic properties of these elements at high temperatures as a function of time. The tested gallium orthophosphate elements proved to possess very stable efficiency and sensing capability when subjected to high temperature. The results are very encouraging for proceeding with development of phased array probes using gallium orthophosphate, for inspection and condition monitoring of high temperature pipelines in power plants at a temperature up to 580°C.

1. Introduction

Today, a number of power plants (PP) are operating beyond their designed life. Rigorous standards, regulations and codes are in place by regulatory bodies to ensure structural health inspection is carried out periodically during planned outages. Since the inspections are performed during outages, the inspection techniques and technologies used are effective at room temperature. However, if a vital part of the PP needs to be monitored in service following an outage, such as a pipeline carrying superheated steam, it may present a problem, particularly for an aging PP. As an example, high temperatures (HT) and pressures experienced in these pipelines, can lead to creep, fatigue and corrosion type defects, which if undetected may have catastrophic consequences. For this reason, *in situ* condition monitoring techniques should be developed to retain reliability and extend the lifetime of aging PP.

Advanced ultrasonic techniques, such as Phased Array (PA) can be used to detect defects at room temperature. However, at HT operating conditions this technique cannot be applied due to lack of availability of HT probes for operation at temperatures up to 580°C in electrical PP [1]. The key challenge is to develop HT PA probes for continuous monitoring of defect growth over time, so that when the defect reaches a critical size the PP can be shut down and maintenance can take place before failure.

These PA probes use piezoelectric elements for generation and reception of ultrasound. Lead Zirconate Titanate (PZT) is the most commonly used piezoelectric for PA probes, but has a maximum operating temperature of around 180°C (half of the Curie temperature for PZT-5A) [2] which is not suitable for this application. Therefore, alternative piezoelectric materials need to be considered. A number of piezoelectric materials that possess high operating temperature have been reported [3], [4]. Piezoelectric nonferroelectric single crystals such as quartz (α-SiO_2), gallium orthophosphate ($GaPO_4$), langasite ($La_3Ga_5SiO_{14}$, LGS) and aluminium nitride (AlN) stand out because they exhibit no Curie temperature and no domain-related aging behaviour while showing good sensitivity and the ability to function over a broad temperature range up to very HT e.g. 800°C for the langasite crystals [4].

Due to excellent thermal stability of most of its material properties up to 970°C (>>580°C) as well as commercial availability of high quality high precision piezoelectric elements, single crystal $GaPO_4$ has been selected for further study [5].

Ideally the PA probes should function continuously in between planned outages. However, material properties of $GaPO_4$ for continuous operation at 580°C were not available in literature. Several methods, such as impedance method, the quasi-static method and the laser interferometry method are used to measure the material properties of piezoelectrics and they all provide high accuracy results [6], [7]. However, due to a fact that an impedance analyser was available for impedance measurements to be carried out on the $GaPO_4$ elements, it was decided to proceed with the impedance method. This method was used to determine the material properties of $GaPO_4$ single crystal elements at HT such as thickness coupling factor, charge coefficient and compliance and stiffness coefficients; these properties were derived from measured values of the resonant (f_r) and anti-resonant (f_a) frequency pair, capacitance (measured at 1 kHz), density and the dimensions of appropriately cut and excited $GaPO_4$ elements as outlined in the European Standard on Piezoelectricity [8]. From the derived material properties it is possible to draw conclusions about the efficiency, sensing capability and mechanical stability of the $GaPO_4$ elements at HT and consequently about the HT operation of the PA probe that would use these elements for the generation and reception of ultrasound.

2. Development of PA Elements

The $GaPO_4$ elements had to be manufactured before the measurements were carried out. The PA probe will use angled compression waves for the inspection of HT pipes. The way in which the piezoelectric element vibrates when stimulated by an electrical pulse depends upon the "cut" in the case of piezoelectric single crystals such as $GaPO_4$. The X-cut (equiv. to thickness extension mode for piezoelectric ceramics) of single crystals can generate and detect compression waves [9]; for this reason raw X-cut $GaPO_4$ plates were purchased from the manufacturer "Piezocryst Advanced Sensorics GmbH", Austria, to be appropriately shaped and coated with electrode material, as shown in Figures 1 a) and b). To build an efficient linear array of piezoelectric elements for the PA probe, the element dimensions must comply with the following: one dimension along the element and perpendicular to the thickness should be narrow for better directivity of the produced ultrasonic beam. The other dimension perpendicular to the above two directions should be large in order to increase the total element radiating area for meeting the power output requirement [10]. $GaPO_4$ elements with dimensions of 3 mm x 12 mm x 1 mm (width, length and thickness) were manufactured to meet those requirements, but at the same time to provide piezoelectric elements robust enough to manipulate with when performing impedance measurements at HT. The dimensions of the $GaPO_4$ elements that will be used in the HT PA probes will be different from the above dimensions, taking into account specification of the defects that want to be measured with the PA probes. Finally, the $GaPO_4$ elements were coated in the standard parallel electrode configuration, with a platinum layer of 100 nm considering the platinum thin film electrodes are used for HT up to 650°C due to their excellent electrical properties, high melting point and outstanding oxidation resistance [11].

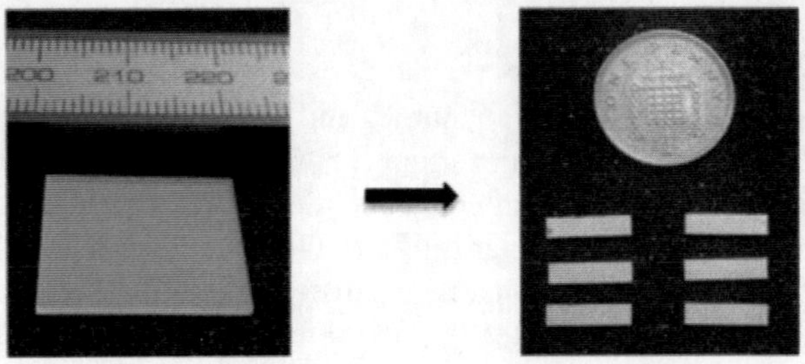

a) b)

Figure 1: a) A raw $GaPO_4$ single crystal plate. **b)** The raw $GaPO_4$ plate was shaped into thin, slender rectangular bars to build an efficient linear array of piezoelectric elements for the HT PA probe.

3. Material Properties

The material properties of piezoelectrics are anisotropic and exhibit various values in different directions, depending on the cut in terms of piezoelectric single crystals. According to the step-by-step procedure for calculating a complete set of material properties for piezoelectric materials found in the European Standard on Piezoelectricity [8], for the X-cut single crystal elements vibrating in the thickness extension mode there are four relevant material properties that has to be derived: (i) thickness coupling factor k_t (electromechanical), (ii) charge coefficient d_{11} (piezoelectric) and (iii) compliance and stiffness coefficients, s^E_{11} and c^E_{11}, respectively (elastic). All of these four properties represent numerical measures of the efficiency, sensing capability and mechanical stability of the piezoelectric elements, as explained below.

3.1. Electromechanical Coupling Factor

The electromechanical coupling factor k_{ij} is a key parameter when describing the behaviour of piezoelectric materials, given that it expresses a numerical measure of efficiency for electro-mechanical conversion of the piezoelectric materials [12]. The first subscript to k denotes the direction along which the electrodes are applied; the second denotes the direction along which the mechanical energy is applied, or developed. The dimensions of a piezoelectric element dictate unique expressions of this factor. For piezoelectric elements such as the X-cut $GaPO_4$ elements shaped into thin, slender rectangular bars, whose surface dimensions are large relative to the thickness, the thickness coupling factor – k_t is the matching factor and it expresses the coupling between an electric field in direction 1 and mechanical vibrations in the same direction. Electromechanical coupling factors are found to range from 8% for quartz crystals to >70% for PZT ceramics [4].

3.2. Piezoelectric Charge Coefficient

The piezoelectric charge coefficient d_{ij} gives the ratio of the charge generated per unit of mechanical stress applied to a piezoelectric material, and conversely is the ratio of the mechanical strain developed by a piezoelectric material per unit of electric field applied. The first subscript to d indicates the direction of polarization generated in the material; the second subscript is the direction of the applied stress or the induced strain, respectively. The sensitivity needs to be sufficiently high so that the generated signal can be detected above the background noise. In practice, the generated signal is small and has to be enhanced by an appropriate charge or voltage amplifier [13]. The matching coefficient for the X-cut $GaPO_4$ elements vibrating in the thickness extension mode is the coefficient d_{11} where the subscript *11* denotes that both, the excitation of the element and the response of the element are in the same direction. Generally, typical values of the charge coefficient for piezoelectric materials range from about 2 pC/N for quartz crystals to about >700 pC/N for PZT ceramics [4].

IOP Conf. Series: Materials Science and Engineering **108** (2016) 012008 doi:10.1088/1757-899X/108/1/012008

3.3. Compliance and Stiffness Coefficients

The elastic compliance and stiffness coefficients, s^k_{ij} and c^k_{ij}, respectively relate the applied stress with the relative deformation of the material – the strain. These two coefficients are numerical measures for the mechanical stability of piezoelectric materials and ideally should not change with temperature. The matching compliance and stiffness coefficients of the X-cut $GaPO_4$ elements vibrating in the thickness extension mode are s^E_{11} and c^E_{11} coefficients, respectively, where the superscript *E* denotes a constant electric field, and the subscript *11* indicates that the direction of strain and the direction of stress are the same. For the *11* direction of piezoelectric materials, the *c* coefficient is the Young's modulus of elasticity.

3.4. Calculation of Material Properties from Measured f_r and f_a Values

Applying suitable modification, the mathematical relations from formulas presented within the European Standard on Piezoelectricity [8] were derived, considering the direction of positive polarization for piezoelectric ceramics vibrating in the thickness extension mode usually is made to coincide with the direction 3 (Z) of a rectangular system of 1 (X), 2 (Y) and 3 (Z) axes and for the piezoelectric single crystals vibrating in the same mode the direction of positive polarization will be the direction 1 (direction X / X-cut of a crystal).

The thickness coupling factor k_t can be derived based on the measured resonance and anti-resonance frequencies of piezoelectric elements, Equation 1:

$$k_t^2 = \frac{\pi}{2}\frac{f_r}{f_a}\cot\left(\frac{\pi}{2}\frac{f_r}{f_a}\right) \tag{1}$$

The piezoelectric charge coefficient d_{11} can be derived from the relation of the thickness coupling factor k_t, the dielectric coefficient ε^T_{11} and the compliance coefficient s^E_{11}, Equation 2:

$$d_{11} = k_t(\varepsilon^T_{11}\, s^E_{11})^{1/2} \tag{2}$$

The dielectric coefficient ε^T_{11} can be calculated from the "free" capacitance C^T measured well below the lowest resonant frequency e.g. at 1 kHz (*T* denotes the constant mechanical stress in the element), the thickness *t* of the piezoelectric element and the electrode area *A*, Equation 3:

$$\varepsilon^T_{11} = C^T\frac{t}{A} \tag{3}$$

The compliance coefficient s^E_{11} can be derived from the thickness coupling factor k_t and the compliance coefficient s^D_{11} where *D* denotes the constant displacement in the element, Equation 4:

$$s^E_{11} = \frac{s^D_{11}}{1-k_t^2} \tag{4}$$

The compliance coefficient s^D_{11} can be calculated directly from the measured anti-resonance value f_a, the density ρ (taken from the datasheet) and the length l (CTE taken into account) of the piezoelectric element, Equation 5:

$$s^D_{11} = \frac{1}{4\rho f_a^2 l^2} \tag{5}$$

Finally, the compliance coefficient c^E_{11} can be derived from the k_t factor and the compliance coefficient c^D_{11}, Equation 6:

$$c_{11}^{E} = c_{11}^{D}(1 - k_t^2) \tag{6}$$

And to calculate c^{D}_{11} one has to measure f_a, ρ and t, Equation 7:

$$c_{11}^{D} = 4\rho f_a^2 t^2 \tag{7}$$

4. Experimental Setup and Procedure

Five $GaPO_4$ elements were used for impedance measurements at HT. The thickness extension mode element has a compression velocity of 4356 m/s and calculated resonant frequency of 2.17 MHz. The dimensions at 25°C were determined using a digital caliper to an accuracy of 0.01 mm, and for HT the dimensions were calculated using temperature dependent coefficients of thermal expansion (CTEs) provided in the material datasheet. Density of the elements was also taken from the datasheet [14].
At room temperature, devices such as "smart tweezers" are usually used for the impedance measurements of piezoelectric elements [15]. However, those commercially available devices are not suitable for measurements at HT. Hence, a special holder for the $GaPO_4$ elements was designed to ensure that all the elements were tested in the same manner. The holder allowed electrical connection between an Agilent 4294A impedance analyser and the $GaPO_4$ elements under test.

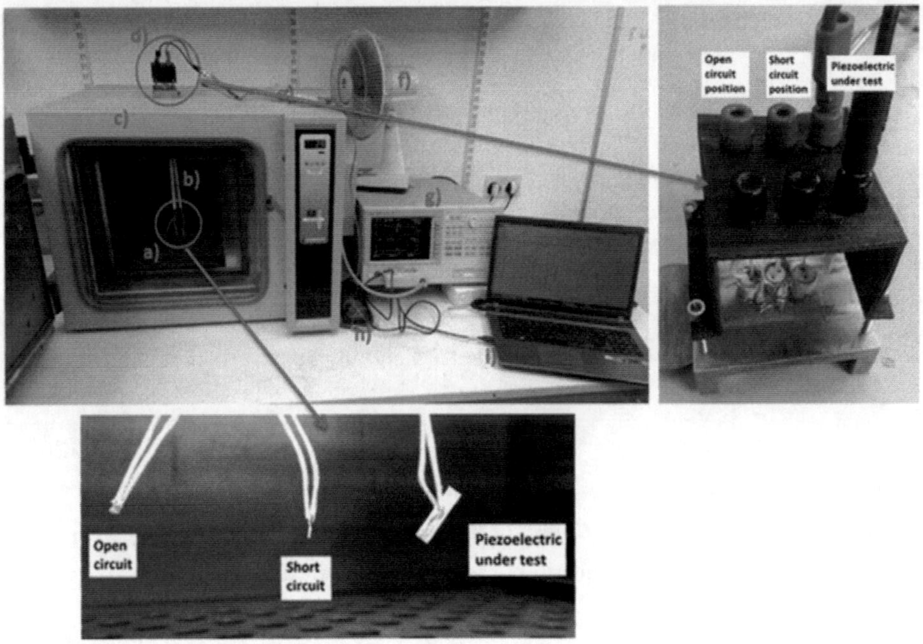

Figure 2: Experimental setup used for the impedance measurements of $GaPO_4$ elements at HT consisted of the following elements: a) piezoelectric element under test affixed to nickel conductors, b) alumina insulation tubing, c) Carbolite LHT6-30 oven, d) plastic panel with banana plug sockets for open and short circuit compensation, e) BNC test leads with Agilent 16048A test fixture, f) fen used for cooling of the outer part of the piezoelectric element holder, g) Agilent 4294A impedance analyser, h) temperature logger with K type thermocouple and i) computer to collect the data.

To establish a baseline, impedance characteristics of the $GaPO_4$ elements were recorded first at 25°C. Then, measurements were taken at 200°C, 400°C and 580°C. Finally, the temperature was kept at a constant value of 580°C for a period of 25 days in order to examine the material properties of the

$GaPO_4$ elements at HT over time. Prior to proceeding with measurements at each temperature, compensation was carried out to remove the effect of parasitic impedance of the holder [16]. The f_r and f_a frequencies taken from the impedance characteristics, capacitance (measured at 1 kHz), density and dimensions of the $GaPO_4$ elements were used to derive material properties of the elements at HT.

5. Results and Discussion

5.1. High Temperature Material Properties

The material properties k_t, d_{11}, s^E_{11} and c^E_{11} (averaged for five $GaPO_4$ elements) were derived and used to assess the effect of increasing temperature on the HT operation of $GaPO_4$ piezoelectric elements, Figure 3 and 4:

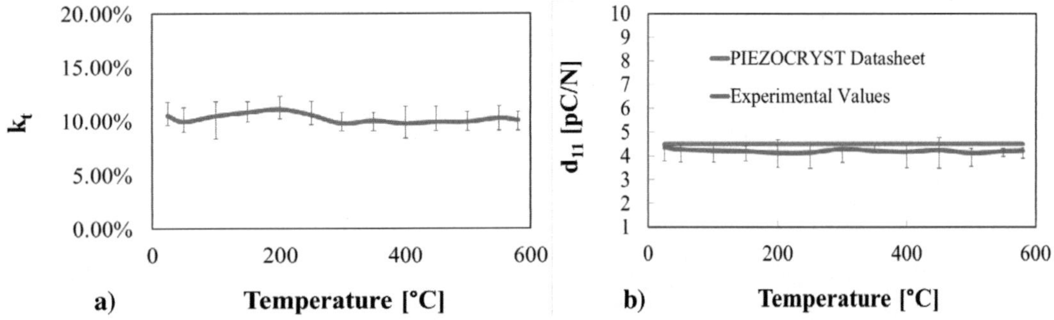

Figure 3: a) Thickness coupling factor k_t and **b)** Piezoelectric charge coefficient d_{11}, averaged for five $GaPO_4$ elements, as a function of temperature from 25°C to 580°C.

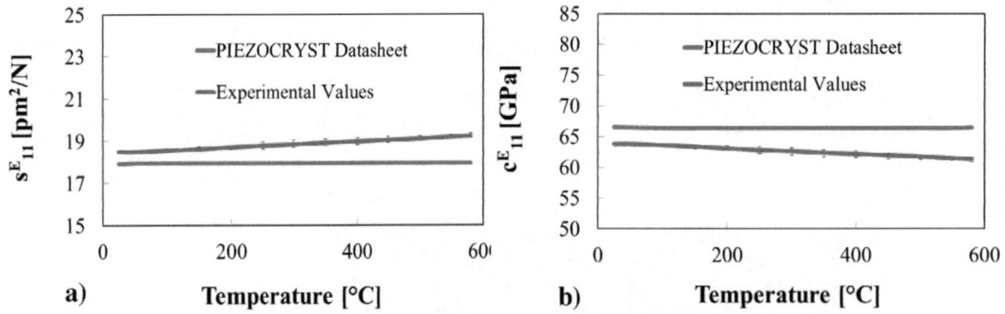

Figure 4: a) Elastic compliance s^E_{11} and **b)** Stiffness c^E_{11} coefficients, averaged for five $GaPO_4$ elements, as a function of temperature from 25°C to 580°C.

From Figure 3 a) it can be seen that the value of k_t remained almost constant at 10% from 25°C up to 580°C. This means that in this temperature range, the tested $GaPO_4$ elements can convert 10% of the energy delivered. In order to design a PA probe with broader bandwidth for better resolution, it is important for each of the $GaPO_4$ element to have high electromechanical coupling coefficient [10]. When comparing $GaPO_4$ elements to PZT ones which display conversion of up to 70% [2], the conversion of 10% seems to be very poor. However, it is important to keep in mind that e.g. PZT-5A elements cannot be used above around 180°C, where $GaPO_4$ elements were tested up to 580°C. From Figure 3 b) it can be seen that the derived d_{11} coefficient stayed constant at 4.2 pC/N in the whole tested temperature range. Same as with the k_t factor, the value of $d_{11} = 4.2$ pC/N when compared with PZT-5A which shows $d_{11} \approx 370$ pC/N is quite low; however the $GaPO_4$ elements showed very stable piezoelectric response far above the upper limitation temperature of PZT piezoelectrics. Finally, the calculated elastic properties proved very good mechanical stability of this piezoelectric material at up to 580°C, Figure 4 a) and b). The calculated s^E_{11} coefficient was 18.5 pm²/N at 25°C and increased up to 19.22 at 580°C, the change was less than 3%. Furthermore, in the same temperature range, the calculated c^E_{11} coefficient deviated less than 4%, with the value of 63.84 GPa at 25°C and 61.42 GPa at 580°C. The comparison of the derived material properties with the datasheet values confirmed that the adopted measurement technique is reliable in determining the material properties of $GaPO_4$ elements at HT; thus, this technique can be used for the characterisation of $GaPO_4$ elements at HT in a long term.

5.2. *Long Term Material Properties at 580°C*

One $GaPO_4$ element was tested at 580°C for 25 days to obtain preliminary results on HT operation of $GaPO_4$ material in a long term in order to evaluate the feasibility of $GaPO_4$ elements for long-term continuous monitoring at HT. The results of the material properties at HT against time are shown in Figures 5 and 6:

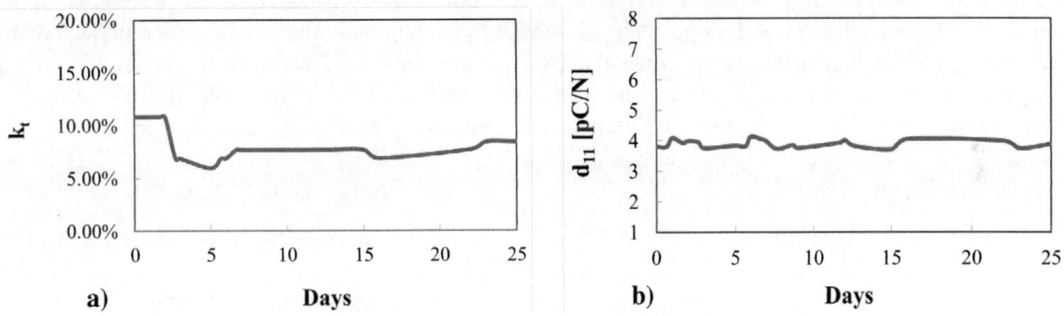

Figure 5: **a)** Thickness coupling factor k_t and **b)** Piezoelectric charge coefficient d_{11}, of a $GaPO_4$ element, as a function of time for 25 days at 580°C.

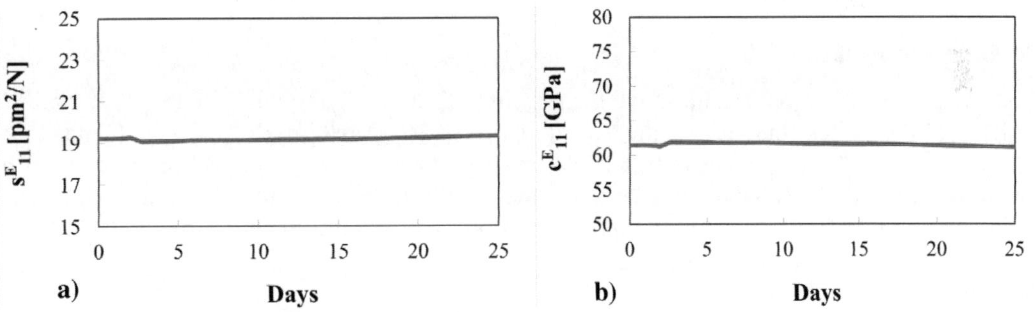

Figure 6: **a)** Elastic compliance s^E_{11} and **b)** Stiffness c^E_{11} coefficients, of a $GaPO_4$ element, as a function of time for 25 days at 580°C.

In Figure 5 a) it can be seen that the value of k_t factor decreased from the initial value of 10% in the first four days of testing and then stabilised at around 7.5% until the end of the test. According to literature, when compared to piezoelectric and elastic properties, the coupling factor is found to be insensitive to effect of temperature in single crystals [4]. This decline of 25% in the conversion of the energy delivered shows the opposite and more $GaPO_4$ elements need to be subjected to HT tests in a long term to validate the initial results. Figure 5 b) shows that the d^E_{11} coefficient remained at around 4 pC/N during the entire duration of the test confirming stable sensing capability of this piezoelectric material at HT. The derived elastic properties s^E_{11} of 19.2 pm²/N and c^E_{11} of 61.6 GPa, stayed virtually unchanged at 580°C for 25 days, Figure 6 a) and b). After the 25-day period, the tested $GaPO_4$ element was cooled down to 25°C and impedance measurement was carried out again. The very pronounced f_r

and f_a peaks after the heat treatment confirmed that this piezoelectric material could potentially be used in development of PA probes for continuous operation at HT.

6. Conclusions

In this study the material properties have been derived for five $GaPO_4$ elements using the impedance method as outlined in the European Standard on piezoelectricity. In order to ensure that all the $GaPO_4$ elements were tested in the same manner, a special holder was designed. The five $GaPO_4$ elements proved to possess very stable material properties from 25°C up to 580 °C and one $GaPO_4$ element stayed very stable when subjected to 580°C for 25 days. However, the very low measured values of k_t factor of 7.5% (PZT $k_t \approx 70\%$) and d^E_{11} constant of 4 pC/N (PZT-5A $d^E_{11} \approx 370$ pC/N) urge caution and further study is needed. One of the proposed solutions to utilise this very low percentage of the converted energy (low efficiency) and low sensing capability when using $GaPO_4$ elements for generation and reception of ultrasound is development of appropriate low-noise signal amplifiers. Also, the measured electrical impedance of the $GaPO_4$ elements was in the order of kΩ which is several orders of magnitude higher than the load on the standard BNC cables (50Ω). The electrical impedance mismatching can affect the ultrasonic probe noise performance, driving response, bandwidth and sensitivity. Appropriate matching circuits need to be developed to overcome this issue.

7. Future Work

More $GaPO_4$ elements will be tested at HT for a period of 25 days in order to provide sufficient evidence to prove the reliability of such elements. Fabrication of $GaPO_4$ elements to achieve higher frequencies up to 5 MHz to be able to detect smaller defects should be possible and will form part of future effort. The next step in this study will be to simulate the ultrasonic propagation in order to assess the transmission and reception quality of the $GaPO_4$ array elements, both at room and HT operating conditions. Finally, PA probes will be built using $GaPO_4$ elements and subjected to laboratory and field trials.

8. Acknowledgements

This research has received funding from the European Commission through the FP7 Programme (FP7-SME-2013-1) under the grant agreement no. 605267, as part of a collaborative project "HotPhasedArray". See project website: http://www.hotphasedarray.eu/project for more information. "HotPhasedArray" is collaboration between the following organisations: iKnowHow Informatics, CeramTec, InnoTecUK, Brunel University, Enkon, Vermon, Tecnitest Ingenieros and INETEC.

References

[1] FP7 Project: "High Temperature Pipe Structural Health Monitoring System utilising PA probes on TOFD configuration" ("HotPhasedArray"). Grant Agree. No.: 605267, 2013.
[2] M. W. Hooker: "Properties of PZT-Based Piezoelectric Ceramics Between -150 and 250°C". NASA / CR- 1998-208708, 1998.
[3] D. Damjanovic: "Materials for high-temperature piezoelectric transducers". Curr. Opin. Solid State Mater. Sci., Vol.: 3, Pg.: 469–473, 1998.
[4] S. Zhang and F. Yu: "Piezoelectric Materials for High Temperature Sensors". J. Am. Ceram. Soc., Vol.: 94 [10], Pg.: 3153-3170, 2011.
[5] P. Krempl: "Quartz Homeotypic Gallium-Orthophosphate – A New High Tech Piezoelectric Material". IEEE Proc., UT Symp., France, Pg.: 949-954, 1994.
[6] W. P. Mason and H. Jaffe: "Methods for Measuring Piezoelectric, Elastic and Dielectric Coefficients of Crystals and Ceramics," Proc. IRE, Vol.: 42, No.: 6, Pg.: 921-930, 1954.
[7] J. Fialka and P. Benes: "Comparison of Methods for the Measurement of Piezoelectric Coefficients". IEEE Trans. Instrum. Meas., Vol.: 62, No.: 5, May 2013.

[8] Active European Standard: “Piezoelectric properties of ceramic materials and components – Part 2: Methods of measurement – Low power”. EN 50324-2:2002.
[9] J. C. Drury: "Ultrasonic Flaw Detection for Technicians". Silverwing Ltd, 3rd Edition, 2004.
[10] M. Kim, J. Kim and W. Cao: “Electromechanical coupling coefficient of an ultrasonic array element”. Journal of Applied Physics, Vol.: 99, 2006.
[11] D. Richter and H. Fritze: “High-temperature Stable Electrodes for Langasite Based Surface Acoustic Wave Devices”. Proc. Sensor+Test Conf., Pg.: 532–537, Germany, 2011.
[12] V. Krstelj: “Ultrasonic control – Selected chapters”. FSB, Zagreb, 2003.
[13] G. Gautschi, Piezoelectric Sensorices. Springer-Verlag, NY, 2002.
[14] Piezocryst Advanced Sensorics GMBH: “$GaPO_4$-Resonators”. $GaPO_4$ Material Datasheet.
[15] Keysight Technologies: “16334A Test Fixture (Tweezers Contacts)”. Available online at: http://www.keysight.com. Accessed on 13/04/2015.
[16] S. Sherrit, X. Bao, Y. Bar-Cohen and Z. Chang: "Resonance Analysis of High Temperature Piezoelectric Materials for Actuation and Sensing". San Diego, CA., SPIE Vol.: 5387, 2004.

A secure and easy-to-implement web-based communication framework for caregiving robot teams

G Tuna[1], R Daş[2], A Tuna[3], H Örenbaş[4], M Baykara[2] and K Gülez[4]
[1] Department of Computer Programming, Trakya University, Edirne, Turkey
[2] Department of Software Engineering, Firat University, Elazig, Turkey
[3] School of Foreign Languages, Trakya University, Edirne, Turkey
[4] Department of Control and Automation Engineering, Yildiz Technical University, Istanbul, Turkey

E-mail: gurkantuna@trakya.edu.tr, rdas@firat.edu.tr, aysetuna@trakya.edu.tr, orenbas@yildiz.edu.tr, mbaykara@firat.edu.tr, gulez@yildiz.edu.tr

Abstract. In recent years, robots have started to become more commonplace in our lives, from factory floors to museums, festivals and shows. They have started to change how we work and play. With an increase in the population of the elderly, they have also been started to be used for caregiving services, and hence many countries have been investing in the robot development. The advancements in robotics and wireless communications has led to the emergence of autonomous caregiving robot teams which cooperate to accomplish a set of tasks assigned by human operators. Although wireless communications and devices are flexible and convenient, they are vulnerable to many risks compared to traditional wired networks. Since robots with wireless communication capability transmit all data types, including sensory, coordination, and control, through radio frequencies, they are open to intruders and attackers unless protected and their openness may lead to many security issues such as data theft, passive listening, and service interruption. In this paper, a secure web-based communication framework is proposed to address potential security threats due to wireless communication in robot-robot and human-robot interaction. The proposed framework is simple and practical, and can be used by caregiving robot teams in the exchange of sensory data as well as coordination and control data.

1. Introduction

Caregiving is hard work because it is tedious and physically and emotionally exhausting. Sometimes it is dangerous or disgusting. Typically on a 24/7 basis, it is an unpaid or low wage job, and has adverse health consequences for those who do it. Therefore, generally women or immigrants work as caregivers. With an increase of elderly among populations around the world, many countries struggle to provide the funds and necessary staff for caregiving services and have acknowledged this reality by investing in robot development [1, 2]. Many countries who realize this reality are investing serious money into the development of caregiving robot technologies. In Japan, the health ministry started to promote the development of nurse robots which can perform certain tasks such as lifting and moving patients in hospitals [3]. In Sweden, a humanoid robot called GiraffPlus was developed. GiraffPlus was programmed to complete a number of household and caregiving tasks [4]. The European Union

[1] To whom any correspondence should be addressed.

started to fund a project called Mobiserv to develop touch-screen humanoid robots that can serve as social companion for the elderly [5]. While these examples and developments in robot technologies are revolutionary and make it possible to replace human caregivers with robots, there are ethical points to take into consideration when replacing real human caregivers with robot caregivers. Although robot caregivers makes sense economically on a macro scale, the fear of isolating the elderly further from real social interaction is an important problem which needs to be addressed.

It is widely accepted that robot caregivers' tasks will emerge according to the improvement in their abilities, even if current robot caregivers are equipped with little interaction skills to accomplish useful tasks that rely on interaction [6]. Therefore while exploring the possible tasks of the robot caregivers, their social abilities must be improved. In the past, researchers tried to develop robots that communicate with people, but they used auditory and visual input for completely different tasks [6, 7]. They used vision to navigate and audio for speech communication. However such robots were not able to perform complex tasks which require interaction skills. Therefore, robot caregivers must be provided with simple social skills and more importantly the ability to infer meaning from incoming sense data.

It is known that multi-robot teams can perform specific tasks more efficiently than a single robot or sometimes can accomplish tasks not executable by a single one. In addition, multi-robot teams have potential advantages such as increasing tolerance to possible robot faults, providing flexibility to the task execution, and taking advantages of distributed sensing and actuation [8]. Communications play an important role in multi-robot teams. Wireless communication is used in multi-robot teams for coordination, cooperation, task management, inter-team communication, and other needs. Even if an efficient and reliable communication network is the main requirement of multi-robot teams, potential information security threats must not be overlooked since wireless communication is broadcast over radio waves, attackers can eavesdrop on the communications and easily pick up unencrypted messages [9-12]. This may lead to numerous security risks in multi-robot teams.

In this paper, an easy-to-implement secure web-based communication framework is proposed to address common information security risks and prevent attacks and its efficiency is proved using a case study. Although, the current networks are assumed to be designed for specific applications, the future ones are envisioned as comprising heterogeneous devices assisting to a large range of applications [13]. To achieve this goal, in this study, a flexible web-services based architecture is used. The rest of the paper is organized as follows. A secure web-based communication framework to enhance robot-robot and human-robot interactions is proposed in Section 2. The paper is concluded in Section 3.

2. Proposed secure web-based communication framework

Caregiving robots are promising since the population of the elderly has been increasing and caregiving is a difficult task in many aspects. Since most robots have wireless communication capability, their inter-team communication is vulnerable against common security threats such as data theft, passive listening, and service interruption. In this regard, in this study, a secure web-based communication framework is proposed. The proposed framework is based on three modules and its flowchart is shown in Figure 1. These modules are an android-based application running on each robot, .asmx web service, developed in C#, running on the server and Microsoft SQL server database running on the server.

As shown in Figure 2, each robot only interacts with the android-based application and the application sends its requests using the addresses of web services and parameters. Web services can be described as a standardized way of integrating web-based applications using open standards over an Internet Protocol (IP) backbone. The open standards used for this integration are Extensible Markup Language (XML), Simple Object Access Protocol (SOAP), Web Services Description Language (WSDL) and Universal Description, Discovery and Integration (UDDI). Web services bring many advantages in terms of usability, accessibility and security.

The parameters are sent to the service by means of HTTP POST method. As shown in Figure 3, the web service returns its responses in JavaScript Object Notation (JSON) [14] format and is responsible for encryption and database connections. To perform the encryption feature of the proposed framework a class is used. Advanced Encryption Standard (AES) [15] is the default encryption technique of the proposed framework and enables to secure message transfers as shown in Figure 4. As shown in Figure 5, the proposed framework secures not only messages but also user profiles, in this case robot identities.

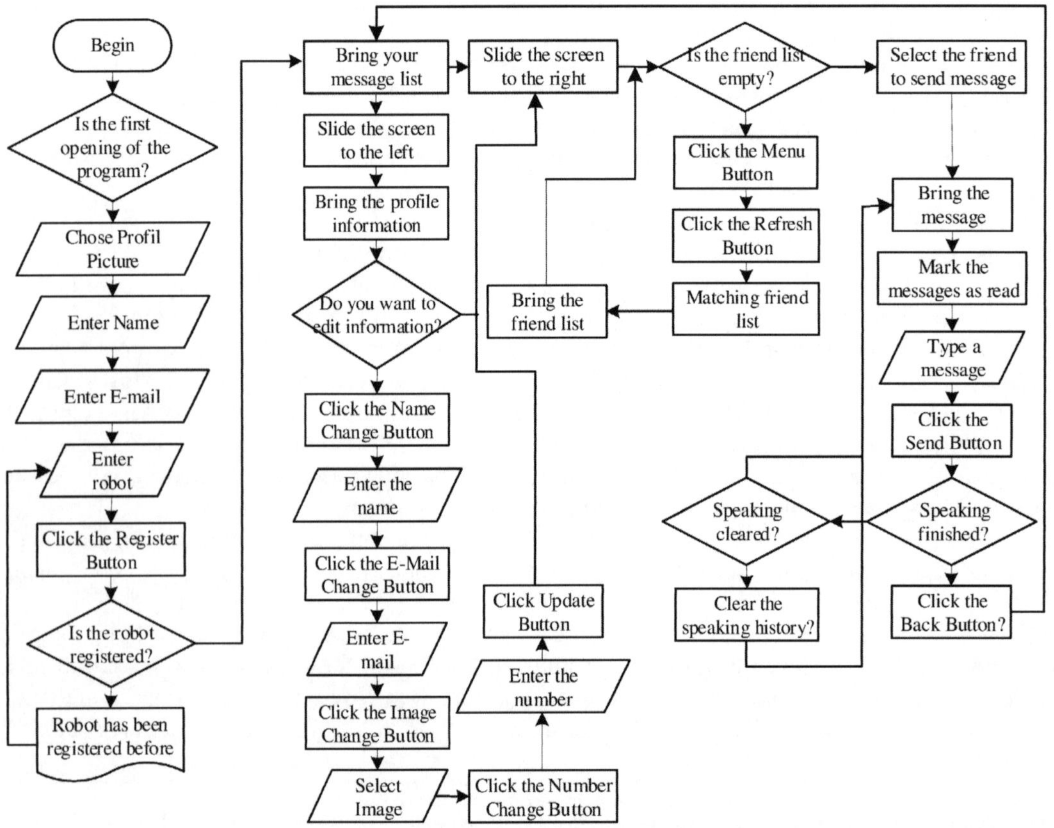

Figure 1. Flowchart of the proposed framework.

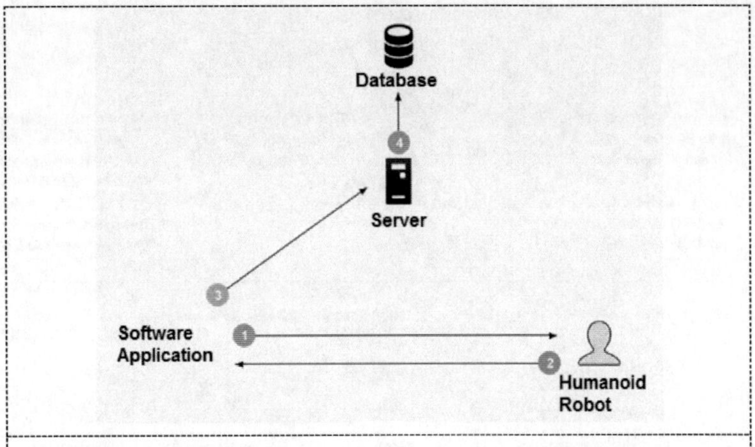

Figure 2. Proposed secure web-based communication framework.

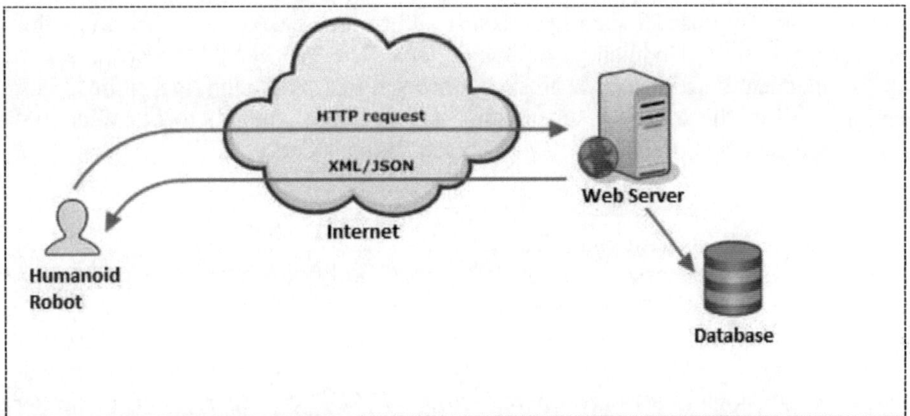

Figure 3. Infrastructure of the proposed framework.

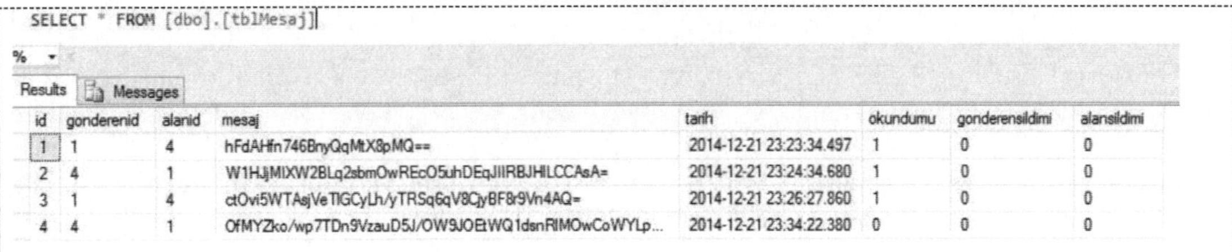

```
SELECT * FROM [dbo].[tblMesaj]
```

Results | Messages

id	gonderenid	alanid	mesaj	tarih	okundumu	gonderensildimi	alansildimi
1	1	4	hFdAHfn746BnyQqMtX8pMQ==	2014-12-21 23:23:34.497	1	0	0
2	4	1	W1HJjMIXW2BLq2sbmOwREcO5uhDEqJIIRBJHILCCAsA=	2014-12-21 23:24:34.680	1	0	0
3	1	4	ctOvi5WTAsjVeTIGCyLh/yTRSq6qV8CjyBF8r9Vn4AQ=	2014-12-21 23:26:27.860	1	0	0
4	4	1	OfMYZko/wp7TDn9VzauD5J/OW9JOEtWQ1dsnRIMOwCoWYLp...	2014-12-21 23:34:22.380	0	0	0

Figure 4. Encrypted messages.

IOP Conf. Series: Materials Science and Engineering **108** (2016) 012009 doi:10.1088/1757-899X/108/1/012009

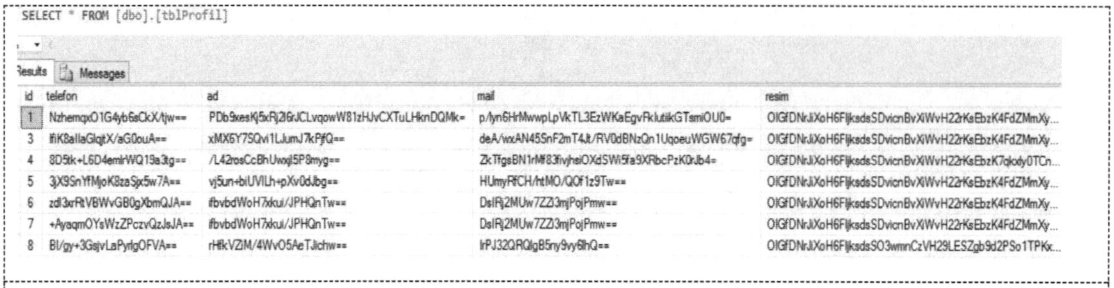

Figure 5. Encrypted profiles.

3. Conclusion

The increase in the population of the elderly and the recent advances in robotics and wireless communications have led to the emergence of autonomous caregiving robot teams. However, due to wireless communications, these robot teams are open to many information security risks. In this paper, to address the information security risks, a secure web-based communication framework is proposed for robot-robot and human-robot interactions. The proposed framework can be used to enhance the communication security of caregiving robot teams. Although, the proposed framework's current version requires the use of a supervisor running one of the Microsoft Windows operating systems, it is a promising solution due to its cloud networking support.

References

[1] Mukai T, Shinya H, Nakashima H, Kato Y, Sakaida Y, Guo S and Hosoe S 2010 Development of a nursing-care assistant robot RIBA that can lift a human in its arms *2010 IEEE/RSJ International Conference on Intelligent Robots and Systems* pp. 5996-6001

[2] Gross H-M, Schröter C, Müller S, Volkhardt M, Einhorn E, Bley A, Martin C, Langner T and Merten M 2011 Progress in developing a socially assistive mobile home robot companion for the elderly with mild cognitive impairment *2011 IEEE/RSJ International Conference on Intelligent Robots and Systems* pp. 2430-2437

[3] http://www.japantimes.co.jp/news/2013/11/26/national/ministry-to-push-nursing-robots-to-aid-[4] caregivers/#.VVBSl_ntmko [Accessed 9 April 2015]

[4] http://www.giraffplus.eu/ [Accessed 10 April 2015]

[5] http://www.mobiserv.info/ [Accessed 10 April 2015]

[6] Asada M, Hosoda K, Kuniyoshi Y, Ishiguro H, Inui T, Yoshikawa Y, Ogino M and Yoshida C 2009 Cognitive developmental robotics: a survey *IEEE Transactions on Autonomous Mental Development* **1(1)** 12-34

[7] Jones K S, Cherry B and Sridharan M 2015 Formative Work Analysis to Design Caregiver Robots *Proceedings of the Tenth Annual ACM/IEEE International Conference on Human-Robot Interaction* pp. 99-100

[8] Long X -L, Jiang J -P, and Xiang K 2004 Towards Multirobot Communication *IEEE International Conference on Robotics and Biomimetics* pp. 307-312

[9] Zhou Y, Fang Y and Zhang Y 2008 Securing Wireless Sensor Networks: A Survey *IEEE Communications Surveys & Tutorials* **10(3)** 6-28

[10] Karlof C and Wagner D 2003 Secure routing in wireless sensor networks: Attacks and Countermeasures *Ad Hoc Networks* **1(2-3)** 293-315

[11] Yang H, Luo H, Ye F, Lu S and Zhang L 2004 Security in mobile ad hoc networks: challenges and solutions *IEEE Wireless Communications* **11(1)** 38-47

[12] Undercoffer J, Avancha S, Joshi A and Pinkston J 2002 Security for sensor networks *CADIP Research Symposium*

[13] Delicato F C, Pires P F, Pirmez L and Carmo L F R C 2003 A Flexible Web Service Based Architecture for Wireless Sensor Networks *The 23rd International Conference on Distributed Computing Systems* p. 730

[14] https://www.ietf.org/rfc/rfc4627.txt [Accessed 11May 2015]

[15] Daemen J and Vincent V R 2003 AES Proposal: Rijndael Available at: http://csrc.nist.gov/archive/aes/rijndael/Rijndael-ammended.pdf [Accessed 11 May 2015]

Information security threats and an easy-to-implement attack detection framework for wireless sensor network-based smart grid applications

G Tuna[1], H Örenbaş[2], R Daş[3], D Kogias[4], M Baykara[3] and K Gülez[2]
[1] Department of Computer Programming, Trakya University, Edirne, Turkey
[2] Department of Control and Automation Engineering, Yildiz Technical University, Istanbul, Turkey
[3] Department of Software Engineering, Firat University, Elazig, Turkey
[4] Department of Electronics Engineering, Piraeus University of Applied Sciences, Aigaleo, Greece

E-mail: gurkantuna@trakya.edu.tr, orenbas@yildiz.edu.tr, rdas@firat.edu.tr, dimikog@teipir.gr, mbaykara@firat.edu.tr, gulez@yildiz.edu.tr

Abstract. Wireless Sensor Networks (WSNs) when combined with various energy harvesting solutions managing to prolong the overall lifetime of the system and enhanced capabilities of the communication protocols used by modern sensor nodes are efficiently used in are efficiently used in Smart Grid (SG), an evolutionary system for the modernization of existing power grids. However, wireless communication technology brings various types of security threats. In this study, firstly the use of WSNs for SG applications is presented. Second, the security related issues and challenges as well as the security threats are presented. In addition, proposed security mechanisms for WSN-based SG applications are discussed. Finally, an easy-to-implement and simple attack detection framework to prevent attacks directed to sink and gateway nodes with web interfaces is proposed and its efficiency is proved using a case study.

1. Introduction

With the addition of renewable energy solutions, existing power grids manage to partially serve the globally increasing demand for electricity, but still have to deal with problems such as equipment failures, blackouts, poor communication and lack of effective monitoring of the infrastructure. Those challenges, along with production instabilities caused by structural or operational characteristics, can easily lead to huge economic losses, inefficient electricity usage, customer dissatisfaction and pollution from a huge amount of CO_2 emissions. Taking, also, under consideration the costly maintenance of the aged existing infrastructures along with the rising costs for building new ones and the declining number of skilled personnel, the need for more dynamic and efficient operation of the system has risen. Real-time monitoring of the various grid components, surveillance and management of distribution and transmission systems become more and more important. In this respect, the grid's efficiency can be improved by the use of various sensors. This approach allows collecting real-time data from various devices in a power grid and spreading it along the infrastructure. In this way, any problems in the system's functionality can be diagnosed proactively and timely generated remedial

[1] To whom any correspondence should be addressed.

actions can be taken in order to prevent any unpleasant dysfunctionalities that might affect the grid's performance. They will, also, reduce person hours to locate and deal with system faults and, consequently, significantly reduce loss of revenue. The sensors can be installed on critical power grid equipment and can be used to monitor many components essential to the grid, such as voltage, current, temperature, system frequency and power quality disturbances [1-4]. This way a next-generation electric power system is created, the Smart Grid (SG) [2, 3].

Recently Wireless Sensor Networks (WSNs) have been widely recognized as a technology promising to enhance various aspects of SGs, especially those that deal with power generation, bidirectional delivery, utilization and seamless monitoring, providing an energy efficient, reliable and low-cost solution for control management [1,5,6]. WSN applications for smart grids, both existing and future ones, range from advanced metering, demand response and dynamic pricing, equipment fault diagnostics, fraud detection, load control, power automation and remote power system monitoring and control. SG is also considered as a data communications network; therefore the communication capabilities among the elements of the electrical power system will play a huge role on the efficient performance of any WSN-based SG application. But the selection for the right communication technology varies, depending on environmental conditions and cost and is usually the result of careful analysis.

Data collection is an essential operation for WSN-based SG applications. However, WSNs are challenged by various constraints such as limited computation and memory resources, limited power supply resource and vulnerability to failures and therefore addressing these challenges plays a key role in improving the efficiency, reliability and robustness of WSN-based SG applications. For this purpose, effective data collection methods [7, 8] to save energy and speed up the data collection process have been investigated in recent years. On the other hand, improving the data collection process alone is not enough for the success of SG applications, as wireless communication technology incurs various types of security threats. In this paper, an easy-to-implement attack detection framework is proposed to address common information security risks and prevent attacks directed to sink and gateway nodes with web interfaces, and its efficiency is proved using a case study. The rest of the paper is organized as follows. The use of wireless sensor networks for data collection in SG environment is explained in Section 2. Section 3 reviews security threats, challenges and proposed security mechanisms, and explains the proposed framework to address security threats in WSN-based SG applications. Finally, the paper is concluded in Section 4.

2. Wireless sensor network-based smart grid applications

In recent years, surveillance and management of transmission and distribution systems as well as real-time monitoring of the various grid components has become more and more important. In this respect, the grid's efficiency can be improved by the use of various sensors. This approach allows collecting real-time data from various devices in a power grid and communicating it along the infrastructure. In this way, a problem in the grid can be diagnosed proactively and timely generated remedial actions can be taken in order to prevent any unpleasant dysfunctionalities that might affect the grid's performance. It also helps to reduce man hours to locate and deal with system faults and, consequently, significantly reduce loss of revenue. The sensors can be installed on the critical power grid equipment and can be used to monitor essential grid components such as voltage, current, temperature, system frequency and power quality disturbances [1-4]. This way a next-generation electric power system is created, the Smart Grid (SG).

A SG is a modernized power transmission and distribution (T&D) network which uses robust two-way data communications, distributed computing technologies and smart sensors to improve reliability, safety and efficiency of power delivery and use [2-4]. Using a sophisticated information processing and communication technology infrastructure, SG will be able to fully use and benefit from its distributed power generation system, while maximizing the whole system's energy efficiency [2, 3]. Consequently, SG is also considered as a data communication network which, by supporting many power management devices, achieves seamless and flexible inter-operation abilities among different advanced system components that leads to efficient performance. The power system infrastructure is comprised of all the devices found on existing electrical grids, as was described above, with the additions of smart meters and sensors spread throughout the grid to detect outages and measure critical performance metrics that should be forwarded to, the back-haul system's detection and decision data centres [2, 4]. It generally includes the power generation, T&D system and the customer premises.

In a SG, there basically are two types of *information flow* [11]. The first flow is from sensors and electrical appliances to smart meters and the second information flow between the smart meters and the data centres of the utility. The first information flow can be accomplished with the use of power-line communication (PLC) [10, 12] or by using wireless communications technologies, such as 6LowPAN, ZigBee, and others [5]. The second information flow can be accomplished by using the internet (public network) or cellular communications. Generally, the expected characteristics of the selected technology would be its simplicity, robustness, mobility and the capability to improve the network's performance, to balance the network's load, to extend the network coverage range and to increase the data transfer capacity and security.

Due to the various components included in the power system infrastructure, SG transformation requires the handling of both the flow from sensors and electrical appliances to smart meters and the flow between the smart meters and the data centres [2]. Generally, the expected characteristics of the selected communication technology for the SG transformation would be its simplicity, robustness, mobility, the capability to improve the network's performance, to balance the network's load, to extend the network coverage range, to increase the data transfer capacity and security, etc. [2, 10]. The choice for the selected technology highly depends on the environmental conditions in the area and should be the result of careful cost versus benefit analysis because not all solutions combine the characteristics mentioned above, so there is no solution that can be considered as a default.

The evolution in WSN, especially on the problems related to energy exhaustion, energy harvesting, mobility, and communication, has open new perspectives for their usage in a SG environment. Various, diverse range power grid applications from home area networks (HANs) to power T&D monitoring have been developed and used [12-14], allowing for robust and energy-efficient monitoring and control in a SG [4]. Although SG transformation brings many benefits to electric utilities, SG deployments create much more data compared to traditional power grid deployments and require the handling of new types of information [15]. In this regard, WSNs play a key role in data collection operation in SGs and enhance the ability of SG operators by creating new opportunities such as improving service reliability and delivering useful information.

3. Proposed attack detection framework and a case study

A WSN is a wireless network consisting of scattered tiny sensor nodes with limited computational capabilities and battery power. It is typically used to monitor physical or environmental conditions. There is no doubt that WSNs bring many advantages to industrial, scientific and military applications due to the sensing technology combined with processing power and wireless communication capability. On the other hand, in most applications, WSNs process sensitive data and face information security threats such as data fraud, data manipulation, and sensor destruction or replacement. This concerns critical monitoring and control applications, and becomes severe in large-scale WSN deployments in which security risks increase highly due to limited physical protection of the sensor nodes and openness of the wireless communication channel [16].

The existing security solutions proposed for traditional wireless networks cannot be used for WSNs due to the constraints inherent to the WSN infrastructure. In WSNs, all the data collected by the inner layer sensor nodes are forwarded to the sink node which is a resource-rich node. The sink is responsible for collecting the sensed data from the sensor nodes and sending it to the operation centre via the Internet or other communication architectures. Therefore, the placement of the sink node has a great impact on the energy consumption and network lifetime [17] and the sink must be protected from unauthorized access since it acts as a gateway to the external world [18]. One of the newest design trends in WSNs is Web Services approach in which sensor nodes are service providers and applications are clients of such services [19]. This approach enables a flexible architecture where sensed data can be accessed by users spread all over the world. Therefore, especially, the sink nodes with also acts as a web service provides must be protected against potential security threats.

WSNs are vulnerable to many attack types including passive information gathering, traffic analysis, message corruption, capture of a sensor node, addition of malicious sensor nodes, node malfunction, node outage, and sophisticated attacks such as selective forwarding, wormhole attacks, hello floods, Sybil attacks, and spoofed, altered or relayed routing information [20, 21]. Sink nodes providing web services are vulnerable to other common attack types such as Denial of Service (DoS) and Distributed Denial of Service (DDoS). In this section, we present an easy-to-implement attack detection framework for the sink nodes in WSN-based SG applications.

A DoS is a type of attack on a network which is designed to make a server or a network resource unavailable to users by flooding it with useless traffic. Its main goal is to temporarily interrupt or suspend the services of a host connected to the Internet. On the other hand, a distributed DoS (DDoS) is a malicious attempt where multiple compromised systems are used to target a single system causing a DoS attack. Resources targeted in DoS attacks can be an entire network, a specific host, a port or service on the targeted system, or one or more components of a given network. DDoS has started to become the preferred method of attack due to its simplicity, potential for major disruption, and ease of distribution. Different from attacks relying on the security vulnerability of operating systems, DDoS attacks are relatively low-tech and easy-to-stage.

Since DDoS are one of the most common attack types, they must be prevented to ensure system availability. In this study, to address the potential security threats of DDoS attacks, a security framework is proposed. The proposed framework's flowchart is shown in Figure 1 and its pseudo code is listed below. To prove the framework's effectiveness and usability, a simulation study was carried out using Visual Studio 2012, Microsoft SQL Server 2014 and LOIC [22], a network stress testing tool. Figure 2 shows the database and tables used by the proposed security framework. After the developed framework run, the system on which the proposed framework was running was tested with LOIC. As shown in Figure 3, when the proposed security framework detected a DoS attack, it stopped the attack by blocking the host's IP address, from which the DoS attack was sourcing.

Pseudo code of the proposed framework

```
Begin
Get source host's IP address
Get total number of records of this IP address (Record_Num) in BlockedIP table
If Record_Num > 0
        Get system time and the previous request's minute field (Time_Difference)
        If Time_Difference<60
                Open "Access Blocked" Page
        If Time_Difference>60
                Clear information belonging to the source IP address
If the source IP address already exists (IPExists_Num)
        If IPExists_Num=0
                Store IP information
        If IPExists_Num>0
                Begin
                Request_Number++
                Save access information
                End
Calculate and save the difference between the last and first requests of the source host
(Time_Difference)
Calculate and save the total number of requests in the calculated time difference (Request_Num)
If Time_Difference>10 and Request_Num>100
```

IOP Conf. Series: Materials Science and Engineering **108** (2016) 012010 doi:10.1088/1757-899X/108/1/012010

```
        Store the source IP address in BlockedIP table
        Notify system administrator via e-mail
End
```

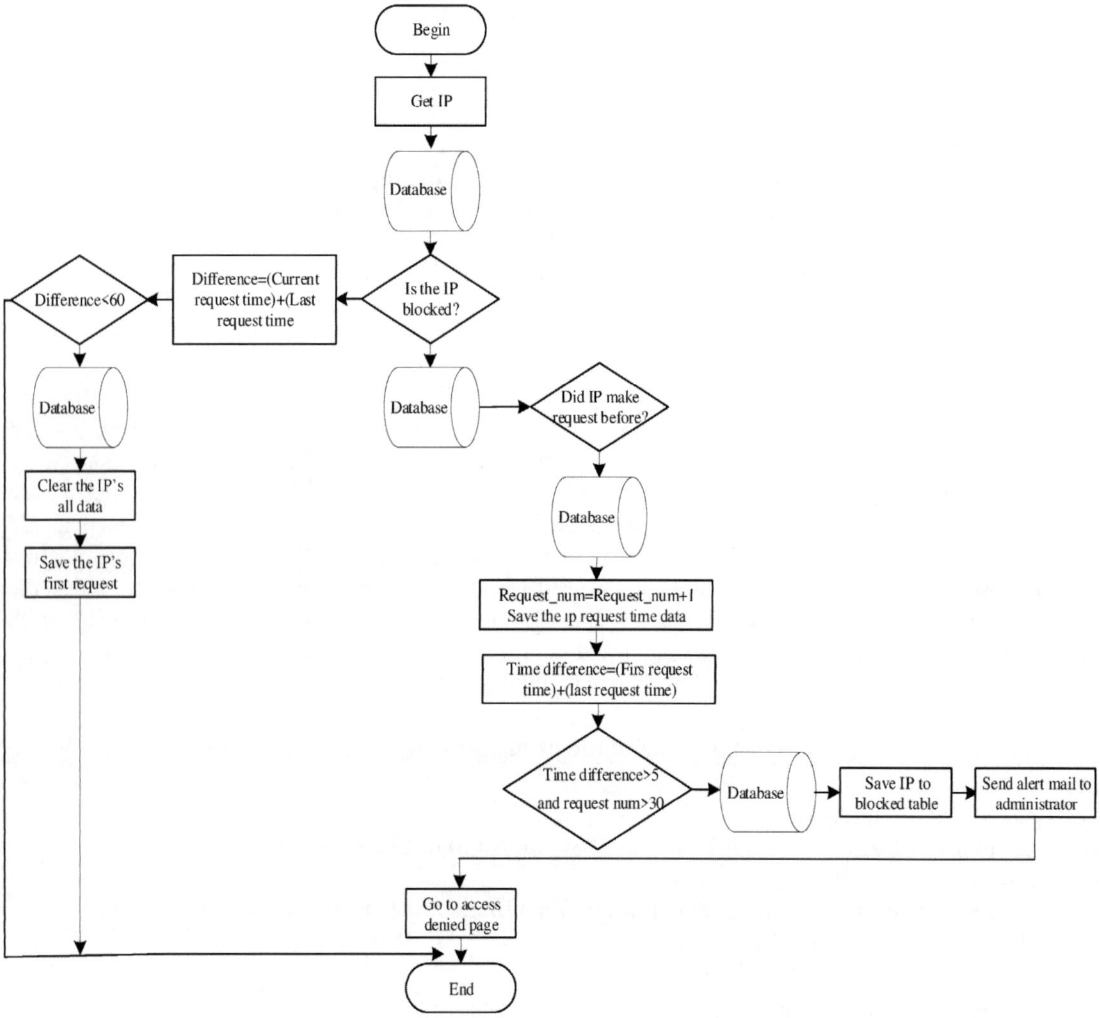

Figure 1. Flowchart of the proposed framework.

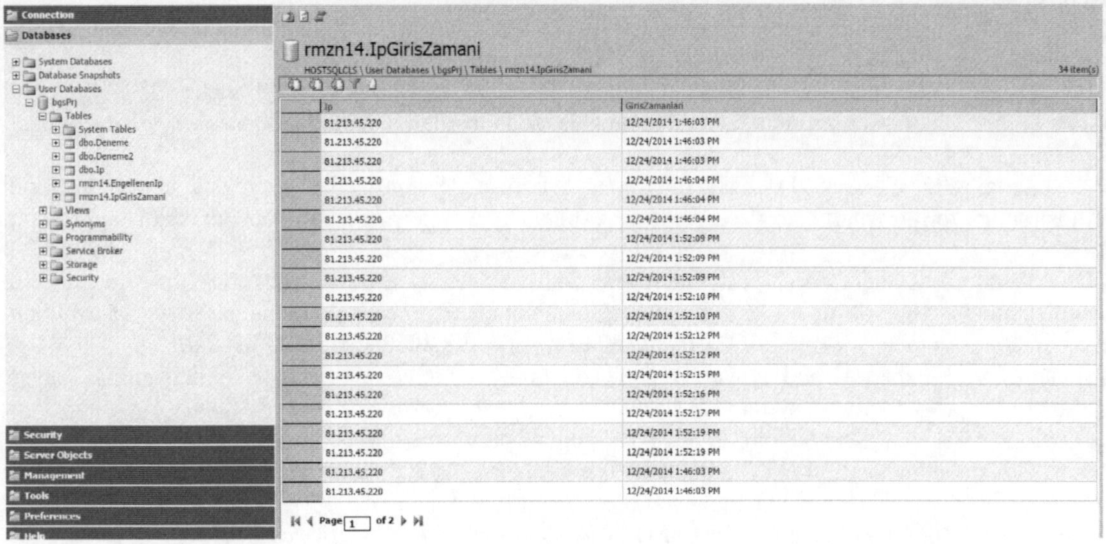

Figure 2. Database structure.

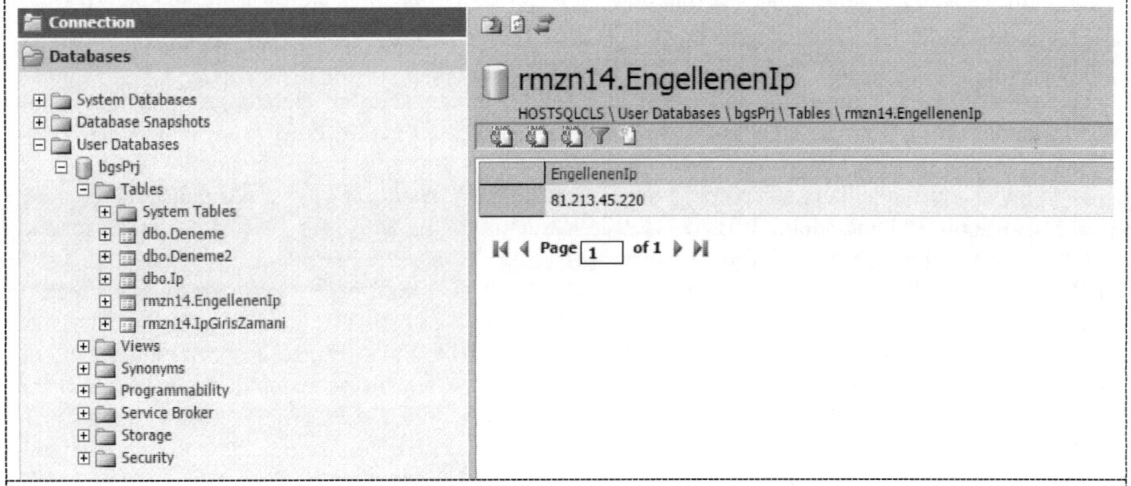

Figure 3. BlockedIP table.

4. Conclusion

Wireless Sensor Networks (WSNs) are commonly used in various monitoring and control applications as well as many other applications. In recent years, WSNs have been commonly implemented in Smart Grid (SG) environment and served for different SG applications. Although many techniques have been proposed to enhance routing, clustering, lifetime and data handling issues of WSNs, security issue somehow has been neglected. However, this may lead to many critical problems in WSN-based SG applications. To address this need, in this study, a practical and easy-to-implement attack detection framework was proposed and its efficiency was proved using a case study. The proposed framework is suitable for all-IP based WSNs.

References

[1] Gungor, VC, Bin, L and Hancke, GP 2010 Opportunities and challenges of wireless sensor networks in smart grid *IEEE Trans. Industrial Electronics* **57(10)** 3557-64

[2] Gungor, VC, Sahin, D, Kocak, T, Ergüt, S, Buccella, C, Cecati C and Hancke, GP 2011Smart Grid Technologies: Communication Technologies and Standards *IEEE Transactions on Industrial Informatics* **7(4)** 529-539

[3] Sood, VK, Fischer, D, Eklund, JM and Brown, T 2009 Developing a communication infrastructure for the smart grid *Proc. IEEE Electrical Power and Energy Conference (EPEC 2009)* pp 1-7

[4] Tuna, G, Gungor, VC and Gulez, K 2013 Wireless Sensor Networks for Smart Grid Applications: A Case Study on Link Reliability and Node Lifetime Evaluations in Power Distribution Systems *International Journal of Distributed Sensor Networks* **2013** Article ID 796248

[5] Yang Y, Lambert F and Divan D 2007 A survey on technologies for implementing sensor networks for power delivery systems *IEEE Power Engineering Society General Meeting* pp. 1-8

[6] Ullo S, Vaccaro A and Velotto G 2010 The role of pervasive and cooperative Sensor Networks in Smart Grids communication *IEEE Mediterranean Electrotechnical Conference (MELECON 2010)* pp. 443-447

[7] Francesco MD, Das SK and Anastasi G 2011 Data Collection in Wireless Sensor Networks with Mobile Elements: A Survey *ACM Trans. Sen. Netw.* **8(1)**

[8] Durmaz Incel O, Ghosh A, Krishnamachari, B and Chintalapudi K 2012 Fast Data Collection in Tree-Based Wireless Sensor Networks *IEEE Transactions on Mobile Computing* **1** 86-99

[9] Wang Y, Zhang J and Xu HS 2012 The Design of Data Collection Methods in Wireless Sensor Networks Based on Formal Concept Analysis *Advances in Intelligent and Soft Computing* **169** 33-38

[10] Yigit M, Gungor VC, Tuna G, Rangoussi M and Fadel E 2014 Power Line Communication Technologies for Smart Grid Applications: a review of advances and challenges *Computer Networks* **70** 366-383

[11] Moghe R, Yi Y, Lambert F and Divan D 2009 A scoping study of electric and magnetic field energy harvesting for wireless sensor networks in power system applications *Proc. of IEEE Energy Conversion Congress and Exposition (ECCE 2009)* pp. 3550-3557

[12] Amin M and Wollenberg BF 2005 Toward a Smart Grid *Power and Energy Mag.* **3(5)** 34-41

[13] Erol-Kantarci M and Mouftah HT 2011 Wireless Multimedia Sensor and Actor Networks for the Next-Generation Power Grid *Ad Hoc Networks* **9(4)** 542-551

[14] Leon RA, Vittal V and Manimaran G 2007 Application of Sensor Network for Secure Electric Energy Infrastructure *IEEE Trans. Power Delivery* 22(2) 1021-28

[15] Kayastha N, Niyato D, Hossain E and Han Z 2014 Smart grid sensor data collection, communication, and networking: a tutorial *Wireless Communications and Mobile Computing* **14(11)** 1055-87

[16] Pathan A K, Lee H -W, and Hong C S 2006 Security in wireless sensor networks: issues and challenges *The 8th International Conference on Advanced Communication Technology* **2** pp. 1043-1048

[17] Zhou Y, Fang Y and Zhang Y 2008 Securing Wireless Sensor Networks: A Survey *IEEE Communications Surveys & Tutorials* **10(3)** 6-28

[18] Shahabuddin M, Furqan Z and Guha R 2005 Wireless sensor network security: a secure sink node architecture *24th IEEE International Performance, Computing, and Communications Conference* pp. 371-376

[19] Delicato F C, Pires P F, Pirmez L and Carmo L F R C 2003 A Flexible Web Service Based Architecture for Wireless Sensor Networks *The 23rd International Conference on Distributed Computing Systems* p. 730

[20] Undercoffer J, Avancha S, Joshi A and Pinkston J 2002 Security for sensor networks *CADIP Research Symposium*

IOP Conf. Series: Materials Science and Engineering **108** (2016) 012010 doi:10.1088/1757-899X/108/1/012010

[21] Karlof C and Wagner D 2003 Secure routing in wireless sensor networks: Attacks and Countermeasures *Ad Hoc Networks* **1(2-3)** 293-315
[22] http://sourceforge.net/projects/loic

IOP Conf. Series: Materials Science and Engineering **108** (2016) 012011 doi:10.1088/1757-899X/108/1/012011

Physiological parameters monitoring of fire-fighters by means of a wearable wireless sensor system

Stelios M Potirakis[1,4], Stelios A Mitilineos[1], Panagiotis Chatzistamatis[1], Savvas Vassiliadis[1], Antonios Primentas[1], Dimitris Kogias[1], Emmanouel T Michailidis[1], Maria Rangoussi[1], Senem Kurşun Bahadır[2], Özgür Atalay[2], Fatma Kalaoğlu[2], and Yusuf Sağlam[3]

[1] Department of Electronics Engineering, Piraeus University of Applied Sciences (TEI of Piraeus), 250 Thivon and P. Ralli, Aigalao, Athens, GR-12244, GREECE
[2] Istanbul Technical University, Faculty of Textile Technologies and Design, Inönü cd. No.65, Gümüşsuyu, Beyoğlu, Istanbul, TR-34437, TURKEY
[3] Kıvanç Group, Ziyagökalp Mh., Başakşehir, Istanbul, TR-34490, TURKEY

E-mail: spoti@teipir.gr

Abstract. Physiological parameter monitoring may be useful in many different groups of the population, such as infants, elderly people, athletes, soldiers, drivers, fire-fighters, police etc. This can provide a variety of information ranging from health status to operational readiness. In this article, we focus on the case of first responders and specifically fire-fighters. Fire-fighters can benefit from a physiological monitoring system that is used to extract multiple indications such as the present position, the possible life risk level, the stress level etc. This work presents a wearable wireless sensor network node, based on low cost, commercial-off-the-self (COTS) electronic modules, which can be easily attached on a standard fire-fighters' uniform. Due to the low frequency wired interface between the selected electronic components, the proposed solution can be used as a basis for a textile system where all wired connections will be implemented by means of conductive yarn routing in the textile structure, while some of the standard sensors can be replaced by textile ones. System architecture is described in detail, while indicative samples of acquired signals are also presented.

1. Introduction

Critical infrastructures represent a nation's material basis for the sustainable welfare of its citizens. Across the world, First Responders (FRs) are called upon every day to protect critical infrastructures against deliberate acts of terrorism, natural disasters, negligence, accidents and malicious behaviour. The occurrence of a crisis (e.g., a fire in a chemical or oil plant, a disastrous earthquake, an incident in a water stemming facility, etc.) requires the presence of a variety of FR teams to restore the situation to a normal (e.g., fire-fighters, civil protection units, police, special forces, etc.)

During the critical phase of a disaster situation, the FRs need to enter a harsh environment to search and rescue human lives, as well as control and mitigate damage. The environment is usually obscured with smoke, noise and rubble, while indoor environments usually pose a risk of collapse or sudden and

[4] To whom any correspondence should be addressed.

life-threating change of the floor plan. Such situations pose the life of FRs and infrastructure users at risk, making orientation and movement difficult. Abundant threats include fire, explosion, toxic gases, emission of unknown substances, falling objects etc. A worst case scenario could be that of a fire at a chemical plant that also includes offices. Reconnaissance is slow, while environment awareness and awareness of the FR team members' whereabouts and physiological parameters monitoring are critical. Another scenario could be that of a bomb disposal mission in populated areas: the combination of the protective suit's weight (~ 40 kg), physical activity, high ambient temperatures, and restricted airflow can cause the operative's temperature to rise to dangerous levels during missions, impairing their physical and mental ability [1]. A body sensor network could be used in order to monitor physiological parameters in such conditions. Other approaches may consist in using low processing power and low consumption tags for aggregate collection of physiological data and delivering pertinent information to FRs [2].

The main concern of FR teams in cases of crises is the lack of useful information regarding the situation at the field and the operational level of the FRs. The local manager needs to have access to timely and accurate information regarding the environment (temperature, hazardous substances, etc.) and the physiological parameters of each FR as well as the FR team as a whole (anxiety level, sufficient respiratory resources / lack of oxygen, heart-pulse rate etc.) Therefore, during the intervention there is a gap between the FRs' situation (positioning, health, etc.) and the overall overview at their mobile headquarter. The proposed platform aims at increasing the effectiveness and safety of FRs by bridging the gap between their side and the higher command level with respect to FR's health situation (acquisition of physiological data and transmission to local headquarters). Another issue of critical concern is the interoperability of communications platform among diverse teams of FRs [3]. Ubiquitous coverage is a necessity and needs to be provided in a seamless manner. In this context, pervasive technologies, such as Wireless Sensor Networks (WSNs), or technologies regarding the vision and navigation of FRs, can be of great help [4-5].

In this paper, we present early results of a wearable wireless sensor system for the timely acquisition of physiological parameters data of fire-fighters, as a characteristic FRs example. The proposed system comprises of WASN nodes based on components of low frequency wired interface, making it suitable for future implementation using textile electronic components and seamlessly incorporating it to existing fire-fighters uniforms. The communications infrastructure to transmit the data to the local headquarters is based on a WSN that can be deployed on the spot independently from any the existing infrastructure in the operations area, since this might be damaged. In the following paragraphs, the system architecture is described in detail, while indicative samples of acquired signals are presented.

2. Wireless sensor network architecture

The proposed system for the monitoring of the physiological parameters of fire-fighters comprises a typical WSN and is intended for use at the operational level. Specifically, the commander of a team of fire-fighters continuously gets updated information on the health status, stress and fatigue level of the team members and based on this information decides which tasks should be assigned to whom. In this sense, the proposed WSN consists of two kinds of wireless nodes: one central "commander" node and a number of "fire-fighter" sensor nodes. Both kinds of nodes are implemented using the same hardware (HW) devices, but with different functionality. Each one of the monitored fire-fighters carries one "fire-fighter" sensor node. Such nodes are equipped with a number of sensors (cf. section 3) that provide a set of physiological parameters' measurements. These measurements are wirelessly sent to the central "commander" node and from there to a personal computer or tablet; the "commander" node is a central point that collects all available information and forwards it to the computer where the received data are stored and processed. Of course, the "commander" node is also equipped with sensors. Note that since both kinds of node are identical in terms of HW, in case of

IOP Conf. Series: Materials Science and Engineering **108** (2016) 012011 doi:10.1088/1757-899X/108/1/012011

emergency, e.g., if the command authority has to be transferred to another person, the system can be rearranged so that new roles are distributed to the wireless nodes. A block diagram of the proposed WSN is shown in Figure 1.

The short range radio frequency (RF) category includes a broader range of technologies and has become synonymous with Bluetooth, Wi-Fi and ZigBee. However the protocol selected for the specific application was Wi-Fi, mainly for reasons of compatibility with a wide range of personal computers and tablets, which provides flexibility in the procurement of this kind of commercial off-the-self equipment (COTS). However, this choice also provides flexibility for future upgrades of the system, either in the form of an increase of the monitored sensors per sensor node (which would increase the required wireless throughput), or in the form of an upgrade of the processing and display means (computer).

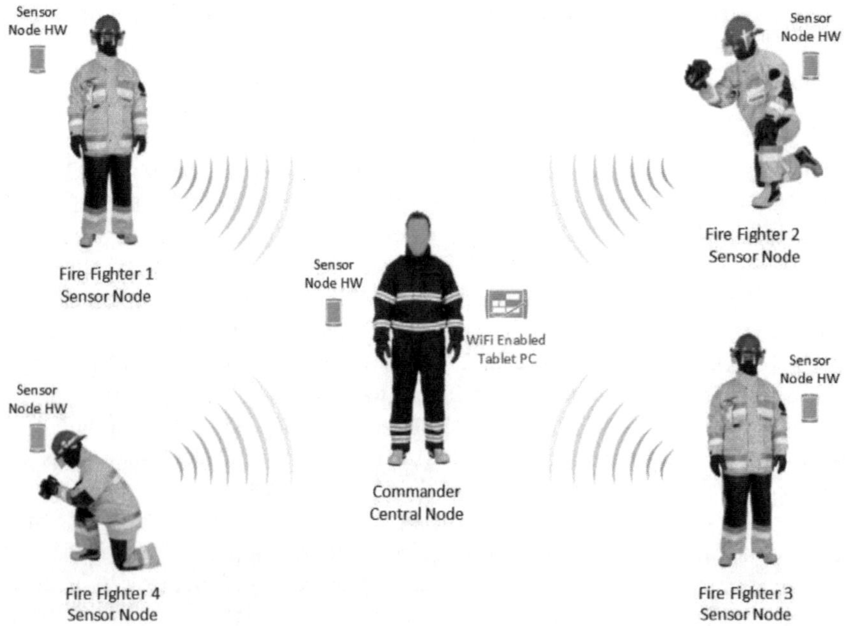

Figure 1. Wireless sensor network for fire-fighters physiological parameters monitoring.

The 802.11 standards, which are widely known with term Wi-Fi, constitute a set of IEEE standards for the creation of wireless local area networks (WLAN) and when initially introduced were destined to expand the wired computer connection (Ethernet) to the wireless area. The 802.11 family contains a series of standards which are: 802.11a, 802.11b, and 802.11g. These standards use the orthogonal frequency division multiplexing method (OFDM) and provide theoretical speeds up to 54Mbps. 802.11a uses the frequency band of 5GHz but does not provide compatibility with the respective network cards used by the other two standards. 802.11b and 802.11g transmit at 2.4GHz and are compatible between them and provide maximum possible speed when communicating between them. Wi-Fi became known in the last years, creating a new method for devices to access the Internet. Table 1 depicts the frequency bands and the theoretical and actual speeds of the various 802.11 standards. [6].

In our application, we use the 802.11 basic services set (BSS). BSS topology is a networking topology in which the wireless network has cluster form. In each cluster there are a number of wireless stations (clients), which communicate between them through a central distributor BS (Base Station), or, more commonly, AP (Access Point). In our case the central "commander" node is the BS [7].

Table 1. The frequency bands of 802.11 standards.

Standard	*Frequency Band*	*Typical Transmission Rate*	*Theoretical Transmission Rate*	*Range Indoors*
802.11a	5GHz	23Mbps	54Mbps	35m
802.11b	2.4GHz	4.3Mbps	11Mbps	38m
802.11g	2.4GHz	19Mbps	54Mbps	38m

3. Wearable sensor node

The WSN nodes of the proposed system are an example of wearable electronic system incorporated in the fire-fighter's suit. Both types of nodes, at this point, carry five kinds of sensors: electrocardiogram (ECG), galvanic skin response, thorax belt sensor, body temperature and tri-axial acceleration sensors for the monitoring of physiological parameters of the firemen, such as heart rate, perspiration, respiration rate, body temperature and motion, respectively.

The wearable sensor node design has been based on the widely spread single-ship control unit Arduino Uno [8]. This is an open source HW and software (SW) platform, providing a low-cost solution, which, on the other hand, provides more than enough processing power and memory for the real-time application of interest. It uses the ATmega328P microcontroller, an Atmel 8-bit RISC processor. It incorporates the following key features: 32KB flash memory, 2KB SRAM, 16MHz clock, 6 channel 10bit analogue to digital converter (ADC) sampling up to 76.9 kSPS, 2-wire serial interface (TWI), and programmable serial interface (USART).

All the sensors used in order to monitor the firemen's physiological parameters, except from the tri-axial accelerometer, are analogue. This means that their signal is digitized through the available ADC. Tri-axial accelerometer directly produces digital output and uses the TWI, which is a serial interface compatible with the widely employed I2C, to send its measurements to the processor. IEEE 802.11 connectivity is obtained through an XBee Wi-Fi module which communicates with the processor over a serial interface. The structure of a sensor node is depicted in Figure 3 in the form of a block diagram.

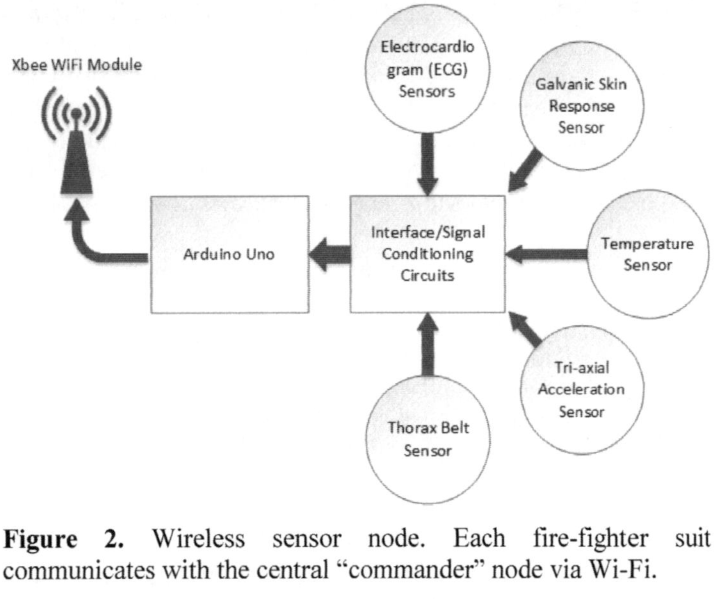

Figure 2. Wireless sensor node. Each fire-fighter suit communicates with the central "commander" node via Wi-Fi.

The current implementation is wearable in a limited degree, in the sense that all sensors, printed circuit boards (PCBs) and electrical connections are implemented using conventional low-cost COTS electronic modules which are integrated into the garment using different types of textile hosting structures like pockets, hemming etc. Figure 3 shows the currently employed electronic components of a wearable sensor node.

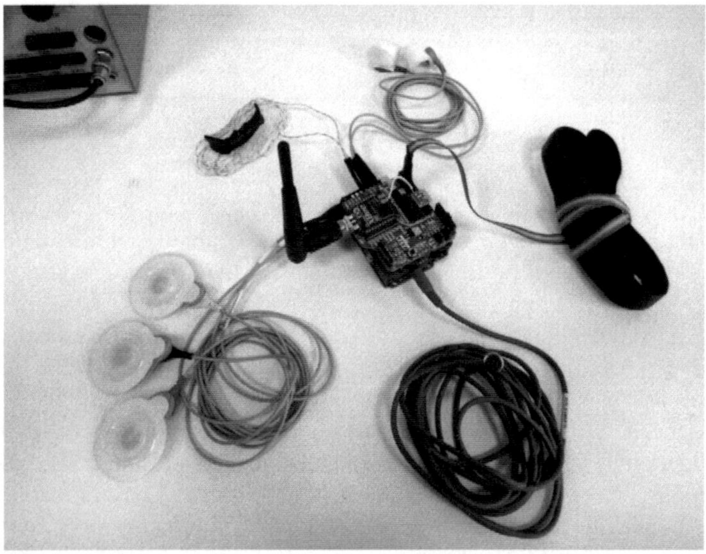

Figure 3. Low-cost COTS HW implementation of the "fire-fighter" sensor node.

In the next steps of our project we intend to replace as much as possible of the electronic modules and the sensors of our system with textile structures or electronics that can be easier incorporated to the garment. For example, Arduino Uno can easily be replaced by a flex-PCB on which the same electronic components as in Arduino Uno will be hosted. On the other hand, our intention is to replace all cables, even the transmission lines carrying the RF signal, with conductive yarn routes on the fire-fighter suit [8-11]. Moreover, we intend to replace ECG sensors with textile electrodes, either based on conductive yarns or screen-printed [12,13], while respiration rate, temperature and skin moisture (perspiration) could be monitored using appropriately designed variations of conductive yarn based knitted sensors [14-17]. Finally, even the antenna could be partially or fully textile [18,19]. Of course, a key aspect of successfully integrating these structures into the textile substrate is the main aim of our project: reliable textile welding. The integration will be considered successful if this results, at the same time, to a reliable electronic system and a functional, as well as comfortable, garment.

4. Physiological parameters acquisition

In the following we present some characteristic examples of signals acquired in real-time by the proposed wearable wireless sensor node. Actually, the signals presented in Figures 4 to 6 were acquired by a "fire-fighter" sensor node, wirelessly sent to the central "commander" node and then to a personal computer where they were stored.

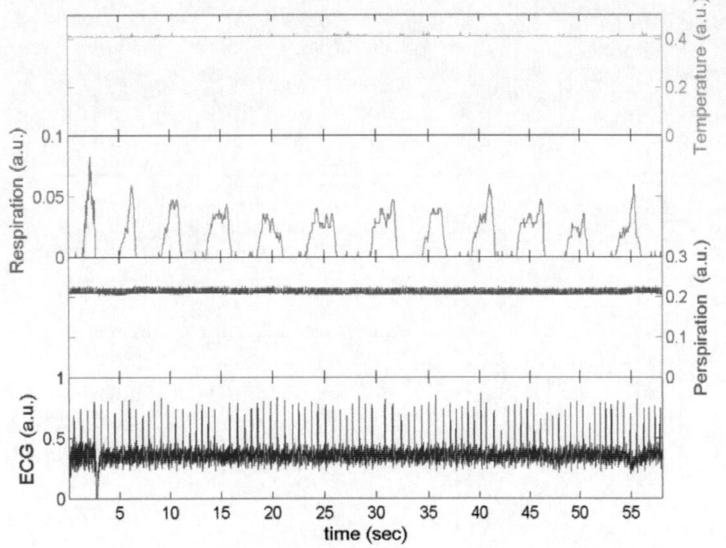

Figure 4. Simultaneous monitoring of all analogue sensor signals. From bottom to top panel: ECG, perspiration, respiration and body temperature. All y-scales are normalized in the interval [0,1].

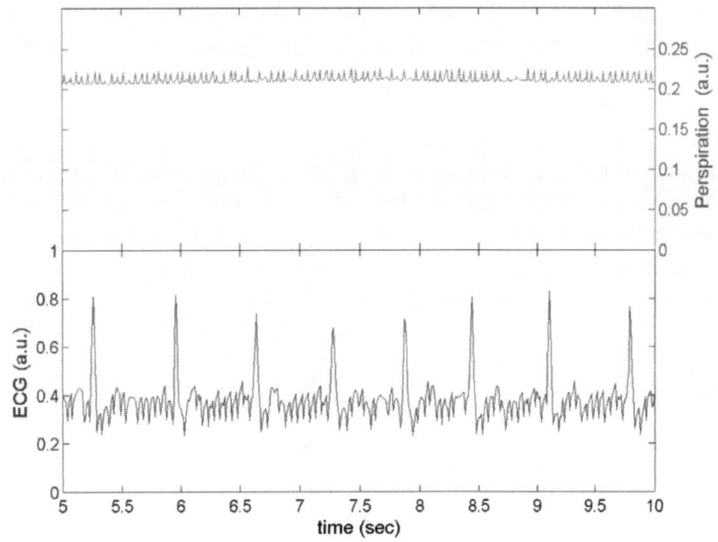

Figure 5. Excerpt of the high sampling rate analogue signals, shown in Figure 4. Bottom panel: ECG, top panel: perspiration. All y-scales are normalized in the interval [0,1].

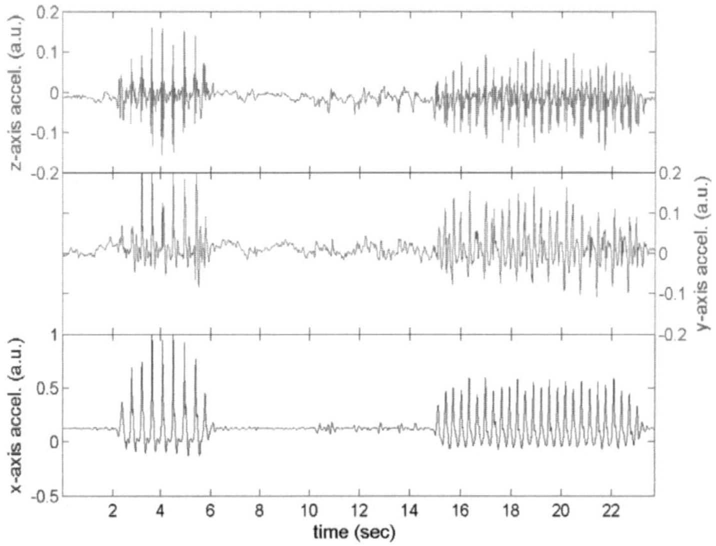

Figure 6. Excerpt of the tri-axial accelerometer time series. First a jumping up and down and then on the spot jogging are presented. All y-scales are normalized in the interval [0,1].

In the next step, which is still under development, this raw physiological signals information will be used to extract useful information about the health status, stress and fatigue level of the fire-fighters. For example, ECG signal will be used to extract the heart rate, tri-axial accelerometer measurements to extract information about motion condition of the firemen, i.e., if a fireman is not moving for a time period longer than a specific threshold, this might indicate that has lost consciousness and his/her life is in immediate danger. On the other hand, respiration, perspiration and body temperature data will be combined with heart rate data to infer stress and fatigue level.

The main idea is that the commander will have a screen depicting each one of his team members in the form of a light indications column vector. Each line of this column vector will correspond to one specific characteristic of firemen's condition; a green light at a specific point will mean that he/she is within the normal levels, whereas a red light will indicate that normal levels have been exceeded for the specific characteristic. Nevertheless, further analysis on the processing of the raw sensor data is out of scope of this work.

5. Conclusion

A wireless sensor network based on wearable sensor node for the monitoring of fire-fighters' physiological parameters has been presented. The current form of the employed wearable wireless sensor node is based on low cost, commercial off-the-self electronic modules, which can easily be attached on a standard fire-fighters' uniform by means of appropriate textile hosting structures like pockets, hemming etc.. System architecture is described in detail, while indicative samples of acquired signals are also presented.

Our future plans include the elimination of all cables. All wired connections will be implemented by means of conductive yarn routing in the textile structure, which is considered feasible due to the low frequency inter-module communications. Even a short-length transmission line based on conductive yarns can be used for the connection between the Wi-Fi transceiver and the antenna, while the antenna itself can be printed on flex-PCB or implemented as fully textile. Moreover, some sensors can be replaced by textile ones. Of course a key aspect of such a high degree of integration into the garment is the textile welding, which is the main scope of our project.

Acknowledgments
Authors wish to acknowledge financial support from Marie Skłodowska-Curie Actions (MSCA) Research and Innovation Staff Exchange (RISE) H2020-MSCA-RISE-2014: Welding of E-Textiles for Interactive Clothing "ETexWeld" (EU Grant Agreement number: 644268).

References
[1] Gaura E I, Brusey J, Kemp J, Thake C D 2009 Increasing safety of bomb disposal missions: a body sensor network approach *IEEE Transactions on Systems, Man and Cybernetics* 39(6), pp. 621-636.
[2] Gao T, et al. 2007 The advanced health and disaster aid network: a light-weight wireless medical system for triage *IEEE Transactions on Biomedical Circuits and Systems* 1(3), pp. 203-216.
[3] Miller H G, Granato R P, Feuerstein J W, Ruffino L 2005 Towards interoperable first response *IT Professional* 7(1), pp. 13-20.
[4] Satyanarayanan M 2004 From the editor-in-chief: when disaster strikes *IEEE Pervasive Computing* 3(4), pp. 2-3.
[5] Lorincz K 2004 Sensor networks for emergency response: challenges and opportunities *IEEE Pervasive Computing* 3(4), pp. 16-23.
[6] IEEE 802.11 WG, IEEE Std 802.11-2007 (Revision of IEEE Std 802.11-1999), *International Standard [for] Information Technology - Telecommunications and information exchange between systems-Local and metropolitan area networks-Specific Requirements - Part 11: Wireless LAN Medium Access Control (MAC) and Physical Layer (PHY) specifications*, ISBN 0-7381-5655-8
[7] Shooshtary S 2008 *Development of a MATLAB Simulation Environment for Vehicle - to – Vehicle and Infrastructure Communication Based on IEEE 802.11p* (Vienna: University of Gavle, Master Thesis.
[8] Arduino, http://www.arduino.cc/ (last access 6/8/2015)
[9] Kursun-Bahadir S, Koncar V, Kalaoglu F, Cristian I, Thomassey S 2011 Assessing the signal quality of an ultrasonic sensor on different conductive yarns usedas transmission lines *Fibres & Textiles in Eastern Europe* 19, 5(88), pp. 75-81.
[10] Kursun-Bahadir S 2015 Effect of textile pretreatment processes on the signal transferring capability of textile transmission lines *Fibres & Textiles in Eastern Europe* 23(2), pp. 55-62.
[11] Kursun-Bahadir S. 2014 Decision of sensor location and best classification method for entrail and muscle disease detection in healthcare smart clothing based on acceleration measurements *Transactions of the Institute of Measurement and Control*, 0142331214552513, doi: 10.1177/0142331214552513.
[12] Rattfält L, Lindén M, Hult P, Berglin L, Ask P 2007 Electrical characteristics of conductive yarns and textile electrodes for medical applications *Medical & Biological Engineering & Computing* 12, pp. 1251-1257.
[13] Rattfält L, Lindén M, Hult P, Ask P, Borga M 2013 Robust heartbeat detector based on weighted correlation and multichannel input. –Implementation on ECG recorded with textile electrodes *International Journal of E-Health and Medical Communications* 4, pp. 61-71.
[14] Atalay O, Kennon W R, Husain M D 2013 Textile-based weft knitted strain sensors: Effect of fabric parameters on sensor properties *Sensors* 13(8), pp. 11114-11127, doi:10.3390/s130811114.

[15] Atalay O, Kennon W R 2014 Knitted strain sensors: Impact of design parameters on sensing properties *Sensors* 14(3), pp. 4712-4730, doi:10.3390/s140304712

[16] Husain M D, Atalay O, Kennon W R 2013 Effect of strain and humidity on the performance of temperature sensing fabric *International Journal of Textile Science* 2(4), pp. 105-112, doi:10.5923/j.textile.20130204.04.

[17] Atalay O, Kennon W R, Demirok E 2015 Weft-knitted strain sensor for monitoring respiratory rate and its electro-mechanical modeling *IEEE Sensors Journal* 15(1), pp. 110 – 122, doi: 10.1109/JSEN.2014.2339739.

[18] Symeonidis S, Vassiliadis S, Mitilineos S A, Potirakis S, Vossou C 2013 On thefeasibility and applications of wearable antennas, *Proc. International Istanbul Textiles Congress: Innovative and Functional Textiles*, pp. 1-5, May 30 – June 1, 2013, Istanbul, Turkey

[19] Mitilineos S A, Vassiliadis S G, Simeonidis S, Potirakis S M, Stathopoulos N A, Savaidis S P 2014 A new wearable antenna designed using genetic algorithms and partially made of textile materials *Applied Researches in Technics, Technologies and Education* 2(3), 239-245.

Sound-maps of environmentally sensitive areas constructed from Wireless Acoustic Sensors Network data

E. T. Michailidis[1,2], J. Liaperdos[3], N.-A. Tatlas[1], S. M. Potirakis[1], and M. Rangoussi[1]

[1] Department of Electronics Engineering, School of Technological Applications, Piraeus University of Applied Sciences, Petrou Ralli & Thivon 250, 12244, Aigaleo, Greece
[2] Department of Digital Systems, School of Information and Communication Technologies, University of Piraeus, 80 Karaoli & Dimitriou St., 18534, Piraeus, Greece
[3] Department of Computer Engineering, School of Applied Technology, Technological Educational Institute of Peloponnese, 23100, Sparta, Greece

E-mail: emichail@unipi.gr, gliaperd@teikal.gr, ntatlas@teipir.gr, spoti@teipir.gr, mariar@teipir.gr

Abstract. "E-SOUNDMAPS" is a distributed microelectronic system for the sound/acoustic monitoring of areas of environmental interest that is based on an appropriately designed wireless acoustic sensor network (WASN). It involves the automated generation of multi-level sound-maps for environmental assessment of areas of interest. This paper focuses on the method and the software application for the construction of sound-maps, which is developed as part of the integrated "E-SOUNDMAPS" system. The software application periodically produces geographically-referenced, accurate environmental sound information, based on real-field measurement data, and integrates them in the geographic map of the area of interest in a concise and comprehensive manner. Following the field recording of sound and the hierarchical recognition/classification of sound events and corresponding sources, the obtained sound sources characterization tags feed the specific software application. The output is a multi-level soundmap, constructed on the basis of the data and published electronically on the Web, for human inspection and assessment. All necessary steps for handling, archiving, monitoring, visualization and retrieval of sound data are also presented.

1. Introduction

In recent years, electronic monitoring of areas of environmental interest has emerged due to the advances in electronics, computers, and radio communications and has attracted considerable interest by academia and industry [1, 2]. In the context of environmental protection and prevention strategies, the sound/acoustic component plays an important role, since it provides critical information related with both the current status of environmentally sensitive areas and the spatio-temporal evolution of anthropogenic activities and/or natural/meteorological/geological phenomena that significantly influence this status. By exploiting sound/acoustic monitoring instead of other alternatives, i.e., optical monitoring by cameras, is time- and cost-efficient, as well as less sophisticated. The exigencies for protection of high quality acoustic environments has been also recognized by EU Directives

2002/49/EC 2003/10/EC on the Assessment and Management of Environmental Noise [3, 4]. This paper considers an integrated system, called "E-SOUNDMAPS", which incorporates a distributed microelectronic networking system (DMNS) for the sound/acoustic monitoring of areas of environmental interest. This subsystem is based on an appropriately designed small-scale wireless network of acoustic sensors, i.e., microphones, and attempts to automatically generate multi-level soundmaps for environmental assessment. In general, sound maps concisely depict the results of recognition and illustrate in a specific geographic point on the map the sound sources information in multiple levels of detail [5–7]. As far as field measurements of sound/acoustic events in a selected area of environmental interest are conducted, a hierarchical recognition/classification scheme for the classification, recognition and characterization of sound/acoustic events and sources (fauna, meteorological, anthropogenic and further detailed analysis into subclasses) is developed and the generation of multi-level soundmaps is automatically generated. In this paper, a valuable software web application for sound/acoustic monitoring fed by the sound sources characterization tags is presented. This application is capable of visualizing the recorded sound/acoustic environmental information based on real-field measurement data (provided by the aforementioned DMNS) on the geographic map of the area of interest and automatically publish and update the critical information for human inspection and assessment. A relational database model and a client-server architecture are exploited to construct the proposed application.

The remainder of the paper is organized as follows. Section 2 describes the system architecture, whereas Section 3 outlines the design and implementation of the database, where the useful information is stored. Section 4 details the web application that visualizes the processed sound data. Finally, Section 5 provides concluding remarks.

2. System Architecture

In this paper, a DMNS equipped with acoustic sensors is considered. The DMNS is capable of recording the sound/acoustic information of a specific area and then transmit this information to the central network site, after conditioning, pre-processing and coding for compression. The field-recording of sound/acoustic sources and events, at a selected set of cartographic points span the area of interest, taken at seasonal, daily and spatial resolution. Signal detection and digital processing methods are employed to localize sound sources and events (onset and endpoint of sources of interest, possibly overlapping) across the sound recording; a novel hierarchical scheme is developed for the classification, recognition and characterization of the sources. Sound sources characterization tags feed the software application that generates multi-level soundmaps and publishes them electronically for human inspection and assessment. A pilot wireless sensors network (WSN) is employed to support the software applications as well as the study of requirements (in hardware/computations, energy, communication resources) for the scaling up of the pilot network into a full-scale deployment. This deployment involves several tasks, such as programming and managing the sensors, programming the gateways, interpreting and processing the data streams, and setting up servers for data archival and retrieval.

Depending on the area coverage schemes, different communication protocols between network units and associated network architectures can be utilized. In particular, for limited area coverage, the Integrated Peripheral Units (IPUs) are distributed over the area of interest and communicate wirelessly by using the IEEE-802.11.x protocol (wireless local area network – WLAN) with the Central Network Unit (CNU). On the contrary, a number of appropriately distributed Intermediate Network Units (INUs) gather a group of IPUs each (local area network – LAN) and communicate with the CNU by using Worldwide Interoperability for Microwave Access (WiMAX) or Global System for Mobile Communications (GSM) technologies (wide area network – WAN), when extended area coverage is required. The hardware of DMNS and the communication network architecture and protocol are designed and realized in a functional prototype form and support long-term operation. The scope of the aforementioned system is the development of a software application for the localization and hierarchical recognition/characterization of sound sources and events within sound recordings. This application superimposes, regularly updates, and electronically publishes the sound data on the digital geographic map of the area in web format and in multiple layers of detail for visual inspection. The primary information produced by the proposed software should be initially processed. This information can be properly visualized and the soundmaps can be composed, as long as the processed information is transferred and stored in a database. Next section handles the design and implementation of this database.

3. Design and Implementation of Database

The database is designed according to the simplified entity-relationship (ER) model [8] shown in Fig. 1, where primary key attributes are underlined. Entities correspond to (a) the sensor

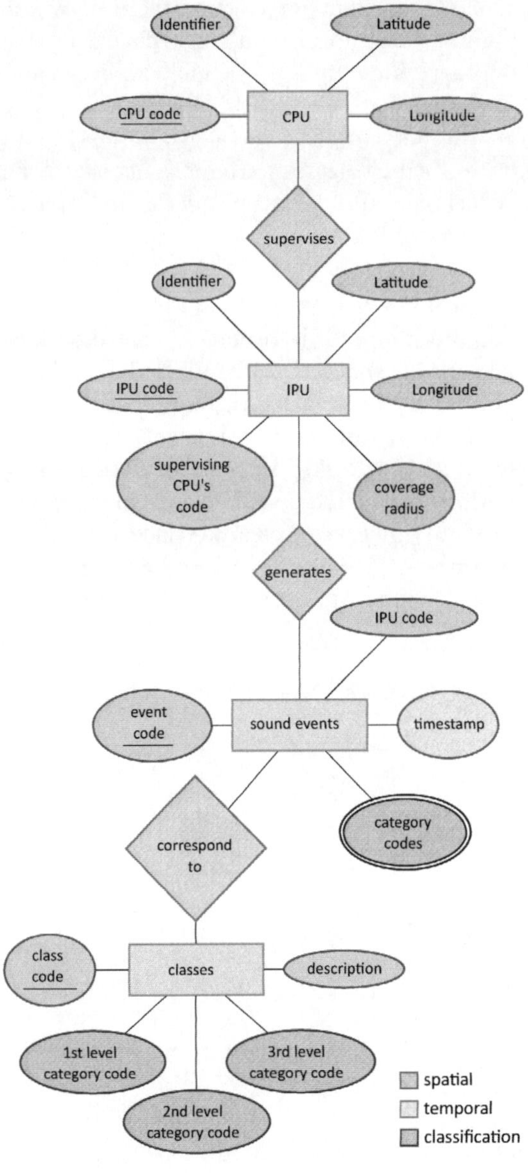

Figure 1. The simplified database entity-relationship model.

network nodes, namely the Central Processing Units (CPU), and the Integrated Peripheral Units (IPU), (b) to the sound events generated by the IPUs, and (c) to the classes to which these events are associated. Network node attributes are selected such as to allow the identification of each specific node, i.e., via the "identifier" attribute, the definition of its precise location, i.e., via its geographical coordinates, together with other spatial information if applicable, i.e., sensor coverage radius, as well as its position within the network hierarchy which is, for instance, determined by the code of the CPU to which a specific IPU is attached. Similarly, sound events inherit the spatial properties of their parent IPU, while temporal information, i.e., a timestamp, is also necessary in order to define an events location on the time line. Incorporating the results of the classification procedure, each sound event is also mapped to a specific class that is explicitly identified by its 1st, 2nd and 3rd level category codes. The structured design of the database allows the generation of alternative soundmap visualizations, based on different spatial and temporal criteria, as well as the composition of various statistical reports, as described in the following section.

The practical implementation of the database follows a client-server architecture, where a MySQL database server is used for central storage of the information, in order to ensure security and compatibility and ease development and integration. MySQL is widely accepted and supported across many platforms and coding languages and capable of describing the network hierarchy, the classified sound events collected from the network nodes (tagged in space and time), as well as the class hierarchy used for sound event categorization. Remote access to the stored information is allowed to soundmap clients, which can apply custom queries in order to retrieve spatiotemporal slots of information, based on user preferences.

4. Design and Implementation of Web Application

In raw data format, a soundmap consists of all the spatio-temporal information that describes the classified sound events being collected from a specific sensor network. While data related to the network topology is stationary requiring update only after network modifications, i.e., when sensor nodes are added or removed sound event data correspond to a rather large volume of information that should be periodically updated. Such a procedure, that involves loading recent classified sound events to the database, is performed by the sound event classification software and is indicated as "soundmap synthesis/update" in Fig. 2 (see interface I1), which illustrates the modules involved in soundmap generation and visualization. In order to provide a soundmap visualization according to user preferences, a web application was designed that

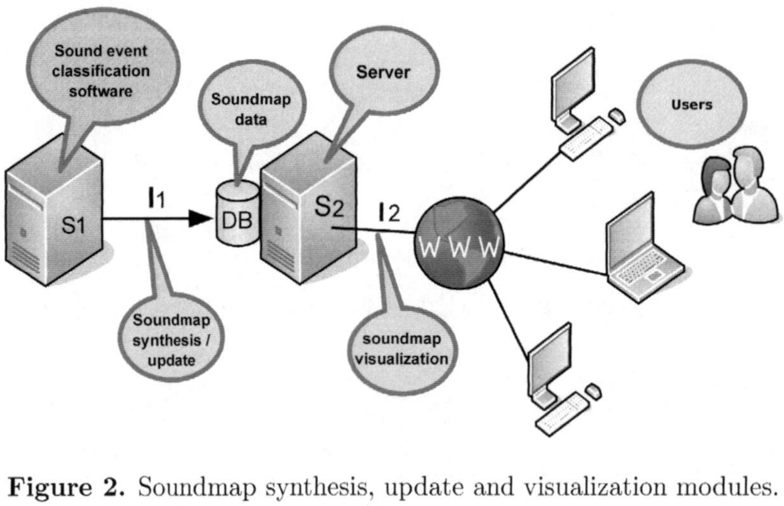

Figure 2. Soundmap synthesis, update and visualization modules.

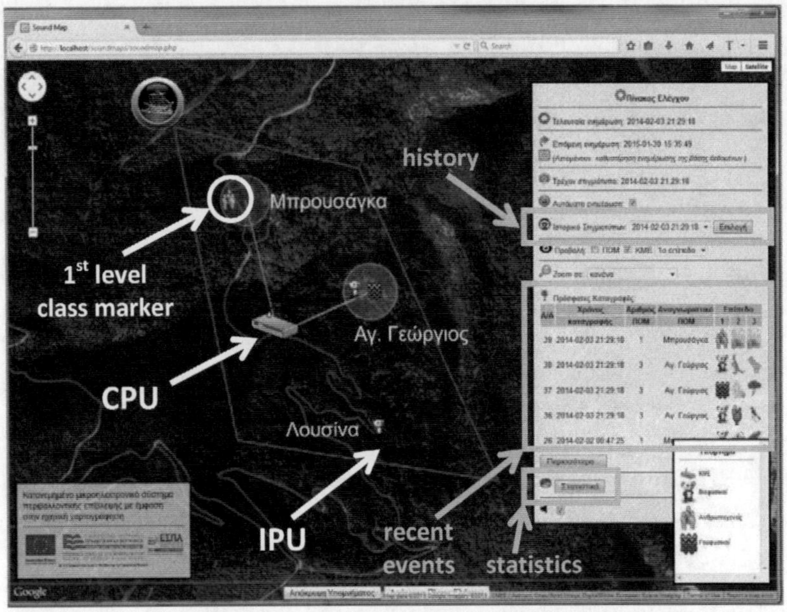

Figure 3. Soundmap visualization and graphical user interface.

handles the interaction between users and the database, and is responsible for querying the appropriate information and for its visual representation in the users web browser (see interface I2 in Fig. 2). This application has been implemented using open-source web technologies, e.g., php, Google maps, javascript, etc.

A typical soundmap visualization is shown in Fig. 3, where the graphical user interface (GUI) is also presented. The soundmap visualization provides a graphical representation of the supervised area of environmental interest, where the sensor network elements are presented as marker overlays. Classified sound events associated to a specific IPU over a certain time interval are also represented by their corresponding class markers, while the classification tree level is user configurable. A control panel provides access to recent event information, while the user can navigate to older soundmap snapshots using the "history" selector. Furthermore, the user can request the generation of various statistical reports by defining several criteria, such as the period of time, specific IPUs, specific sound events/classes etc. Based on similar criteria, the GUI also allows the definition of specific alerts which are automatically activated when the user-defined rules are reached or violated.

While soundmap synthesis and update are rather simple procedures that involve loading the tagged sound event data into the database, soundmap visualization is a complex procedure that involves several interoperating software modules, as shown in Fig. 4 and described below. A client-server architecture is adopted, where the servers functionality is implemented in system S2 (see also Fig. 2) to which client systems have access through the World Wide Web. This system hosts the database (DB) and also incorporates web server and dynamic web-page generation software modules. Client communication with the server is exclusively performed through the clients web browser software. Apart from the main soundmap server (S2), client machines also retrieve spatial and terrain-specific information from the geographical map providers server, i.e., Google Maps, and may also connect to other external servers in order to maintain access to specific services, i.e., statistics visualization.

Fig. 5 illustrates a simplified sequence diagram [9] that describes the phases involved in soundmap visualization. The corresponding steps (indicated by the circled numbers shown in

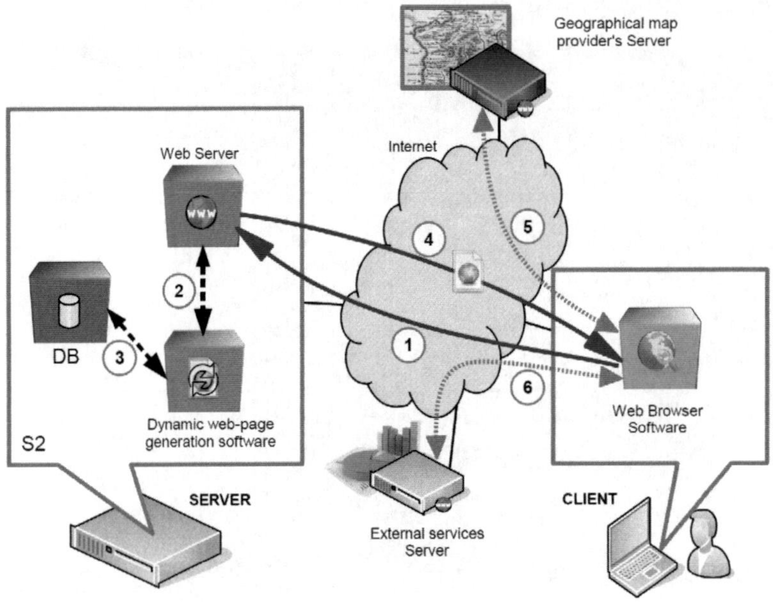

Figure 4. Simplified software interoperability diagram for soundmap visualization (numbers correspond to the sequence diagram shown in Fig. 5).

Figs. 4 and 5) are analyzed as follows:

(i) Clients web browser software sends a soundmap web-page request to the server (1) .

(ii) Web server requests the main soundmap web-page from the dynamic web-page generation software (2) .

(iii) Dynamic web-page generation software connects to the database, from which the current digital soundmap representation is retrieved in order to synthesize the soundmaps main dynamic web-page that incorporates all specified functionality provided by its graphical user interface (3) .

(iv) Main dynamic soundmap web-page is forwarded from the dynamic web-page generation software to the web server (2) .

(v) Main dynamic soundmap web-page is forwarded from the web server to the clients web browser (4) .

(vi) By execution of the client-side software code that has been incorporated to the soundmaps web-page, the client may establish connections to external servers, in order to complete soundmap visualization, namely to retrieve and display the underlying geophysical map and also to provide statistics visualization features (5) ,(6) .

Note that the soundmap dynamic web-page retrieval sequence shown in Fig. 5 is repeated periodically in an automatic manner, with a frequency that is related to the database update frequency. However, external services call is a user-initiated procedure.

Due to the wide employment of MySQL during the last years, many software packages and interfaces have been developed to facilitate the communication between the database and several programming language and software applications. Among these applications, Microsoft Excel is a simple and user-friendly but operational way of interconnection with a MySQL database. Alternatively, MATLAB software package can be used, which represents a complicated and

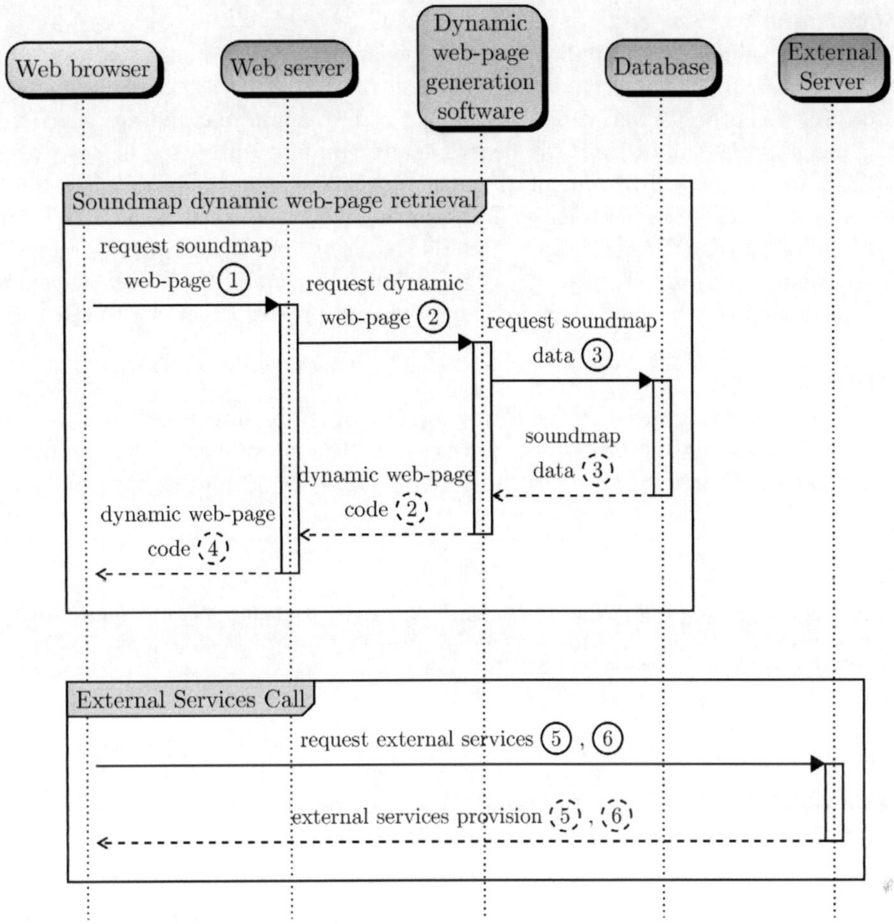

Figure 5. Simplified sequence diagram for soundmap visualization.

approach. Microsoft Excel enables the organization of data in rows and columns, the accurate and efficient process of data, the automatic realization of mathematical operations, and the visualization of data through graphs. To enable the automatic interaction between the MySQL database and spreadsheet files of Microsoft Excel, the "MySQL for Excel" Add-in is used [10]. This simple, flexible, friendly, and accurate tool allows for real-time data import/export. In particular, the MySQL data can be imported into Excel, Excel data can be exported into MySQL as a new table or appended to a current table, and MySQL for Excel enables the edit of the MySQL data directly though Excel. MySQL for Excel is loaded and executed by selecting the "Data" menu tab in Excel, and choosing the "MySQL for Excel" Database icon. Then, a new Excel sidebar with the available MySQL for Excel options is opened. To access the data obtained from previous layers of the proposed web application, the desired information is sent to a gateway computer and exported into Comma Separated Values (CSV) files using specialized algorithms, which analyze and separate the raw byte stream [11]. The CSV format files store tabular data, i.e., numbers and text, in plain text. Hence, the file is interpreted as a sequence of characters and each line of the file is a data record, which consists of one or more fields, separated by commas. Indeed, these files represent sets or sequences of records in which each record has an identical list of fields and can be directly read by various software packages, including Microsoft Excel, Matlab, Open Office Calc, Google Docs, etc.

5. Conclusions

In this paper, a valuable web application that registers and accurately and periodically records and visualizes sound/acoustic environmental information on the geographic map of the area of interest has been presented. The periodic update of soundmaps is designed on the basis of real-field measurement data that are electronically communicated to the software application. Using this application, environmentally sensitive areas can be monitored and the impact of anthropogenic activities/uses as well as the evolution of natural factors (fauna, climate) can be demonstrated. Target groups for the exploitation of the proposed web application are environmental/ecological groups, organizations or agencies as well as central and local governmental agencies involved with development and protection of the physical environment.

Acknowledgments

This work has been co-financed by the European Union (European Social Fund - ESF) and Greek national funds through the Operational Program "Education and Lifelong Learning" of the National Strategic Reference Framework (NSRF) - Research Funding Program ARCHIMEDES III.

References

[1] Mainwaring A, Polastre J, Szewczyk R, Culler D and Anderson J 2002 Wireless sensor networks for habitat monitoring *Proc. ACM Int. Conf. WSNA02 (September 2002, Atlanta, Georgia, USA)*

[2] Boonma P and Suzuki J 2006 A biologically-inspired sensor network architecture for autonomous ecological observation *Proc. 5th Elsevier/ISEI Int. Conf. on Ecological Informatics (ISEI5) (December 2006, Santa Barbara, CA, USA)*

[3] Wrightson K 1999 An introduction to acoustic ecology *J. of Electroacoustic Music* **12**

[4] World Forum for Acoustic Ecology (WFAE), website: `http://interact.uoregon.edu/MediaLit/wfae/home/` and "Soundscape: The Journal of Acoustic Ecology".

[5] `http://riarevolution.com/2008/11/26/google-sound-maps/`

[6] `http://www.soundseeker.org/`

[7] `http://woices.com/`

[8] Chen P 1976 The entity-relationship model – Toward a unified view of data *ACM Tran. on Database Systems* **1** 9

[9] Unified Modeling Language v2.4.1, Open Management Group. `http://www.omg.org/spec/UML/2.4.1/Superstructure/PDF/`

[10] `https://www.mysql.com/why-mysql/windows/excel/`

[11] da Silva Campos B, Rodrigues J, Mendes L, Nakamura E and Figueiredo C 2011 Design and construction of wireless sensor network gateway with IPv4/IPv6 support *Proc. IEEE International Conference on Communications (ICC) (June 5–9 2011)* p 1

Stress state evaluation in low carbon and TRIP steels by magnetic permeability

M.-E. Kouli[1] **M Giannakis**[2]
Dpt. of Mining & Metallurgical Engineering,
National Technical University of Athens, Zografou Campus, Athens, 15780, Greece

[1]E-mail: mariliza@mail.ntua.gr, [2]E-mail: giannakismic@gmail.com

Abstract. Magnetic permeability is an indicative factor for the steel health monitoring. The measurements of magnetic permeability lead to the evaluation of the stress state of any ferromagnetic steel. The magnetic permeability measurements were conducted on low carbon and TRIP steel samples, which were subjected to both tensile and compressive stresses. The results indicated a direct correlation of the magnetic permeability with the mechanical properties, the stress state and the microstructural features of the examined samples.

1. Introduction

Steel is described as an industrial iron - carbon alloy, where the weight content of carbon ranges from 0.025 %w/w to 2.1 %w/w. In steels, various phases are shown, due to the allotropic behaviour of iron and to the complex heat treatments, during the manufacture of steel. Some of these phases, such as ferrite, perlite, martensite, and bainite, exhibit ferromagnetic behaviour [1].

It is clear that steel is a composite ferrous metal alloy with varying mechanical properties, which impart a wide range of applications. However, the different parts of a structure made of steel need to be systematically checked, in order to determine their condition. This control is carried out by various techniques, either destructive or non-destructive [2-11]. The magnetic permeability measurement technique is included among the non-destructive techniques; however, this method is still in laboratory level [11-22].

Conclusions about the characteristics of the material and its structure can be derived by the analysis and study of the output signal of this magnetic sensor. This method can detect defects on the material surface, or throughout the mass, without requiring special preparation of the specimen. Of course, as a method of magnetic non-destructive testing, it is limited to ferromagnetic materials with relatively smooth surfaces, preferably without paint or other coatings.

In recent years, the interest of researchers has grown around the application of magnetic methods in determining the stress condition of steels. However, there is a limited number of published articles on the determination of residual stress using the magnetic permeability method [12-13, 16-17].

For this reason, the magnetic permeability measurements were conducted on low carbon and TRIP steel samples, which were subjected to both tensile and compressive stresses, and thereby there is a direct correlation of the magnetic response values to those deriving from the mechanical stresses inherent in the material.

[1] **Tel.:** +30 210 772 2183, **E-mail :** mariliza@mail.ntua.gr
[2] **Tel.:** +30 210 772 2183, **E-mail :** giannakismic@gmail.com

2. Experimental procedure

2.1. As received materials

In the present article two types of steel are studied: the first was low carbon steel type St37- 2 and the second TRIP. The dimensions of the samples were $20cm^L$ x $15cm^W$ x $0,15cm^T$. The chemical composition of each steel type is presented on table 1.

Table 1. Chemical composition of tested steels.

	C	Mn	Si	P	S	Fe
St37- 2	0.090	0.520	0.010	0.019	0.007	Balanced
TRIP 800	0.195	1.64	1.58	-	-	Balanced

2.2. Magnetic measurements

Two sensors for the magnetic permeability measurement: one was identifying the surface magnetic response in St-37 steel, while the second was identifying the bulk magnetic permeability of the TRIP 800 steel.

The magnetic sensor for the determination of the surface magnetic permeability used in the laboratory is shown in Figure 1(a). The sensor is excited by a sinusoidal signal of 1KHz frequency by a function generator. The excitation and reception signal is received by a NI PXI system data acquisition card (DAQ), in order to be processed (filtering output signal, graph display) in LabView environment. The receiving coil is wounded around an electromagnet, resulting in obtaining the surface's output signal. The maximum value of this voltage is the representative value of the surface magnetic permeability, at the point of measurement in St-37 grade of steel.

The laboratory magnetic sensor, which is used in order to determine the bulk magnetic permeability is illustrated in Figure 1(b). An alternating sinusoidal voltage pulse (amplitude 1V, frequency 1Hz), is applied to the excitation coil. The change in magnetic flux, due to the presence steel, is measured by a receiving coil located on the one leg of the electromagnet. The output voltage of the magnetic sensor is proportional to the magnetic permeability of the TRIP 800 steel. The absolute value of the output signal's total maximum is the examined magnetic parameter.

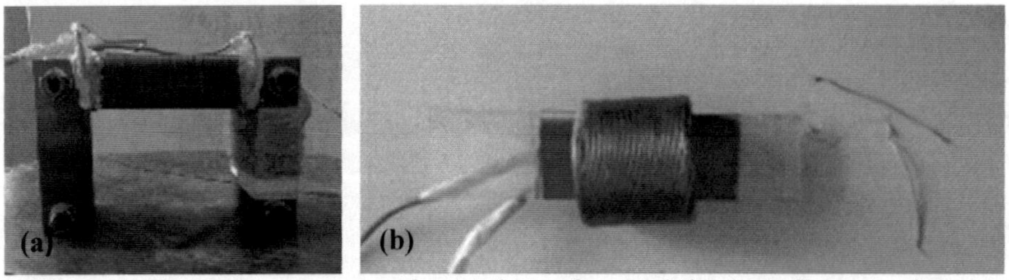

Figure 1. Magnetic measurement sensors for (a) the surface and (b) the bulk magnetic permeability.

The samples have to be calibrated, in order to determine their residual stress values. Therefore, tensile and compressive tests, at preselected deformation steps within the elastic region, took place. The direction loading tests is parallel to the rolling direction, in both samples. Dog-boned samples are cut from the as-received ones, according to the ASTM E8 standard.

At each pre-defined deformation step, five magnetic values were recorded and averaged. These averaged values represent the corresponding magnetic parameter. During the experiment, the stress values were recoded. The completion of the above process led to the determination of the calibration curve of the samples. Through this curve, it was possible to assign the magnetic parameter to the values of residual stresses in the studied material.

IOP Conf. Series: Materials Science and Engineering **108** (2016) 012013 doi:10.1088/1757-899X/108/1/012013

3. Results

The correlation between the applied stresses (either tensile or compressive) is possible, taking as parameter the total maximum of the magnetic sensor's output waveform. This is illustrated as a calibration curve (figure 2 and 3).

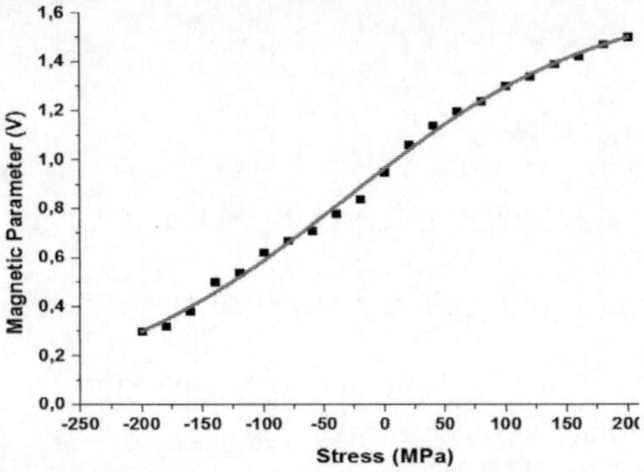

Figure 2. Calibration curves of St 37-2 steel grade.

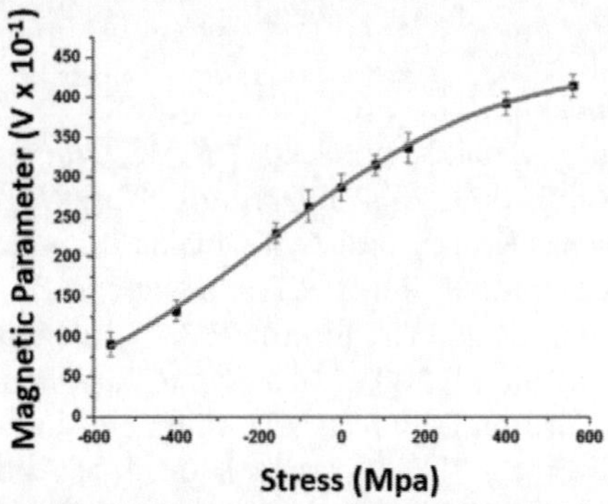

Figure 3. Calibration curves of TRIP 800 steel grade.

Steels are positively magnetostrictive. The presence or absence of applied stresses cause the rearrangement of the magnetic domains, in such a way as to minimize the magnetoelastic and the anisotropy energy. When the direction of the applied stress differs from the magnetization orientation or the number of the 180° domain walls is increased the total energy of the magnetic configuration is minimized [23-27].

Thus, in stressed samples, the initial easy axis magnetization rotates, in order to become parallel to the direction of the applied stress. Therefore, the magnetization vectors of a ferromagnetic material under tensile stresses, which are considered as positive stresses, is aligned towards the direction of the external stress, resulting in the increment of the 180° domain walls. On the contrary, compressive stresses, which are considered as negative stresses, tend to align the magnetization vector perpendicular to the stress axis. As a result, regarding a positively magnetostrictive ferromagnetic alloy, the tensile stresses increase the magnetic response, while the compressive ones reduce it (figure 4 and 5).

The differences of the calibration curves of St-37 low carbon steel and the TRIP steel, within the elastic region, are interpreted on the basis of the changes in the microstructures. The St-37 is a common ferrite-pearlite steel (figure 4), while the Transformation Induced Plasticity steel represent a family of multiphase steels (figure 5). TRIP steel consists of a mixture of ferrite and bainite and what seems to retained austenite.

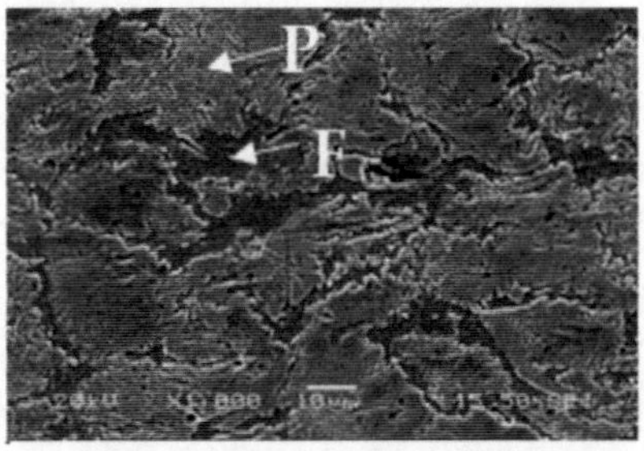

Figure 4. SEM micrograph of St-37 steel grade.

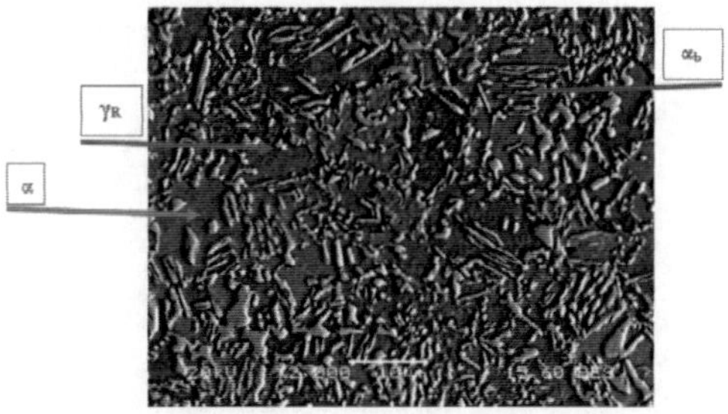

Figure 5. SEM micrograph of TRIP 800 steel grade.

4. Conclusions

Surface and bulk magnetic permeability measurements potentially can be used as non-destructive evaluating methods for estimating the stress state in a ferromagnetic materials. The variations of the magnetic permeability values are sensitive to the microstructural characteristic of the ferromagnetic steels.

5. References

[1] Vourna P and Ktena A 2011 *Key Engineering Materials* **495** 257
[2] Hristoforou E 2002 *J. Opt. Adv. Mat.* **4** 245
[3] Kosmas K, Sargentis C, Tsamakis D and Hristoforou E 2005 *J. Mat.Proc.Tech.* **161** 359
[4] Hristoforou E, Reilly R E and Niarchos D 1993 *IEEE Trans. Magn.* **29** 3171
[5] Kosmas K and Hristoforou E 2007 International J. of App. Electr. and Mech. **25** 319
[6] Hristoforou E 2003 *Meas. Sci. & Technol.* **14** R15
[7] Hristoforou E, Niarchos D, Chiriac H and Neagu M 2001 *Sensors & Actuators A* **92** 132
[8] Hristoforou E and Kosmas K Int. J. of Appl. Electrom. and Mechanics **25** 287
[9] Hristoforou E and Reilly RE 1993 *J. Magn. Magn. Mat.*, **119** 247
[10] Hristoforou E, Kosmas K and Kollar M 2008 *J. Electr. Eng.* **59** 90
[11] Augustyniak B, Piotrowski L, Chmielewski M, Kosmas K and Hristoforou E 2010 *IEEE Trans. Magn.* **46** 544
[12] Vourna P, Ktena A, Tsakiridis P E and Hristoforou E 2015 *Measurements* **71** 31
[13] Vourna P, Ktena A and Hristoforou E 2014 *IEEE Trans. Mag.* **50** 1
[14] Vértesy G, Mészáros I and Tomáš I 2013 *NDT & E Int.* **54** 107
[15] Franco FA, González MFR, De Campos MF and Padovese LR 2013 *J. Nondestruct. Eval.***32** 93
[16] Vourna P, Ktena A, Tsakiridis P E and Hristoforou E 2015 *NDT & E Int.* **71** 33
[17] Vourna P, Hervoches C, Vrána M, Ktena A and Hristoforou E 2015 *IEEE Trans.* **51** 6200104
[18] Ktena A, Hristoforou E, Gerhardt GJL, Missell FP, Landgraf FJG, Rodrigues Jr DL and Alberteris-Campos M 2014 *Physica B: Coned Matter* **435** 109
[19] Hristoforou E and Reilly R E 1990 *Trans. Mag.* **26** 1563-1565
[20] Hristoforou E and Niarchos D 1992 *IEEE Trans. Mag.* **28** 2190-2192
[21] Dimitropoulos P D, Avaritsiotis J N, Hristoforou E 2003 *Sensors & Actuators A* **107** 238-247
[22] Giouroudi I, Ktena A and Hristoforou E 2004 *J. Opt. Adv. Mat.* **6** 45-50
[23] Capó Sánchez J, Campos M F and Padovese L R 2012 *J. of Mag. and Mag. Mat.* **324** 11-14
[24] Lindgren M and Lepistö T 2004 *NDT & E Int.* **37** 403-410
[25] Moorthy V, Shaw B A and Hopkins P 2006 *J. of Mag. and Mag. Mat.* **299** 362-375
[26] Pérez-Benítez J A, Espina-Hernández J H, Martínez-Ortiz P, Chávez-González A F and Rosa J M 2013 *J. of Mag. and Mag. Mat.* **347** 51-60
[27] Blaow M, Evans J T and Shaw B A 2005 *Acta Mat.* **53** 279-287

Design and development of a new magnetic sensor for stress measurements

S Aggelopoulos[1]
Dpt. of Mining & Metallurgical Engineering,
National Technical University of Athens, Zografou Campus, Athens, 15780, Greece

E-mail: spyrosag@gmail.com

Abstract. This paper describes the design and the development of a new magnetic sensor for stress measurements using the magnetic Barkhausen noise and the magnetic permeability techniques in ferromagnetic steels. Both techniques together, become an important non-destructive technique, due to its exceptional material and stress characterization capabilities. The correlation of the two methods was investigated. Conclusions were derived based on the experimental results.

1. Introduction

The Barkhausen effect has become an important Non-Destructive Technique, which can be used in order to evaluate ferromagnetic materials. It was discovered in 1919 by the German physicist Heinrich Barkhausen, who wrapped a wire around a ferromagnetic material. He noticed that, under the influence of a varying magnetic field, a sound was heard from a speaker, which was connected to the coil. The sound was similar to static noise, which led to the name "Barkhausen noise" [1]. The Barkhausen signal can be used as a method of Non-Destructive Testing of materials, since it can provide useful information about the structure and the properties of the ferromagnetic materials [2-4].

2. Theoretical background

The magnetic materials consist of magnetic regions, called domains, the boundaries of which are called domain walls. Each domain is characterised by a magnetization vector. Under the influence of an external magnetic field, the magnetization vector of each of the domains rotates, in order to point in the same direction with the vector of the external magnetic field. As a result, the domain walls move and the magnetic domains change their size.

If we place a coil near or around the sample during the magnetization process, we will be able to record an electrical signal. The signal is abrupt and it is caused by the small and not continuous movement of the domain walls, due to the external magnetic field. This signal is called "Barkhausen Noise" and is affected by the properties and the various impurities and structural imperfections of the sample [5-9].

3. Measuring procedure

A basic concept of the measuring procedure includes the transfer of an alternating electrical signal from a source to an excitation coil, the conversion of the electrical signal to a magnetic flux, which

[1] Tel: +30 2107722183, E-mail: spyrosag@gmail.com

propagates through the tested material, the reconversion of the magnetic signal to an electrical signal, the amplification and the filtering of the output signal and finally, the recording of the processed signal.

The coils are usually wrapped around a U-shaped core [10-28]. The excitation coil is wrapped around the core's body, while the pickup coil is placed in the free space between the two legs of the electromagnet. As a result, when the electromagnet is placed on the tested material, a closed circuit is created, which forces the signal to travel through the sample and be affected by its properties. The electromagnet described above is called "single yoke arrangement" (figure 1).

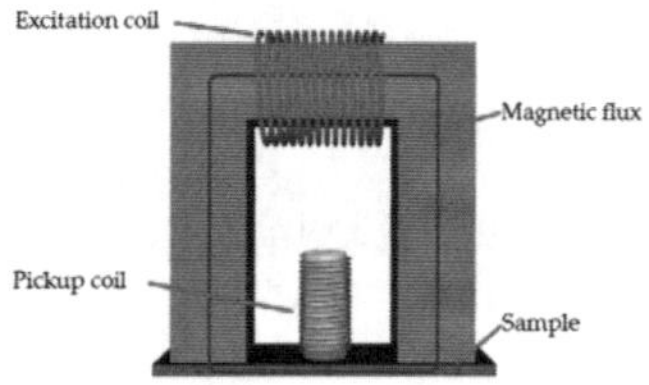

Figure 1. The single-yoke arrangement.

Using this information, it is possible to detect deformations induced by mechanical stresses [29], monitor the fatigue of a material [30] and inspect its structure [31].

4. Development of the magnetic Barkhausen sensor

4.1. The arrangement

The chosen arrangement includes a signal generator which produces the excitation signal, an electromagnet with the suitable excitation and pickup coils, electronic amplification and filtering circuits, a power supply for the operation of the electronic components and an oscilloscope to display the output signal. In order to construct the sensor, we chose the single-yoke design.

4.2. The electronic circuit design

The signal amplification and filtering was possible due to a printed circuit board (PCB), which was designed and constructed in our laboratory. This electronic circuit includes a preamplifier, a low pass and a high pass filter (figure 2).

The preamplifier circuit amplifies the output signal of the pickup coil. The Texas Instruments DRV134 and INA163 differential amplifiers were used for this purpose, due to their low noise performance. The amplification factor (gain) of the input signal can be set from 2 to 200 times.

Two Texas Instruments OPA209 operational amplifiers were also used, in order to construct the high pass and the low pass filters. The high pass filter was designed to block the signal frequencies which are lower than 2 kHz and additionally, to amplify the signal 10 times. Similarly, the low pass filter blocks the frequencies which are higher than 100 kHz and amplifies the signal 10 more times. The filters were designed as second-order Butterworth filters.

As a result, the output signal was amplified by a total of 20000 times and it was limited in the range of 2-100 kHz, which is commonly chosen, in order to eliminate the ambient noise [32,33].

4.3. Development of the sensor

A printed circuit board was designed in order to include the electronic circuit above. The electronic circuit was printed on a photosensitive board, which was exposed to ultraviolet radiation.

The electromagnet was made using a U-shaped electrical steel core. Two connected in series excitation coils of 200 windings were constructed, using a Ø0.5 mm copper wire. Between the two legs of the electromagnet, the pickup coil was placed, which consisted of 600 windings, made of a

IOP Conf. Series: Materials Science and Engineering **108** (2016) 012014 doi:10.1088/1757-899X/108/1/012014

Ø0.1 mm copper wire. In order to achieve the optimum contact between the pickup coil and the various test samples, a spring was placed between the yoke and the pickup coil.

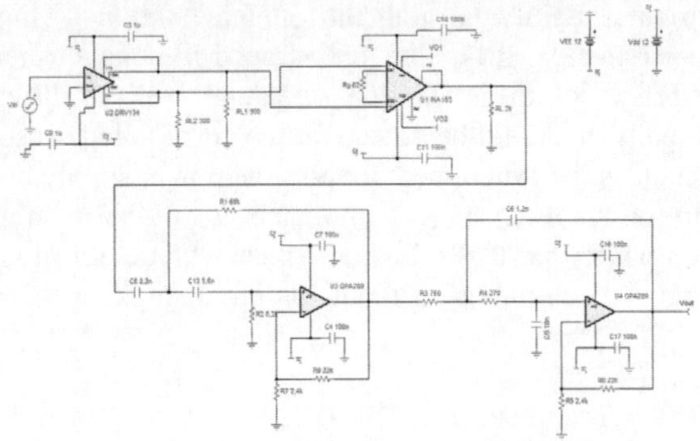

Figure 2. The electronic circuit design.

5. Experimental results

In order to test the constructed arrangement, we used a low carbon steel plate as a sample. Using a triangular 13 V, 2 Hz waveform as an excitation signal, we were able to record the Barkhausen noise signal obtained by the pickup coil, after its amplification and filtering (figure 3). We were, also, able to plot the output/input signal ratio versus input (figure 4), which indicates the effect of the induced magnetic field to the magnetic permeability of the tested material.

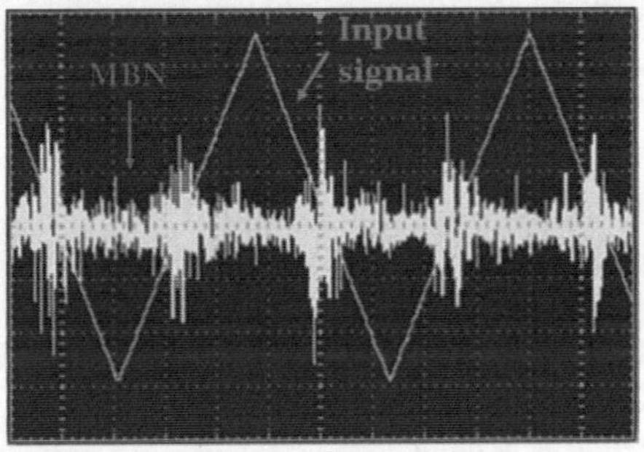

Figure 3. The input signal and the Barkhausen noise signal of a low carbon steel plate versus time.

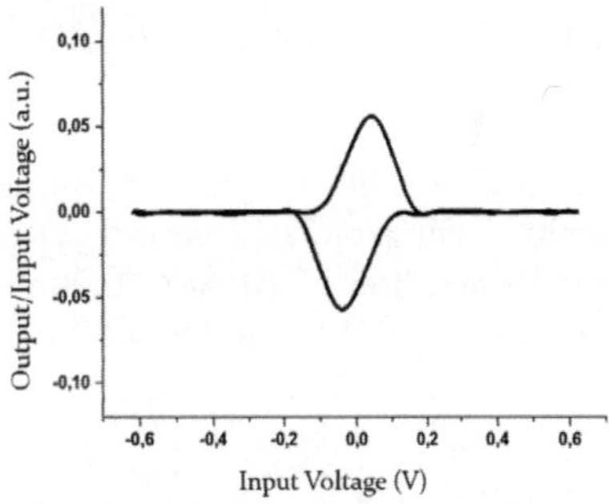

Figure 4. Correlation between the magnetic permeability and the induced magnetic field.

6. Discussion - Conclusions

The results mentioned above indicate that the developed sensor can be used effectively in order to obtain the magnetic Barkhausen noise signal of a tested material. Using the acquired data, it is possible to correlate the magnetic permeability of the sample with the induced magnetic field. As a result, we can use the magnetic Barkhausen sensor in order to study both the Barkhausen noise signal and the magnetic permeability of a material.

Furthermore, we conclude that it is possible to develop a magnetic Barkhausen sensor and the necessary electronic components with relatively low cost. However, the sensor's very weak output signal in conjunction with the ambient noise, make it too difficult to capture and process the data. Thus, it is necessary to design and construct the appropriate electronic circuits, in order to amplify the signal and eliminate the noise through its filtering.

The selection of the electronic components and the design of the electronic circuit was based on the elimination of the ambient noise and the best possible signal amplification, without adding extra noise to the processed signal, through the components. As a result, it would be useful to further improve the electronic circuits, mainly through the implementation of higher cost components, in order to maximize their performance. It is, also, encouraged to conduct measurements using various amplitude and frequency ranges of the excitation signal, as well as different types of samples of ferromagnetic materials. Finally, it should be noted that the constructed sensor arrangement consists of some laboratory devices, such as the function generator, the power supply and the oscilloscope, which help in error reduction, but, on the other hand, complicate its outdoor operation. As a result, it would be useful to develop a completely portable device, which will be able to include all the necessary electronic and electromechanical equipment, without compromising its performance.

7. References

[1] Barkhausen H 1919 *Physische Zeitschrift* **20** 401
[2] Vourna P, Ktena A, Tsakiridis P E and Hristoforou E 2015 *Measurements* **71** 31
[3] Vourna P, Ktena A and Hristoforou E 2014 *IEEE Trans. Mag.* **50** 1
[4] Vértesy G, Mészáros I and Tomáš I 2013 *NDT & E Int.* **54** 107
[5] Vourna P and Ktena A 2011 *Key Engineering Materials* **495** 257
[6] Franco FA, González MFR, De Campos MF and Padovese LR 2013 *J. Nondestruct. Eval.***32** 93
[7] Vourna P, Ktena A, Tsakiridis P E and Hristoforou E 2015 *NDT & E Int.* **71** 33
[8] Vourna P, Hervoches C, Vrána M, Ktena A and Hristoforou E 2015 *IEEE Trans.* **51** 7029219
[9] Ktena A, Hristoforou E, Gerhardt GJL, Missell FP, Landgraf FJG, Rodrigues Jr DL and Alberteris-Campos M 2014 *Physica B: Coned Matter* **435** 109
[10] White S, Krause T and Clapham L 2007 *Meas. Sci. Technol.* **18** 3501
[11] Capó Sánchez J, Campos M F and Padovese L R 2012 *J. of Mag. and Mag. Mat.* **324** 11-14
[12] Stupakov O, Tomas I and Kadlecova J 2006 *J. Phys. D: Appl. Phys.* **39** 248-54
[13] Piotrowski L, Augustyniak A, Chmielewski M, Hristoforou E and Kosmas K 2010 *IEEE Trans. Magn.* **46** 239
[14] Hristoforou E 2002 *J. Opt. Adv. Mat.* **4** 245
[15] Kosmas K, Sargentis C, Tsamakis D and Hristoforou E 2005 *J. Mat.Proc.Tech.* **161** 359
[16] Hristoforou E, Reilly R E and Niarchos D 1993 *IEEE Trans. Magn.* **29** 3171
[17] Kosmas K and Hristoforou E 2007 International J. of App. Electr. and Mech. **25** 319
[18] Hristoforou E 2003 *Meas. Sci. & Technol.* **14** R15
[19] Hristoforou E, Niarchos D, Chiriac H and Neagu M 2001 *Sensors & Actuators A* **92** 132
[20] Hristoforou E and Kosmas K Int. J. of Appl. Electrom. and Mechanics **25** 287
[21] Hristoforou E and Reilly RE 1993 *J. Magn. Magn. Mat.*, **119** 247
[22] Hristoforou E, Kosmas K and Kollar M 2008 *J. Electr. Eng.* **59** 90
[23] Augustyniak B, Piotrowski L, Chmielewski M, Kosmas K and Hristoforou E 2010 *IEEE Trans. Magn.* **46** 544
[24] Kasai N, Koshino H, Sekine K, Kihira H and Takahashi M 2013 *J. Nondestruct. Eval.* **32** 277
[25] Hristoforou E and Reilly R E 1990 *Trans. Mag.* **26** 1563-1565
[26] Hristoforou E and Niarchos D 1992 *IEEE Trans. Mag.* **28** 2190-2192
[27] Dimitropoulos P D, Avaritsiotis J N, Hristoforou E 2003 *Sensors & Actuators A* **107** 238-247
[28] Giouroudi I, Ktena A and Hristoforou E 2004 *J. Opt. Adv. Mat.* **6** 45-50
[29] Lindgren M and Lepistö T 2004 *NDT & E Int.* **37** 403-410
[30] Vincent A, Pasco L, Morin M, Kleber X and Delnondedieu M 2005 *Acta Mat.* **53** 4579-4591
[31] Moorthy V, Shaw B A and Hopkins P 2006 *J. of Mag. and Mag. Mat.* **299** 362-375
[32] Pérez-Benítez J A, Espina-Hernández J H, Martínez-Ortiz P, Chávez-González A F and Rosa J M 2013 *J. of Mag. and Mag. Mat.* **347** 51-60
[33] Blaow M, Evans J T and Shaw B A 2005 *Acta Mat.* **53** 279-287

Correlation of magnetic properties with deformation in electrical steels

S Papadopoulou [1]
Laboratory of Physical Metallurgy, NTUA, Zografos, 15780, Greece

E-mail: mcrsofaki@gmail.com

Abstract. This paper investigates the utilization of magnetic Barkhausen Noise (MBN) and hysteresis loops methods for the non-destructive characterization of deformed electrical steel samples. For this reason electrical steel samples were subjected to uniaxial tensile tests on elastic and plastic region of deformations. Both the MBN and hysteresis loops were measured. The results shown a strong degradation of the magnetic properties on plastically strains. This was attributed to the irreversible movement of the magnetic domain walls, due to the presence of high dislocation density. The resulting magnetic properties were further evaluated by examining the microstructure of the deformed samples by using scanning electron microscopy.

1. Introduction

Non-destructive testing (NDT) are techniques, which are used either in laboratory or in industry, in order to evaluate the properties of a material, without causing any damage. Common NDT methods include eddy currents, magnetic particles, magnetic leakage (MDL), hysteresis loops (B, M – H loops), magnetic permeability (μ – H loops), magnetoacoustic (MAE), Barkhausen noise (MBN) [1-9].

It is well known that the magnetic non-destructive methods are influenced by the microstructural features of the examined ferromagnetic materials [10-23]. It has been also verified that, when a ferromagnetic material is subjected to uniaxial tensile or compressive tests, it undergoes a reconfiguration of its structure [24-27]. Thus, the deformation influence the final microstructure of the material, resulting in variations of the output magnetic signals.

The effect of the elastic and plastic deformation on the magnetic properties of electrical steels has been investigated in this paper. The utilization of both the MBN and hysteresis loops method for the non-destructive evaluation of the deformed samples. The resulting magnetic properties were further evaluated by examining the samples' microstructure by using scanning electron microscopy.

2. Experimental procedure

The studied alloy was Non-Oriented Electrical Steel samples. Their chemical composition, as given by the manufacturer, is given in Table 1.

Table 1. Chemical composition, in %wt, of the as-received NOES.

Si	Mn	Al	P	S	C	Fe
2.18	0.12	0.35	0.00009	0.00009	0.0018	Balanced

[1] Phone number : +30 210 772 2183, E-mail : xvourna@metal.ntua.gr

Dog-boned samples were cut from the as-received samples, according to the ASTM E8 Standard. These samples were subjected to uniaxial tensile strain at preselected deformation steps. The elastic region was divided into 6 deformation states, while the plastic one into 20 deformation states. As can be clearly seen in the stress – strain curve of the examined NOES (figure 1), the sample undergoes the Lüders band propagation. For this reason, 2 strain values were defined around the yield point.

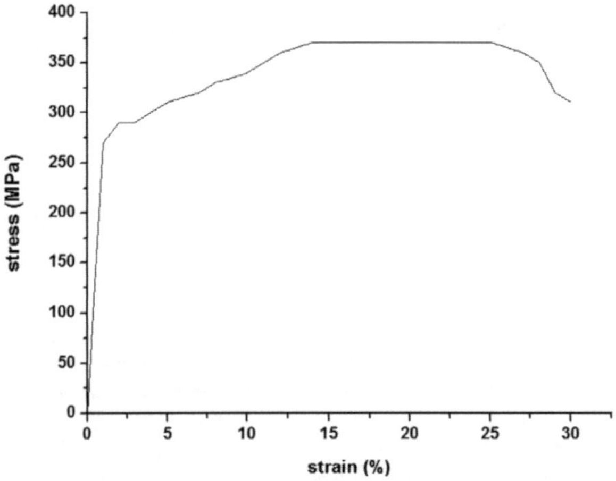

Figure 1. Stress- strain curve of the NOES.

During the tensile tests both the magnetic Barkhausen noise and the magnetic permeability were recorded. The magnetic BN sensor consisted of an excitation coil, which was wounded around the long axis of a U-shaped electromagnet. A triangular signal of 10 Hz was applied on this coil. The output signal was received by a sensing coil, which was wound around a ferrite core and placed at the centre of the electromagnet's leg. The receiving signal was further amplified and send, via a acquisition card, to a computer, for further analysis. The rms value of the output signal was the recorded magnetic Barkhausen parameter.

The permeability magnetic sensor consisted of a double U- shape electromagnet. In this apparatus the receiving coil was wound around the examined sample. The output voltage was proportional to the magnetic permeability of the examined sample.

3. Results

Both the MBN (figure 2) and the magnetic permeability (figure 3) values were increased in the elastic deformation region. During the uniaxial tensile test, the initial easy magnetization axis rotated, in order to become parallel to the direction of the applied stress, resulting in the increment of the 180° domain walls. Thus, the tensile stresses increase the magnetic responses [28-29].

However, in the plastic deformation region it is evident a strong decrease of the magnetic permeability. Additionally, the magnetic Barkhausen rms values were also diminished. Thus, the demarcation between elastic and plastic region was evidenced by the drop of both magnetic output signals.

When the stress passes the yield strength, plastic deformation will start to occur inside the grains of the polycrystalline material. As a result, a dramatic increment of the dislocation density occurred, at the beginning of the plastic region. After the slip system reaches the required critical resolved shear stress value, the dislocation start to slip. Thus, the formation and movement of dislocations will play a crucial role in absorbing the plastic deformation. Moreover, the high density of dislocations formed as tangles for higher plastic strain rates. These tangles acted as strong pinning sites during the magnetization procedure and led to a slight but progressive deterioration of the magnetic behaviour.

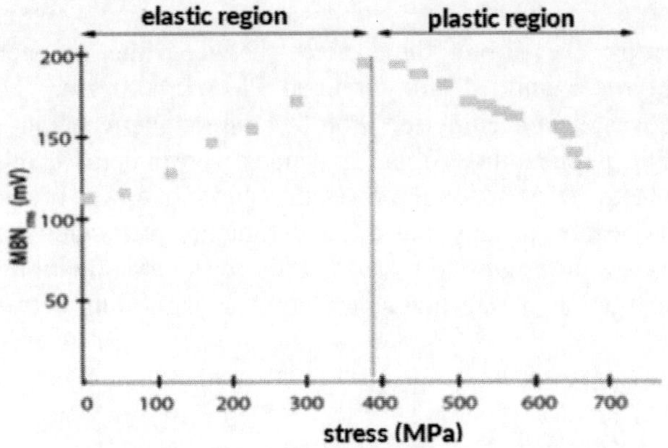

Figure 2. Variation of the rms values of the MBN during the uniaxial tensile test.

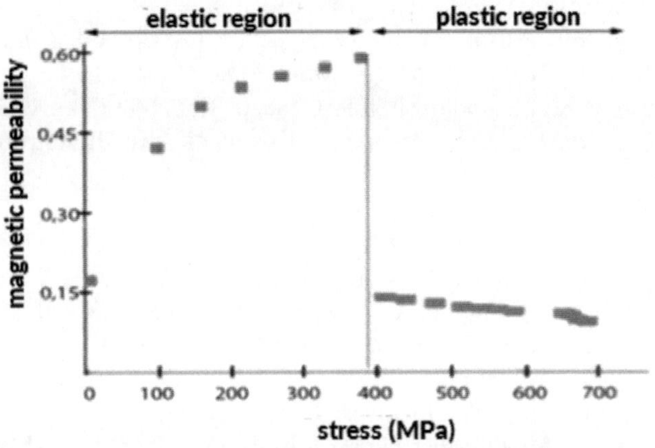

Figure 3. Variations of the magnetic permeability during the uniaxial tensile test.

Microstructural studies revealed that electrical steel consisted of polygonal and equiaxed ferrite grains in the elastic region (figure 4). In the plastic region, the ferrite grains were elongated along the direction of the applied tensile stress (figure 5). The high dislocation density and the increment in the gain boundary area are pinning sites of higher energy, resulting in the pronounced reduction of both the MBN and permeability responses.

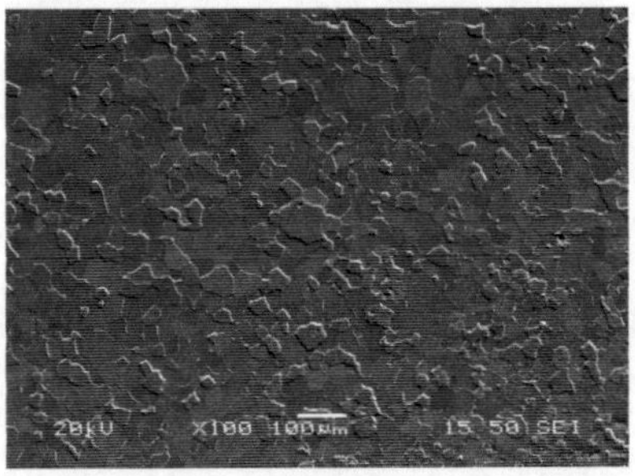

Figure 4. SEM micrograph of NOES in elastic region.

Figure 5. SEM micrograph of NOES in plastic region.

4. Conclusions

Generally, the results attest the strong dependence between the magnetic properties and the deformation behaviour of the examined Non-Oriented Electrical Steels. The demarcation between elastic and plastic region was evidenced by the drop of both magnetic output signals. Both the MBN (figure 2) and the magnetic permeability (figure 3) values were increased in the elastic deformation region. However, in the plastic deformation the formation and movement of dislocations play a crucial role, because the high dislocation density and the development of tangles deteriorated the magnetic properties. Further studies are in progress directed mainly to the examination of the samples in TEM and a more precise investigation of the magnetic response variations between the elastic and the plastic region.

5. References

[1] Hristoforou E 2002 *J. Opt. Adv. Mat.* **4** 245
[2] Kosmas K, Sargentis C, Tsamakis D and Hristoforou E 2005 *J. Mat.Proc.Tech.* **161** 359
[3] Hristoforou E, Reilly R E and Niarchos D 1993 *IEEE Trans. Magn.* **29** 3171
[4] Kosmas K and Hristoforou E 2007 *Int. J. of App. Electr. and Mech.* **25** 319
[5] Hristoforou E 2003 *Meas. Sci. & Technol.* **14** R15
[6] Hristoforou E, Niarchos D, Chiriac H and Neagu M 2001 *Sensors & Actuators A* **92** 132
[7] Hristoforou E and Reilly RE 1993 *J. Magn. Magn. Mater.*, 119 247
[8] Hristoforou E, Kosmas K and Kollar M 2008 J. Electr. Eng. 59 90
[9] Augustyniak B, Piotrowski L, Chmielewski M, Kosmas K and Hristoforou E 2010 *IEEE Trans. Magn.* **46** 544
[10] Vértesy G, Mészáros I and Tomáš I 2013 *NDT & E Int.* **54** 107
[11] Vourna P and Ktena A 2011 *Key Eng. Mat.* **495** 257
[12] Franco FA, González MFR, De Campos MF and Padovese LR 2013 *J. Nondestruct. Eval.* **32** 93
[13] Ktena A, Hristoforou E, Gerhardt GJL, Missell FP, Landgraf FJG, Rodrigues Jr DL and Alberteris-Campos M 2014 *Physica B: Coned Matter* **435** 109
[14] White S, Krause T and Clapham L 2007 *Meas. Sci. Technol.* **18** 3501
[15] Capó Sánchez J, Campos M F and Padovese L R 2012 *J. Magn. Magn. Mater.* **324** 11-14
[16] Stupakov O, Tomas I and Kadlecova J 2006 *J. Phys. D: Appl. Phys.* **39** 248-54
[17] Piotrowski L, Augustyniak A, Chmielewski M, Hristoforou E and Kosmas K 2010 *IEEE Trans. Magn.* **46** 239
[18] Kasai N, Koshino H, Sekine K, Kihira H and Takahashi M 2013 *J. Nondestruct. Eval.* **32** 277
[19] Lindgren M and Lepistö T 2004 *NDT & E Int.* **37** 403-410
[20] Vincent A, Pasco L, Morin M, Kleber X and Delnondedieu M 2005 *Acta Mat.* **53** 4579-4591
[21] Moorthy V, Shaw B A and Hopkins P 2006 *J. Magn. Magn. Mater.* **299** 362-375
[22] Pérez-Benítez J A, Espina-Hernández J H, Martínez-Ortiz P, Chávez-González A F and Rosa J M 2013 *J. Magn. Magn. Mater.* **347** 51-60
[23] Blaow M, Evans J T and Shaw B A 2005 *Acta Mat.* **53** 279-287
[24] Vourna P, Ktena A, Tsakiridis P E and Hristoforou E 2015 *NDT & E Int.* **71** 33
[25] Vourna P, Hervoches C, Vrána M, Ktena A and Hristoforou E 2015 *IEEE Trans.* **51** 6200104
[26] Vourna P, Ktena A, Tsakiridis P E and Hristoforou E 2015 *Measurements* **71** 31
[27] Vourna P, Ktena A and Hristoforou E 2014 *IEEE Trans. Mag.* **50** 1
[28] Stefanita C G, Clapham L and Atherton D L 2000 *J. Mater. Sci.* **35** 2675
[29] Moorthy V, Vaidyanathan S, Jayakumar T, Raj B, Kashyap B P 1999 *Acta Mater.* **47** 1869
[30] Hristoforou E and Reilly R E 1990 *Trans. Mag.* **26** 1563-1565
[31] Hristoforou E and Niarchos D 1992 *IEEE Trans. Mag.* **28** 2190-2192
[32] Dimitropoulos P D, Avaritsiotis J N, Hristoforou E 2003 *Sensors & Actuators A* **107** 238-247
[33] Giouroudi I, Ktena A and Hristoforou E 2004 *J. Opt. Adv. Mat.* **6** 45-50

IOP Conf. Series: Materials Science and Engineering **108** (2016) 012016 doi:10.1088/1757-899X/108/1/012016

Correlation of grain growth phenomena with magnetic properties in non – oriented electrical steels

E Mangiorou[1]
Laboratory of Physical Metallurgy, NTUA, Zografos, 15780, Greece

E-mail: mangiorou.eleni@gmail.com

Abstract. This paper presents a combination of two types of method targeted to investigate the stages of the microstructure evolution in annealed non-oriented electrical steels by means of magnetic measurements and metallographic analysis. The indirect magnetic testing, carried out by Barkhausen noise was associated with the direct structural investigation by Scanning Electron Microscopy measurements. The goal of this work was to study the influence of heat transport phenomena on grain growth processes in non-oriented electrical steels, which were subjected to different annealing conditions. The results determined from the magnetic measurements and predicted from micrograph observations show a relatively good concordance.

1. Introduction

Non – Oriented Electrical Steels (NOES) are soft polycrystalline magnetic materials and are so called because it contains a grain structure with no preferred orientation [1-10]. For this reason they have found extensive applications in domestic appliances that require isotropic magnetic properties in the plane of the sheet [3-6].

The variety of NOES in general depends on the chemical content of Si. Silicon is a ferrite stabilizer, which increase the resistivity and decreases hysteresis loss [4-6, 10]. In order to promote the favourable texture development of (100) plane, which is favourable for magnetization, it is important to control the final microstructure of NOES, in terms of grain size and crystallographic texture [6, 11-12]. Additionally, the rolling in the ferrite region below the austenitization temperature is required [6, 11-12]. Therefore, NOES magnetic properties are controlled by the microstructure.

For this reason, NOES are subjected to various thermally activated process, which initiates recrystallization and grain growth phenomena. During annealing the grain size rapidly increases, because of the large amount of stored energy in the grains, released at elevated temperatures [10-12]. However, the non-linear migration of the grain boundaries and the abnormal growth of certain types of grains at the expense of others, resulting in the final microstructural inhomogeneity, which deteriorates the final magnetic properties of the NOES [4-6, 10-12]. Thus, grain growth kinetics significantly affects the evolution of the texture and, consequently, the final magnetic properties [6].

It is well known that magnetic methods are influenced by the microstructural features of the examined ferromagnetic materials [13-34]. As a result, the annealing influence the final microstructure of the material, resulting in variations of the output magnetic signals. Although, the non-destructive

1 Phone number : +30 210 772 2183, E-mail : mangiorou.eleni@gmail.com

testing using various methods has been of great interest in recent years [35-46], a quantitative or qualitative evaluation on how the microstructure affect the magnetic properties has not predicted yet.

Hence, the present paper presents a combination of two types of method targeted to investigate the stages of the microstructure evolution in annealed non-oriented electrical steels by means of magnetic measurements and metallographic analysis. The indirect magnetic testing, carried out by Barkhausen noise was associated with the direct structural investigation by Scanning Electron Microscopy measurements. The goal of this work was to study the influence of heat transport phenomena on grain growth processes in non-oriented electrical steels, which were subjected to different annealing conditions.

2. Experimental procedure

Commercial samples of NOES sheets were investigated. The cold rolled specimens were subjected to annealing under laboratory conditions. Before the heat treatment each specimen was cut into a strip of 10 cm^L × 3 cm^W with the longest side parallel to the rolling direction. The surface of these specimens was cleaned, degreased and dried. Each specimen was heated to a different temperature within the range of 350 °C – 950 °C at a heating rate of at least 10 °C/s. The specific temperature selected was kept constant for 12 min (Table 1).

Table 1. Annealed samples.

	Annealing temperature (°C)	Holding time (min)	Cooling conditions
NOES 1	350	12	Air cooled
NOES 2	550	12	Air cooled
NOES 3	600	12	Air cooled
NOES 4	700	12	Air cooled
NOES 5	800	12	Air cooled
NOES 6	920	12	Air cooled
NOES 7	950	12	Air cooled

The microstructure of the specimens were examined along the longitudinal cross-section using Scanning Electron Microscopy (SEM). The samples were polished and subsequently etched using a 2% Nital solution for 10-30s until the grain boundaries were revealed. The JEOL JSM 6100 electron microscope was connected with a Noran TS 5500 Microanalyzer for EDS analysis on cross – sections of steel sheets.

Magnetic Barkhausen noise (MBN) was measured using the MEB-2C device. A triangular shaped magnetic field produced by a generator was applied on the surface of the sample by an electromagnet in the shape of a probe. The discrete changes in the resulting local magnetization (MBN) were sensed at the ends of a receiving coil as voltage pulses. All MBN measurements were taken at a constant threshold at different positions of the annealed sample. The rms value of MBN measurements were recorded.

3. Results

All the annealed NOES samples presented a Body Centered Cubic (BCC) ferritic matrix which is characterized by micro-segregation of silicon, strongly in grain boundaries (figure 1). The commencement of abnormal grain growth is represented in figure 1e. At this point, the abnormally growing grain seemed to consume the primary recrystallized grain matrix (figure 1d). The microstructure obtained after heat treatment at 920 °C for 12 min represented a totally abnormal grain growth (figure 1g). Hence, the abnormal grain growth can be obtained during annealing in a very narrow temperature range, within very short time frames. The optimum set of conditions for the complete development of abnormal grain growth in the investigated NOES samples can be annealing at 950 °C for 12min and an extremely high, for this grade of steel, heating rate of 10 °C/s.

It has to be noted, that the microstructure observed by the SEM did not provide all the information required concerning recovery the process during recrystallization, for example the decrement in the dislocations density.

Figure 2 summarizes the average grain size distribution in the investigated NOES samples. It is obvious that last process initiated a comparatively fast and abnormal grain growth mechanism. The MBN results were plotted in figure 3 as a function of the annealing temperature. It was evident, the decrement of the magnetic response, as the annealing temperature is increased. Because the ferrite grains became coarser, as the annealing temperature is increasing, the MBN values of the investigated specimens was decreasing. Grain boundaries act as strong pinning centers for the magnetic measurements. As a result, finer samples have higher magnetic output signal. In a region of annealing temperatures between 550°C and 650°C a slight increment of the magnetic signal was observed, as a result of the strong microsegragation of Si in grain boundaries

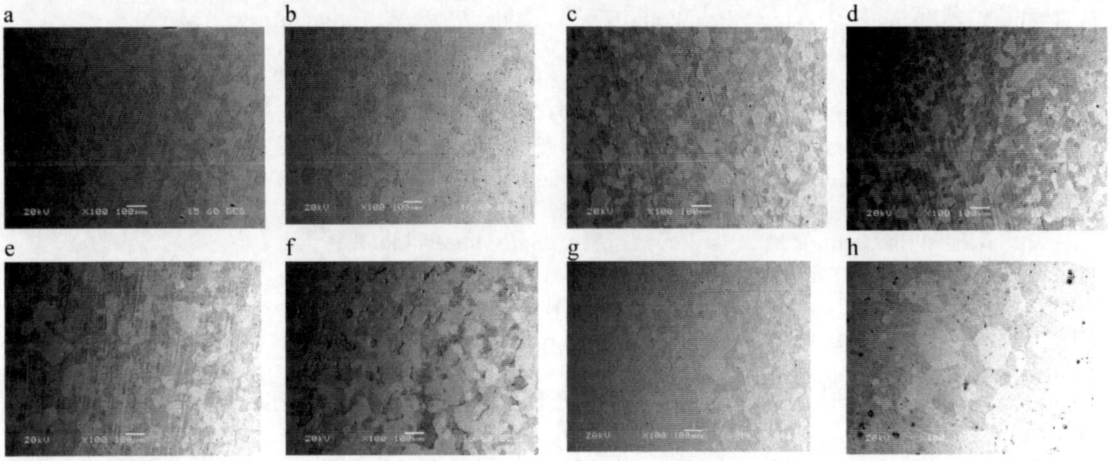

Figure 1. Microstructure developed in the (a) as received sample and after annealing at (b) 350 °C, (c) 550 °C, (d) 600 °C, e) 700 °C , f) 800 °C , g) 920 °C and h) 950 °C for 12min.

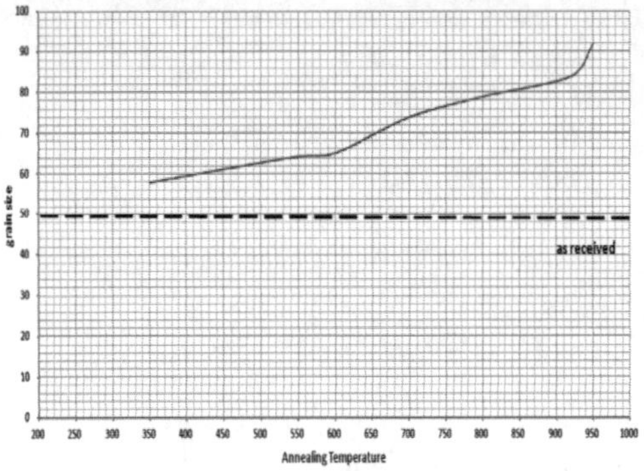

Figure 2. Average grain size of the as received and annealed samples.

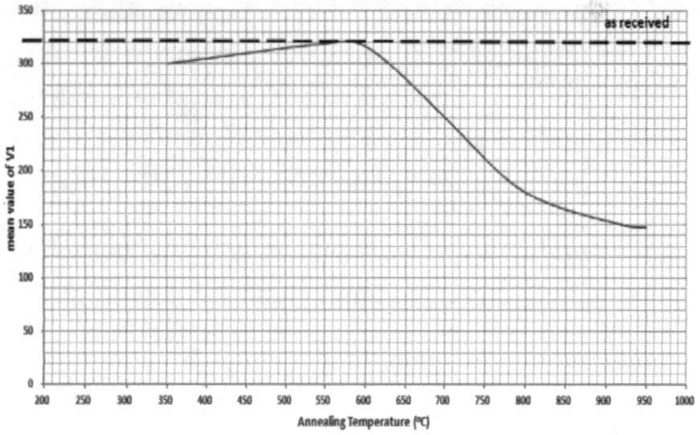

Figure 3. Variations of the MBN rms values at various at various temperatures.

4. Conclusions

A combination of metallographic analyses with magnetic measurements on the annealed NOES specimens allowed the distinguish of the microstructure evolution stages. The anisotropic behavior of the grain boundaries migration under non-equilibrium conditions, leads to a strong decrease of the MBN values.

5. References

[1] Shimazu T, Shiozaki M and Kawasaki KJ, 1994 *J. Magn. Magn. Mater.* **133** 147
[2] Fisher O and Schneider J 2003 *J. Magn.Magn. Mater.* **302** 254
[3] Oshihiko O, Masaaki Y K and Atsuhito H 2008 *J. Magn. Magn. Mater.* **320** 2430
[4] Kumar PR, Samajdar I, Viswanathan NN, Singal V and Seshadri V 2003 *J. Magn. Magn. Mater.* **264** 75
[5] Bertotti G 2008 *J. Magn. Magn. Mater.* **320** 2436
[6] Cunha MA and Paolinelli SC 2008 *J. Magn. Magn. Mater.* **320** 2485
[7] Vourna P, Ktena A, Tsakiridis P E and Hristoforou E 2015 *Measurements* **71** 31
[8] Vourna P, Ktena A, Tsakiridis P E and Hristoforou E 2015 *NDT & E Int.* **71** 33
[9] Vourna P, Ktena A and Hristoforou E 2014 *IEEE Trans. Mag.* **50** 1
[10] Moses AJ, 1990 *J. Magn. Magn. Mater.* **112** 150
[11] Sidor Y, Kovac F and Kvackaj T 2007 *Acta Mater.* **55** 1711
[12] Sha YH, Zhang F, Zhou SC, Pei W and Zuo L, 2008 *J. Magn. Magn. Mater.* **320** 393
[13] Augustyniak B, Piotrowski L, Chmielewski M, Kosmas K and Hristoforou E 2010 *IEEE Trans. Magn.* **46** 544
[14] Vértesy G, Mészáros I and Tomáš I 2013 *NDT & E Int.* **54** 107
[15] Vourna P and Ktena A 2011 *Key Eng. Mat.* **495** 257
[16] Franco FA, González MFR, De Campos MF and Padovese LR 2013 *J. Nondestruct. Eval.* **32** 93
[17] Ktena A, Hristoforou E, Gerhardt GJL, Missell FP, Landgraf FJG, Rodrigues Jr DL and Alberteris-Campos M 2014 *Physica B: Coned Matter* **435** 109
[18] White S, Krause T and Clapham L 2007 *Meas. Sci. Technol.* **18** 3501
[19] Capó Sánchez J, Campos MF and Padovese L R 2012 *J. Magn. Magn. Mater.* **324** 11-14
[20] Piotrowski L, Augustyniak A, Chmielewski M, Hristoforou E and Kosmas K 2010 *IEEE Trans. Magn.* **46** 239
[21] Kasai N, Koshino H, Sekine K, Kihira H and Takahashi M 2013 *J. Nondestruct. Eval.* **32** 277
[22] Lindgren M and Lepistö T 2004 *NDT & E Int.* **37** 403-410
[23] Vincent A, Pasco L, Morin M, Kleber X and Delnondedieu M 2005 *Acta Mat.* **53** 4579-4591
[24] Moorthy V, Shaw B A and Hopkins P 2006 *J. Magn. Magn. Mater.* **299** 362-375
[25] Blaow M, Evans JT and Shaw BA 2005 *Acta Mat.* **53** 279-287
[26] Vourna P, Hervoches C, Vrána M, Ktena A and Hristoforou E 2015 *IEEE Trans.* **51** 6200104
[27] Stefanita CG, Clapham L and Atherton DL 2000 *J. Mater. Sci.* **35** 2675
[28] Bae BK, Woo JS and Kim JK 2003 *J. Magn. Magn. Mater.* **373** 254
[29] Baudouina P, Belhadj A and Houbaert Y 2002 *J. Magn. Magn. Mater.* **238** 221
[30] Cunha MA and Paolinelli C P 2003 *J. Magn. Magn. Mater.* **379** 254
[31] Guerenu AM, Arizti F, Fuentes MD and Gutierrez I 2004 *Acta Mater.* **52** 3657
[32] Park JT and Szpunar JA 2003 *Acta Mater.* **51** 3037
[33] Ray RK, Jonas JJ and Hook RE 1994 *Int. Mater. Rev.* **39** 129
[34] Reis ACC, Bracke L, Petrov R, Kaluba WJ and Kestens L 2003 *ISIJ Int.* **43** 1260
[35] Hristoforou E 2002 *J. Opt. Adv. Mat.* **4** 245
[36] Kosmas K, Sargentis C, Tsamakis D and Hristoforou E 2005 *J. Mat.Proc.Tech.* **161** 359
[37] Hristoforou E, Reilly R E and Niarchos D 1993 *IEEE Trans. Magn.* **29** 3171
[38] Kosmas K and Hristoforou E 2007 *Int. J. of App. Electr. and Mech.* **25** 319
[39] Hristoforou E 2003 *Meas. Sci. & Technol.* **14** R15
[40] Hristoforou E, Niarchos D, Chiriac H and Neagu M 2001 *Sensors & Actuators A* **92** 132
[41] Hristoforou E and Reilly RE 1993 *J. Magn. Magn. Mater.*, **119** 247
[42] Hristoforou E, Kosmas K and Kollar M 2008 J. Electr. Eng. 59 90
[43] Hristoforou E and Reilly R E 1990 *Trans. Mag.* **26** 1563
[44] Hristoforou E and Niarchos D 1992 *IEEE Trans. Mag.* **28** 2190
[45] Dimitropoulos P D, Avaritsiotis J N, Hristoforou E 2003 *Sensors & Actuators A* **107** 238
[46] Giouroudi I, Ktena A and Hristoforou E 2004 *J. Opt. Adv. Mat.* **6** 45

An accurate method for determining residual stresses with magnetic non-destructive techniques in welded ferromagnetic steels

P Vourna [1]
Laboratory of Physical Metallurgy, NTUA, Zografos, 15780, Greece

E-mail: xvourna@metal.ntua.gr

Abstract. The scope of the present research work was to investigate the proper selection criteria for developing a suitable methodology for the accurate determination of residual stresses existing in welded parts. Magnetic non-destructive testing took place by the use of two magnetic non-destructive techniques: by the measurement of the magnetic Barkhausen noise and by the evaluation of the magnetic hysteresis loop parameters. The spatial distribution of residual stresses in welded metal parts by both non-destructive magnetic methods and two diffraction methods was determined. The conduction of magnetic measurements required an initial calibration of ferromagnetic steels. Based on the examined volume of the sample, all methods used were divided into two large categories: the first one was related to the determination of surface residual stress, whereas the second one was related to bulk residual stress determination. The first category included the magnetic Barkhausen noise and the X-ray diffraction measurements, while the second one included the magnetic permeability and the neutron diffraction data. The residual stresses determined by the magnetic techniques were in a good agreement with the diffraction ones.

1. Introduction

Residual stresses can be defined as those stresses existing within a structural material or component, in the absence of any external loading or thermal gradients [1]. Among the factors that are known to cause residual stresses are the development of deformations, due to the application of mechanical loads or to the presence of thermal gradients. Moreover, the changes in volume during solidification or due to solid state transformation introduced residual stresses in a material [1-2].

A considerable amount of residual stresses is introduced during common manufacturing processes. Additionally, during welding the material is subjected to temperatures ranging from the material's melting point to room temperature [3]. During welding, the temperature range varies from the material's melting point to the room temperature. Additionally, the mechanical properties of the joint are temperature dependant and therefore, these are often degraded due to the presence of thermal gradients. Cooling to room temperature invokes stresses, which are inevitably incorporated to the material's residual stress [1-5]. Therefore, the quantitative determination of the residual stresses is important for quality, integrity and performance of the welding joints.

Although there are many scientific articles describing the correlation of magnetic and material's properties [6-18], there are only few reports describing the determination of residual stresses via

[1] Phone number : +30 210 772 2183, E-mail : xvourna@metal.ntua.gr

magnetic techniques [19-35]. In addition, reports on predictions of the spatial distribution of the residual stresses in welded ferromagnetic samples are rather limited [1, 4-5, 21].

The scope of the present research work was to investigate the proper selection criteria for developing a suitable methodology for the accurate determination of residual stresses existing in welded parts. Magnetic non-destructive testing took place by the use of two magnetic non-destructive techniques: by the measurement of the magnetic Barkhausen noise and by the evaluation of the magnetic hysteresis loop parameters. The spatial distribution of residual stresses in welded metal parts by both non-destructive magnetic methods and two diffraction methods was determined. The conduction of magnetic measurements required an initial calibration of ferromagnetic steels. Based on the examined volume of the sample, all methods used were divided into two large categories: the first one was related to the determination of surface residual stress, whereas the second one was related to bulk residual stress determination. The first category included the magnetic Barkhausen noise and the X-ray diffraction measurements, while the second one included the magnetic permeability and the neutron diffraction data.

2. Experimental procedure

Three different grades of ferromagnetic steel were tested: non-oriented electrical steel (NOES), low carbon steel (AISI 1008) and low-alloyed carbon steel (AISI 4130). For each grade of steel, two identical sheets were welded together in a butt joint configuration by different welding techniques. The welding parameters were carefully selected to produce full penetration welds free of defects, such as porosity and cracking.

As far as the magnetic measurement set-up is concerned, two laboratory methods were used: the magnetic Barkhausen noise (MBN) and the magnetic permeability (MP). The apparatus used for the magnetic measurement has been described elsewhere [1, 3-5].

In order to evaluate residual stresses using the magnetic techniques, a two-step procedure is proposed. In the first step the as-received samples have to be cut in rectangular dog-bone-shaped flat specimens and then subjected both in tensile and compressive deformation, in a step-wise manner by a loading machine (INSTRON). For strain measurements, a commercially available strain gauge is adhered on the surface of the testing sample, at the centre of the gauge length. The measured strain values are converted to stress values through elasticity equations. While stain is measured, magnetic values is also recorded, using the appropriate magnetic sensor device. The root mean square (rms) value of the measured MBN signal, as well as the MP from the hysteresis loops are considered as estimated parameters. During the second step the distribution of the magnetic values is measured on the sample, taking measurements at pre-defined intervals along the length of the sample's surface. The temporal fluctuations of the magnetic values yielded information about the spatial distribution of the residual stresses on the surface and bulk of the sample, with the aid of the calibration curve.

3. Results

In the NOES samples, which were welded together using TIG (figure 1) and Plasma (figure 2) as the welding techniques, the residual stresses determined by the MBN and the XRD method followed a similar trend. However, the residual stresses determined by the magnetic permeability method represented a deviation from the results of the XRD method. This was attributed to the different penetration depth of each method.

In order to confirm the accuracy of the aforementioned magnetic techniques in determining residual stresses, AISI 1008 samples were welded using electron beam (EB) as the welding technique welded (figures 5, 6). The residual stresses determined by all four methods presented only minor deviations. To further evaluate the effectiveness of the magnetic techniques in determining residual stresses, AISI 4130 samples were welded by either TIG (figure 7) or EB (figure 8) as welding techniques. It was confirmed that, with regards to the residual stresses the MBN followed a similar trend to the XRD, in both welded samples. Furthermore, the results from the permeability method were in a good agreement with the ND ones, in both welded samples.

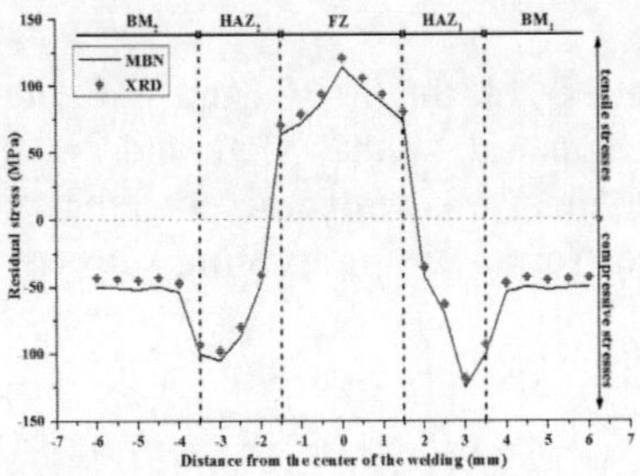

Figure 1. Determination of residual stress with MBN and XRD methods in TIG welded NOES samples.

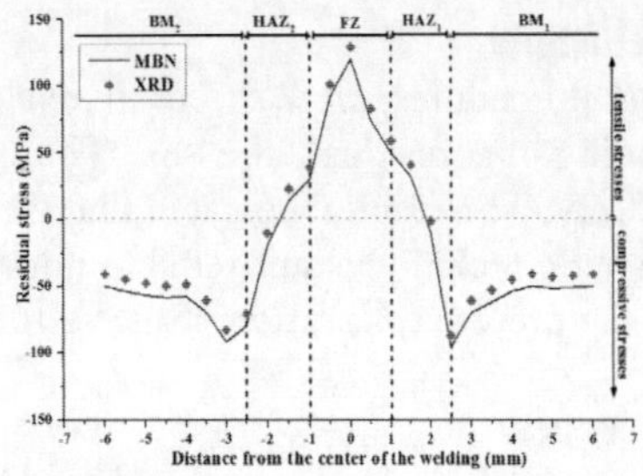

Figure 2. Determination of residual stress with MBN and XRD methods in Plasma welded NOES samples.

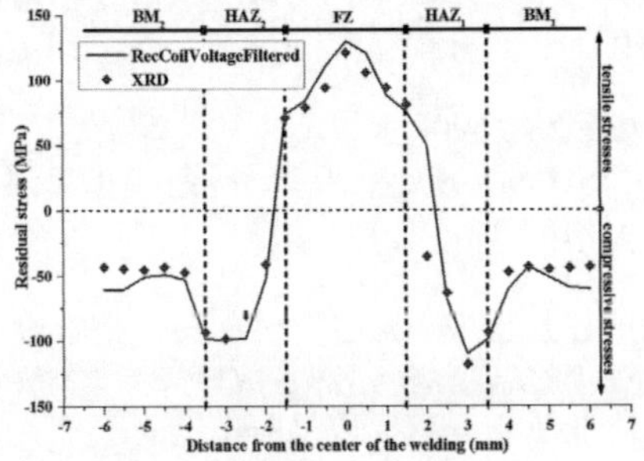

Figure 3. Determination of residual stress with MP and XRD methods in TIG welded NOES samples.

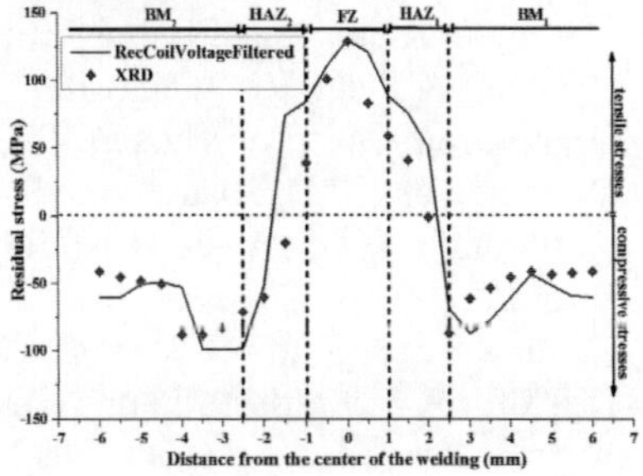

Figure 4. Determination of residual stress with MP and XRD methods in Plasma welded NOES samples.

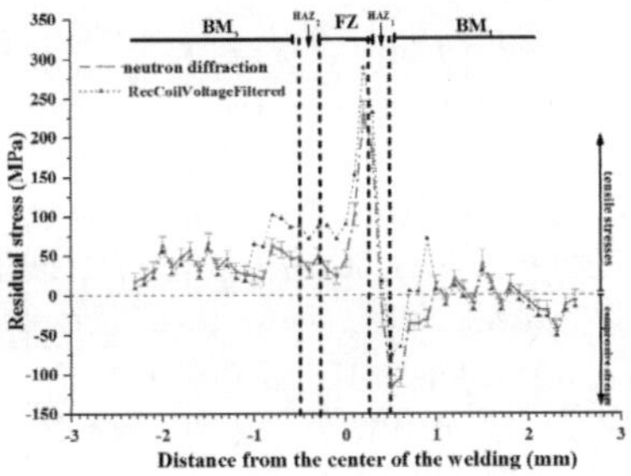

Figure 5. Determination of residual stress with MBN and XRD methods in EB welded AISI 1008 sample.

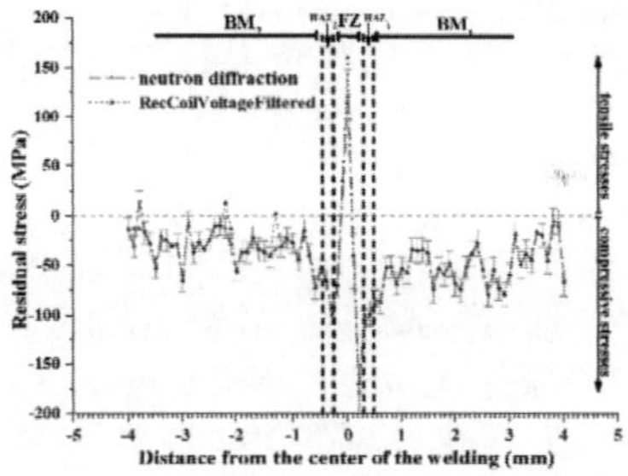

Figure 6. Determination of residual stress with MP and ND methods in EB welded AISI 1008 sample.

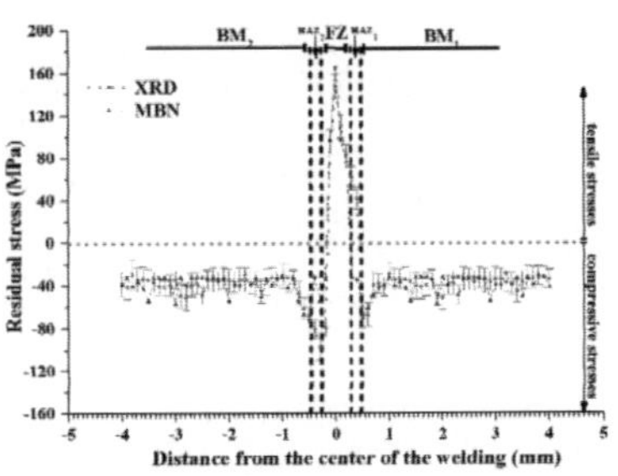

Figure 7. Determination of residual stress with MBN and XRD methods in TIG welded AISI 4130.

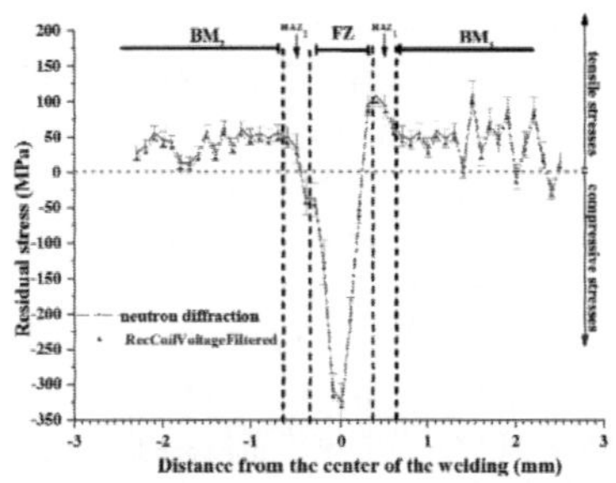

Figure 8. Determination of residual stress with MP and ND methods in EB welded AISI 4130 samples.

4. Conclusions

Welding procedures were used to evaluate the effectiveness of the MBN and MP methods in determining the residual stresses. These methods were compared to the XRD and ND methods, respectively. The results indicated that the deviation between them was within acceptable limits. Thus, the magnetic techniques are reliable non-destructive methods for analysing existing stresses, but they require the precise calibration of the ferromagnetic sample.

5. References

[1] Vourna P, Ktena A, Tsakiridis P E and Hristoforou E 2015 *NDT & E Int.* **71** 33
[2] Rossini NS, Dassisti M, Benyounis KY and Olabi AG 2012 *Mater. Des.,* **35** 572
[3] Vourna P, Hervoches C, Vrána M, Ktena A and Hristoforou E 2015 *IEEE Trans.* **51** 6200104
[4] Vourna P, Ktena A, Tsakiridis P E and Hristoforou E 2015 *Measurements* **71** 31
[5] Vourna P, Ktena A and Hristoforou E 2014 *IEEE Trans. Mag.* **50** 1
[6] Hristoforou E and Reilly R E 1990 *Trans. Mag.* **26** 1563
[7] Hristoforou E and Niarchos D 1992 *IEEE Trans. Mag.* **28** 2190
[8] Dimitropoulos P D, Avaritsiotis J N, Hristoforou E 2003 *Sensors & Actuators A* **107** 238
[9] Giouroudi I, Ktena A and Hristoforou E 2004 *J. Opt. Adv. Mat.* **6** 45
[10] Hristoforou E 2002 *J. Opt. Adv. Mat.* **4** 245
[11] Kosmas K, Sargentis C, Tsamakis D and Hristoforou E 2005 *J. Mat.Proc.Tech.* **161** 359
[12] Hristoforou E, Reilly R E and Niarchos D 1993 *IEEE Trans. Magn.* **29** 3171
[13] Kosmas K and Hristoforou E 2007 International *J. App. Electr. Mech.* **25** 319
[14] Hristoforou E 2003 *Meas. Sci. & Technol.* **14** R15
[15] Hristoforou E, Niarchos D, Chiriac H and Neagu M 2001 *Sensors & Actuators A* **92** 132
[26] Hristoforou E and Kosmas K *Int. J. Appl. Electrom. Mechanics* **25** 287
[17] Hristoforou E and Reilly RE 1993 *J. Magn. Magn. Mater.* **119** 247
[18] Hristoforou E, Kosmas K and Kollar M 2008 *J. Electr. Eng.* **59** 90
[20] Vértesy G, Mészáros I and Tomáš I 2013 *NDT & E Int.* **54** 107
[21] Vourna P and Ktena A 2011 *Key Eng. Mater.* **495** 257
[22] Franco FA, González MFR, De Campos MF and Padovese LR 2013 *J. Nondestruct. Eval.***32** 93
[23] Ktena A, Hristoforou E, Gerhardt GJL, Missell FP, Landgraf FJG, Rodrigues Jr DL and Alberteris-Campos M 2014 *Physica B: Coned. Matter.* **435** 109
[24] White S, Krause T and Clapham L 2007 *Meas. Sci. Technol.* **18** 3501
[25] Capó Sánchez J, Campos M F and Padovese L R 2012 *J. Magn. Magn. Mater.* **324** 11
[26] Stupakov O, Tomas I and Kadlecova J 2006 J. *Phys. D: Appl. Phys.* **39** 248
[27] Piotrowski L, Augustyniak A, Chmielewski M, Hristoforou E and Kosmas K 2010 *IEEE Trans. Magn.* **46** 239
[28] Augustyniak B, Piotrowski L, Chmielewski M, Kosmas K and Hristoforou E 2010 *IEEE Trans. Magn.* **46** 544
[29] Kasai N, Koshino H, Sekine K, Kihira H and Takahashi M 2013 *J. Nondestruct. Eval.* **32** 277
[30] Lindgren M and Lepistö T 2004 *NDT & E Int.* **37** 403
[31] Vincent A, Pasco L, Morin M, Kleber X and Delnondedieu M 2005 *Acta Mater.* **53** 4579
[32] Moorthy V, Shaw B A and Hopkins P 2006 *J. Magn. Magn. Mater.* **299** 362
[33] Pérez-Benítez J A, Espina-Hernández J H, Martínez-Ortiz P, Chávez-González A F and Rosa J M 2013 *J. Magn. Magn. Mater.* **347** 51
[34] Blaow M, Evans J T and Shaw B A 2005 *Acta Mater.* **53** 279
[35] Gür CH and Çam İ 2007 *Mater. Charact.* **58** 447

Acknowledgments

This work was supported in part by the European Union, in part by Greek national funds through the operational program "Education and Lifelong Learning" of the National Strategic Reference

Framework-Research Funding Program: ARCHIMEDES III, and in part by the Investing in Knowledge Society through the European Social Fund.

Tin Oxide Microheater for Chemical Sensors

Mohsen Gharesi[1a] and Mohammad Ansari[1b]

[1]Electronic Materials Laboratory, Electrical Engineering Department, K.N. Toosi University of Technology, Tehran 16317-14191, Iran.

[a]mohsengharesi@ee.kntu.ac.ir, [b]mohammad.ansari1371@gmail.com

Abstract. Tin oxide is the main material utilized for the fabrication of chemical sensing pellets which operate at elevated temperatures. The heating is commonly carried out with ruthenium dioxide resistors. Here, a tin oxide-based microheater is developed for microsensor applications. These microheaters are fabricated on 0.5 mm thick alumina substrates using spray pyrolysis technique. The optimum SnO_2 heaters have a sheet resistivity in the 40-70 Ω/□ range. Ohmic Ag/SnO_2 contacts are formed by silver paste printing followed by an appropriate thermal annealing, which provide connections to the external circuitry. Durability tests are carried out on several samples; the long-term performance of the fabricated devices is satisfactory. The method allows the elimination of the expensive ruthenium dioxide from the structure of generic gas sensors.

Keywords: Tin oxide, Microheater, Gas sensor, Thin film, Spray pyrolysis.

1. Introduction

Owing to vast applications in the fields of food safety, health care, environmental science, and many others, gas analysis is gaining importance. Gas sensors are simple, cheap, and fast tools used for quantitative and qualitative assessment of contaminants in different atmospheres. Today, analysis of simple and complex odors is possible using a single generic gas sensor [1-5]. Metal oxides are the class of materials which have been vastly used as sensing element of gas sensors [6-9]. Precious properties such as high thermal stability, wide variety of materials with different chemical, physical and optical properties, low fabrication cost, and the ability to function in harsh environmental conditions, have led to widespread applications of metal oxides in the fields of optoelectronics and chemical sensors [10-12]. The role of metal oxides in sensor technology is not limited to being used as the sensing element; they have also been successfully used as electrodes and heaters [13-15].

Establishing ohmic contacts to metal oxides is an unavoidable necessity in fabricating various electronic and ionic devices. Metal/metal oxide contacts have been vastly studied in recent years and these studies have led to innovative methods for forming contacts of ohmic quality to these materials [16-20]. Schottky contacts are of remarkable importance in device fabrication technology. It's been reported that Schottky-type metal/metal oxide contacts are more suitable for gas sensor applications due to their enhanced sensitivity [21-24].

In order to achieve high levels of sensitivity and short response times, the majority of gas sensors operate at elevated temperatures; so the presence of a heater in such devices is unavoidable. Durability, temperature uniformity, and the fabrication cost are the parameters which should be considered in choosing the proper heater for gas sensors. Wire heaters made of Ni-Cr alloy are common in experimental gas sensors; although, temperature non-uniformity and relatively high power consumption are the problems which these kind of heaters face. Platinum films are also utilized as heating element of gas sensors. In order to obtain appropriate heater resistance, these films should be deposited in the shape of meanders which causes difficulties in the fabrication of the devices which are very small in size. Furthermore, catalytic interaction of gas molecules with platinum heaters can cause instabilities in sensor response [25]. Ruthenium dioxide resistors are the most common kind of heaters that can be paste printed or sputtered on sensors substrate. RuO_2 resistance has negligible dependence on temperature, but this kind of heater has some problems with its stability [26]. For example, it has been shown that ruthenium dioxide tends to reduce to ruthenium in hydrogen atmosphere at temperatures between 150 and 250°C [27].

In sensor technology, tin oxide has been vastly used for the fabrication of chemical sensing pellets. Here, a tin oxide microheater for chemical sensors is proposed. Since the microheater and the sensing pellet can be deposited using the same technique, the fabrication cost of the gas sensor decreases significantly. In this research work, tin oxide thin films are produced using a homemade spray pyrolysis deposition system. Ohmic Ag/SnO_2 contacts provide connection to the external circuitry. Durability and the microstructure of the deposited heaters are investigated.

2. Experimental

Tin oxide layers are deposited on 5 mm × 5 mm × 0.5 mm alumina substrates. The schematic diagram of the ultrasonic spray pyrolysis system utilized for the deposition of SnO_2 layers is shown in figure 1. Alumina substrates are washed with acetone (96% v/v), rinsed with deionized water, and dried on a hot plate at 150°C. The precursor is 0.1 M solution of $SnCl_2.2H_2O$ (Merck, 107815) in absolute ethanol. Aerosol droplets, generated from the precursor using a piezoelectric transducer (900 kHz), are transported by air as a carrier gas with a flow rate of 0.1 lit/min to the deposition chamber. The piezoelectric source is separated from the precursor container by a thin polymer sheet and maintained in a water filled tank in order to prevent its precursor caused corrosion [28]. The aerosol strikes the substrate surface heated to 450°C where film deposition occurs. The surface temperature is provided using a Ni-Cr alloy wire heater embedded under the substrate and monitored using a fine S-type thermocouple. It takes about an hour to produce layers with sheet resistivity in the range of 50 Ω/□. The samples are annealed at 700°C in air to achieve stable heater resistance.

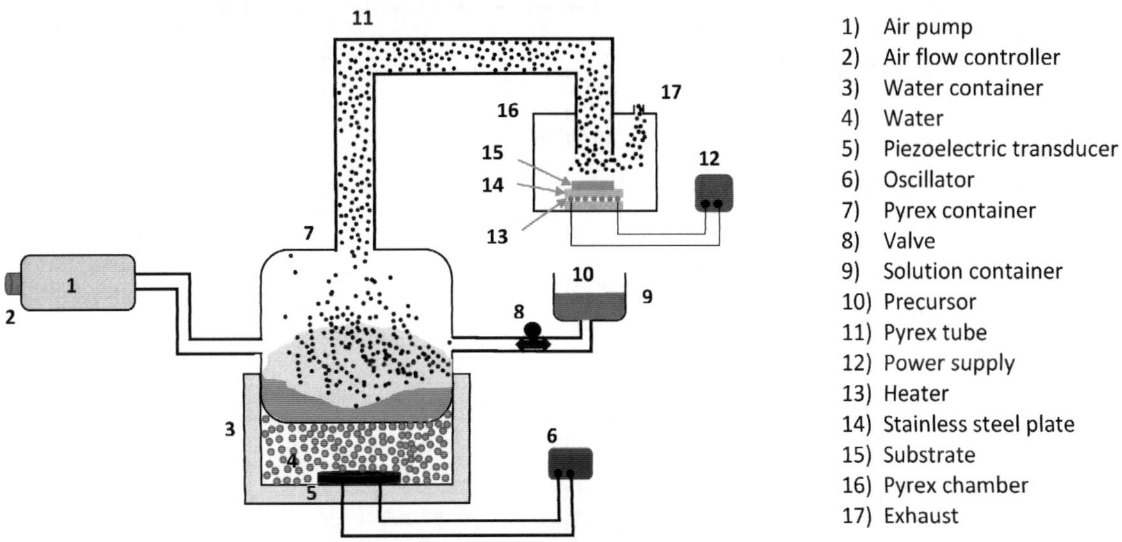

Figure 1. The schematic diagram of the homemade spray pyrolysis system used for tin oxide layer deposition on alumina substrates.

IOP Conf. Series: Materials Science and Engineering **108** (2016) 012018 doi:10.1088/1757-899X/108/1/012018

Two parallel silver electrodes are deposited on tin oxide layer utilizing silver paste printing. Fine Ni-Cr alloy wires attached to the deposited silver layers provide connection to the external circuitry. The forward biased device is heat treated at 200°C for 10 minutes to achieve ohmic Ag contacts [16]. Figure 2.a shows the schematic diagram of the fabricated sample. A fine S-type thermocouple is used for continuous monitoring of the temperature on the device surface. Temperature measurements are carried out with an accuracy of ±1°C. Electronic properties of the device are investigated using a two point probe measurement system. Figure 2.b shows the experimental setup used for these measurements.

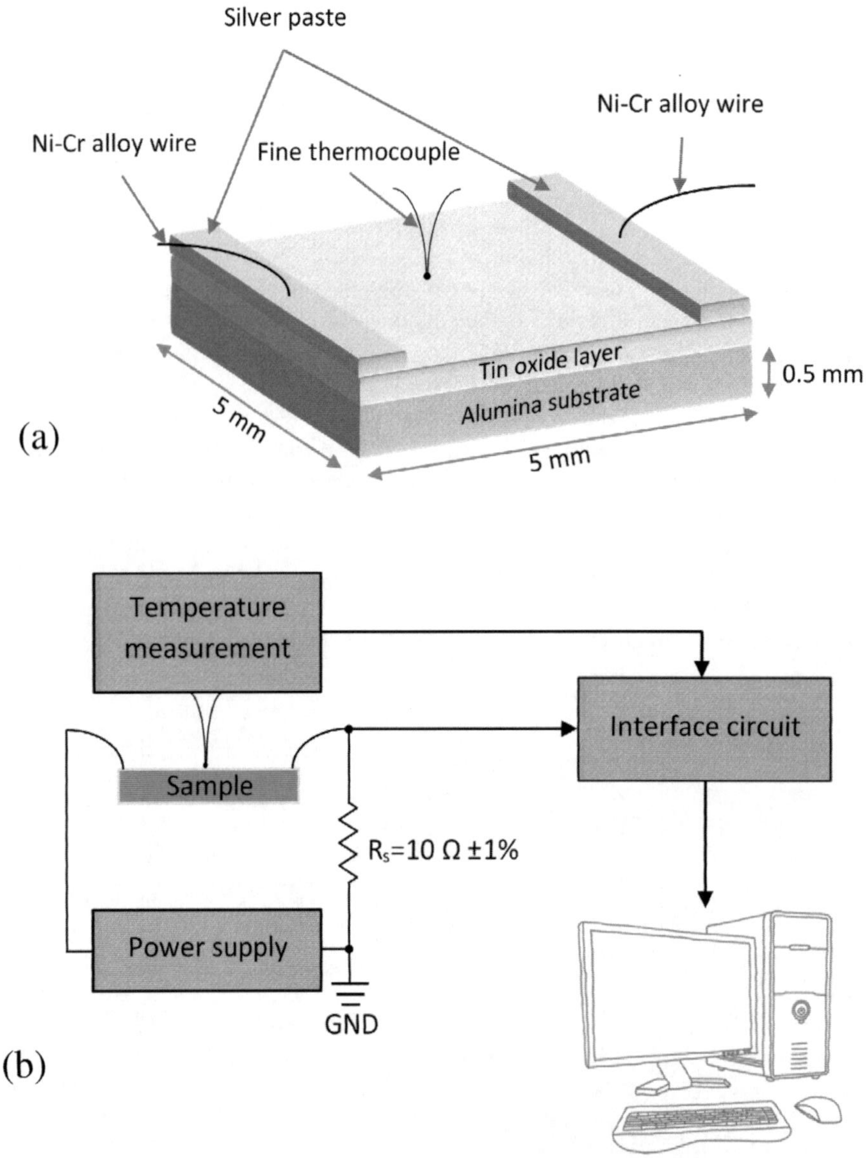

Figure 2. (a) The schematic presentation of the fabricated sample and (b) the experimental setup used for electrical measurements.

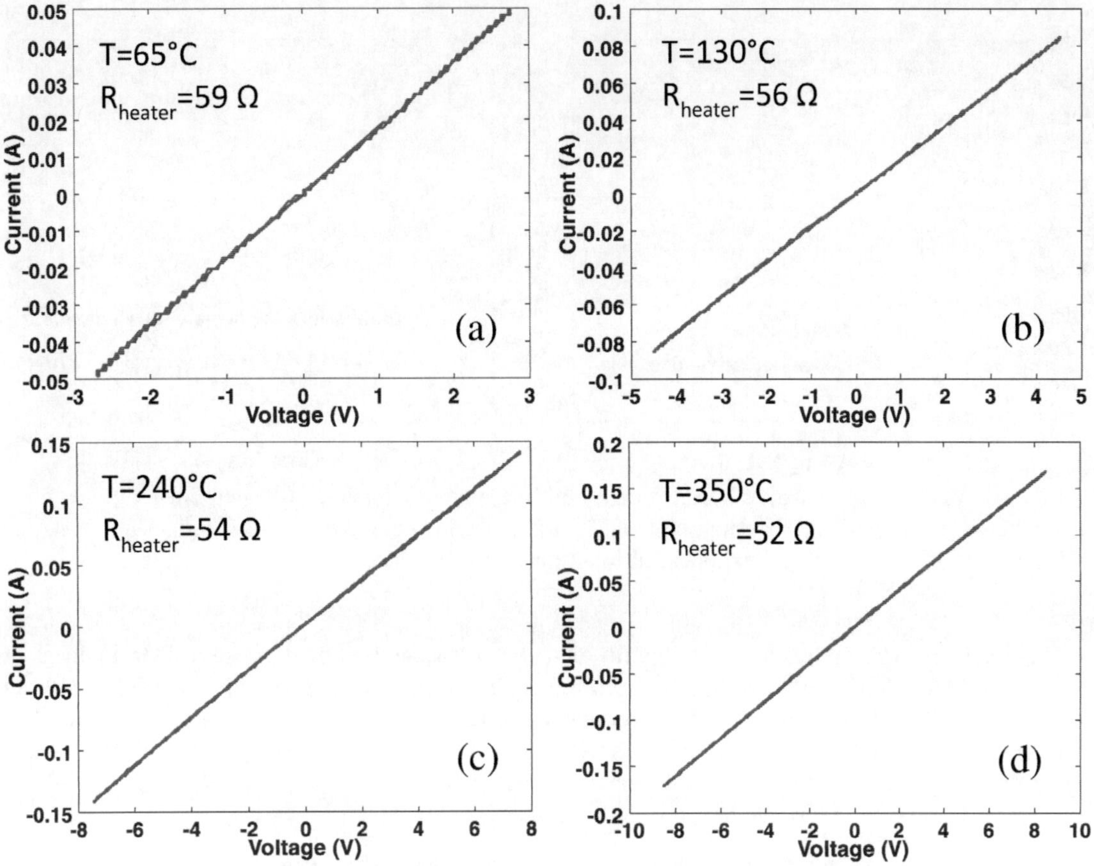

Figure 3. I-V diagrams for a heater sample at (a) T=65°C, (b) T=130°C, (c) T=240°C, and (d) T=350°C. Heater resistances are obtained using the slope of the diagrams.

3. Results

Sinusoidal waveforms with different amplitudes are applied to the fabricated sample and the resulting I-V characteristics are recorded at four different temperatures (figure 3). Linear I-V diagrams show that Ag-SnO_2 contacts are ohmic at operating temperatures in the 65-350°C range. Using the slope of the I-V diagrams, heater resistances are calculated. It can be seen that resistance variation of SnO_2 heater is only about 10% in a 300°C temperature change. For comparison, the resistance of a platinum heater doubles when it's operating at temperatures in 350-400°C range [25].

The effect of thermal annealing has been investigated by recording heater resistance of annealed and not-annealed devices continuously operating at 350°C for 2 weeks (figure 4.a). It's obvious that no significant change occurs in the resistance of the device which has been annealed at 700°C for 30 minutes; while for the not-annealed device, the resistance experiences an increase of 44% in the first 6 days and then reaches its stable value. This change can be expressed by an exponential function of time with a time constant of 1.3 days (figure 4.b). Durability tests have been carried out for different samples continuously operating at 350°C in a 1-month time period. The heater resistance remained almost unchanged in this period and only a 2% fluctuation was observed in the measured values which can be caused by the measurement errors.

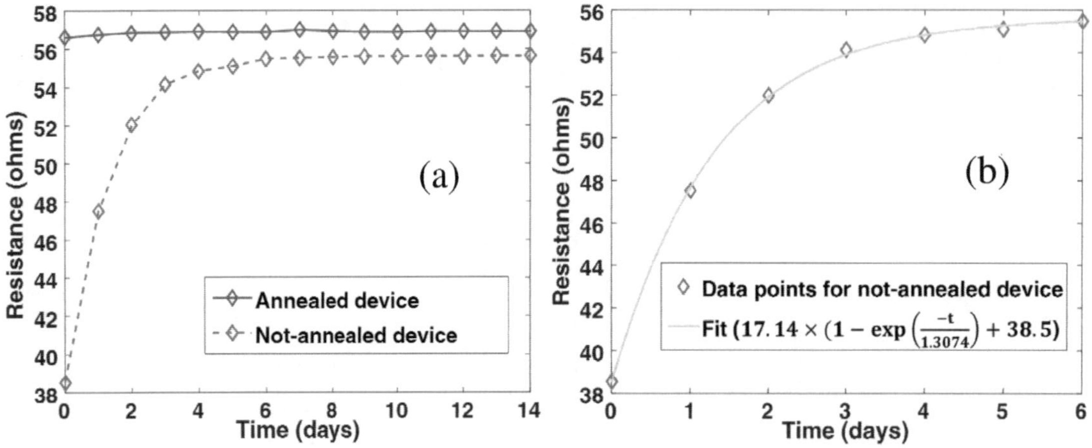

Figure 4. (a) Heater resistance of annealed (solid line) and not-annealed (dashed line) samples for a 2-week time period and (b) data points of the not-annealed device for the first 6 days and the exponential function which fits them.

The SEM micrograph obtained from the tin oxide layer surface is shown in figure 5. Roughness of the polycrystalline SnO_2 layer is obvious from the figure. The deposited layer has particles with an average size of about 500 nm, while the aggregated particles have much smaller size (~50 nm). The crystalline SnO_2 particles show a pyramidal habit which is in agreement with the reported results [29, 30].

Figure 5. SEM micrograph of the SnO_2 layer surface.

IOP Conf. Series: Materials Science and Engineering **108** (2016) 012018 doi:10.1088/1757-899X/108/1/012018

4. Conclusion

A tin oxide-based microheater was fabricated using a homemade ultrasonic spray pyrolysis system. The I-V characteristics of the device at different temperatures were studied and it was shown that the formed Ag/SnO_2 contacts are ohmic in a wide range of operating temperatures. It was shown that the thermally annealed heaters are stable over time. The microstructure of the deposited tin oxide layer was investigated and it was seen that the layer has a polycrystalline structure with an average crystallite size of 500 nm and pyramidal habit. Owing to its negligible characteristics dependence on temperature, and durability, the proposed microheater can effectively function in chemical sensors. Furthermore, this microheater enables the fabrication of gas sensors sensing pellets and microheaters with the same technology reducing the fabrication cost.

References

[1] Hossein-Babaei, Faramarz, and Amir Amini. "A breakthrough in gas diagnosis with a temperature-modulated generic metal oxide gas sensor." *Sensors and Actuators B: Chemical* 166 (2012): 419-425.

[2] Gosangi, Rakesh, and Ricardo Gutierrez-Osuna. "Active temperature modulation of metal-oxide sensors for quantitative analysis of gas mixtures."*Sensors and Actuators B: Chemical* 185 (2013): 201-210.

[3] Hossein-Babaei, Faramarz, Mohammad Paknahad, and Vahid Ghafarinia. "A miniature gas analyzer made by integrating a chemoresistor with a microchannel." *Lab on a Chip* 12.10 (2012): 1874-1880.

[4] Martinelli, Eugenio, et al. "Self-adapted temperature modulation in metal-oxide semiconductor gas sensors." *Sensors and Actuators B: Chemical* 161.1 (2012): 534-541.

[5] Hossein-Babaei, Faramarz, and Amir Amini. "Recognition of complex odors with a single generic tin oxide gas sensor." *Sensors and Actuators B: Chemical*194 (2014): 156-163.

[6] Wetchakun, K., et al. "Semiconducting metal oxides as sensors for environmentally hazardous gases." *Sensors and Actuators B: Chemical* 160.1 (2011): 580-591.

[7] Miller, Derek R., Sheikh A. Akbar, and Patricia A. Morris. "Nanoscale metal oxide-based heterojunctions for gas sensing: a review." *Sensors and Actuators B: Chemical* 204 (2014): 250-272.

[8] Kim, Il-Doo, and Avner Rothschild. "Nanostructured metal oxide gas sensors prepared by electrospinning." *Polymers for Advanced Technologies* 22.3 (2011): 318-325.

[9] Kim, Hyo-Joong, and Jong-Heun Lee. "Highly sensitive and selective gas sensors using p-type oxide semiconductors: Overview." *Sensors and Actuators B: Chemical* 192 (2014): 607-627.

[10] Clavero, César. "Plasmon-induced hot-electron generation at nanoparticle/metal-oxide interfaces for photovoltaic and photocatalytic devices."*Nature Photonics* 8.2 (2014): 95-103.

[11] Hossein-Babaei, Faramarz, Mehdi Mohamadzade Lajvardi, and Farhad Akbari Boroumand. "Large area Ag–TiO_2 UV radiation sensor fabricated on a thermally oxidized titanium chip." *Sensors and Actuators A: Physical* 173.1 (2012): 116-121.

[12] Comini, Elisabetta, et al. "Metal oxide nanoscience and nanotechnology for chemical sensors." *Sensors and Actuators B: Chemical* 179 (2013): 3-20.

[13] Zhu, Lina, Yangong Zheng, and Jiawen Jian. "Effect of palladium oxide electrode on potentiometric sensor response to carbon monoxide." *Ionics*(2015): 1-8.

[14] Chen, Xin, Xiaoyu Cheng, and J. Justin Gooding. "Detection of trace nitroaromatic isomers using indium tin oxide electrodes modified using β-cyclodextrin and silver nanoparticles." Analytical chemistry 84.20 (2012): 8557-8563.

[15] Bai, Zikui, Aihua Wang, and Changsheng Xie. "Laser grooving of Al_2O_3 plate by a pulsed Nd: YAG laser: characteristics and application to the manufacture of gas sensors array heater." *Materials Science and Engineering: A* 435 (2006): 418-424.

[16] Hossein-Babaei, Faramarz, Saeed Moghadam, and Saeed Masoumi. "Forming ohmic Ag/SnO_2 contacts." Materials Letters 141 (2015): 141-144.

[17] Greiner, Mark T., et al. "Metal/Metal-Oxide Interfaces: How Metal Contacts Affect the Work Function and Band Structure of MoO_3." *Advanced Functional Materials* 23.2 (2013): 215-226.

[18] Hossein-Babaei, F., Mehdi M. Lajvardi, and Navid Alaei-Sheini. "The energy barrier at noble metal/TiO_2 junctions." *Applied Physics Letters* 106.8 (2015): 083503.

[19] Lee, Jeong Eun, et al. "Thermal stability of metal Ohmic contacts in indium gallium zinc oxide transistors using a graphene barrier layer." *Applied Physics Letters* 102.11 (2013): 113112.

[20] Hossein-Babaei, Faramarz, and Mohammad Orvatinia. "Thickness dependence of sensitivity in thin film tin oxide gas sensors deposited by vapor pyrolysis."*International Journal of Engineering Transaction B: Applications* 16.1 (2003): 33-40.

[21] Wei, Te-Yu, et al. "Gigantic enhancement in sensitivity using Schottky contacted nanowire nanosensor." *Journal of the American Chemical Society*131.48 (2009): 17690-17695.

[22] Hossein-Babaei, F., and S. Rahbarpour. "Separate assessment of chemoresistivity and Schottky-type gas sensitivity in M–metal oxide–M′ structures." Sensors and Actuators B: Chemical 160.1 (2011): 174-180.

[23] Hu, Youfan, et al. "Supersensitive, Fast-Response Nanowire Sensors by Using Schottky Contacts." *Advanced Materials* 22.30 (2010): 3327-3332.

[24] Hossein-Babaei, F., and S. Rahbarpour. "Titanium and silver contacts on thermally oxidized titanium chip: electrical and gas sensing properties." *Solid-State Electronics* 56.1 (2011): 185-190.

[25] Korotcenkov, Ghenadii. *Handbook of Gas Sensor Materials: Properties, Advantages and Shortcomings for Applications Volume 1: Conventional Approaches*. Springer Science & Business Media, 2013. 266-267.

[26] Jelenkovic, Emil V., et al. "Degradation of RuO_2 thin films in hydrogen atmosphere at temperatures between 150 and 250°C." *Microelectronics Reliability* 43.1 (2003): 49-55.

[27] Kaga, Yukinao, et al. "Thermal stability of RuO_2 thin films and effects of annealing ambient on their reduction process." *Japanese journal of applied physics* 38.6R (1999): 3689.

[28] Moghadam, Saeed, Sepideh Zendehdell, and Ahmad Rouhollahi. "Sensitivity enhancemenet in SnO_2 based gas sensors by surface decoration with platinum nanoparticles." *Electrical Engineering (ICEE), 2014 22nd Iranian Conference on*. IEEE, 2014.

[29] Brinzari, V., et al. "Morphological rank of nano-scale tin dioxide films deposited by spray pyrolysis from $SnCl_4 \cdot 5H_2O$ water solution." *Thin Solid Films* 408.1 (2002): 51-58.

[30] Korotcenkov, G., et al. "Faceting characterization of tin dioxide nanocrystals deposited by spray pyrolysis from stannic chloride water solution." *Thin solid films* 471.1 (2005): 310-319.

IOP Conf. Series: Materials Science and Engineering **108** (2016) 012019 doi:10.1088/1757-899X/108/1/012019

Discriminating among different tea leaves using an operating temperature-modulated tin oxide gas sensor

Ali Rastkhadiv[1,a], Amin Jenabi[1,b], Asma Souri[1,c]

[1]Electronic Materials Laboratory, Electrical Engineering Department, K. N. Toosi University of Technology, Tehran 16317-14191, Iran

[a]ali.rastkhadiv@gmail.com, [b]aminjenabi@yahoo.com, [c]asma_souri@yahoo.com

Abstract. We report distinguishing different types of tea leaves from each other based on their aroma using a thermal shock-induced generic tin oxide gas sensor. The sensor used in this work consists of a microheater and a tin oxide pellet, both connected to outside circuitry with noble metal contacts. The heater is powered with a series of narrow high magnitude voltage impulses of predetermined thermal impacts adjusted to produce step-like temperature rises of different magnitudes on the gas sensitive pellet. The sensor is exposed to aromas collected from various types of tea leaves at different concentrations. Within 4.5 s, nine 500 ms-wide voltage pulses, each as high as 9.3 V in magnitude, are applied to the microheater. Each pulse causes a step-like temperature jump on the pellet temperature. The transient responses recorded for different tea leaves look different even after amplitude normalization. The sensor profiles are recorded, digitized, and compared with the database of previous experiences. A heuristically defined high dimensional feature vector is automatically generated for each analyte. Classifications are graphically achieved in a 3-D feature space after applying principle component analysis for dimension reduction.
Keywords: Artificial olfaction; Metal oxide gas sensor; Thermal shock; Microheater; Tea leaves.

1. Introduction

Successful classification of single component gases and complex odours with a single thermal shock-induced metal oxide gas sensor has recently been demonstrated [1, 2]. Artificial olfaction systems are seriously considered for many applications including dangerous gas identification [3, 4], fire detection [5, 6], disease recognition [7, 8] and food stuff quality control [9, 10]. Gas identification systems such as mass spectroscopy and gas chromatography are massive, expensive and difficult to operate [11, 12]. Moreover, the sensor array-based electronic noses suffer from the drifts of the array components and require expensive array replacements and cumbersome system calibrations [13, 14]. Single sensor electronic noses are proven alternatives for sensor arrays which significantly reduce required calibration activities and maintenance costs [15-19]. A thermally modulated gas sensor can provide sufficient discriminative information for the recognition of different analytes [20-23]. It's also been reported that thermal modulation of a gas sensor reduces aging drifts [24]. Furthermore, environmental fluctuations and gradual microstructural changes cause drift terms in the responses of a thermally modulated gas sensor, which may be eliminated by the means of mathematical compensation methods [25].

IOP Conf. Series: Materials Science and Engineering **108** (2016) 012019 doi:10.1088/1757-899X/108/1/012019

It has been shown that the response patterns obtained by an array made virtually by the thermal shock induction of a single generic tin oxide gas sensor contain analytical information on the nature of the odour enough for its discrimination from a number of other target analytes [1, 2]. Here, we are reporting fabrication of an olfaction system based on this concept. The performance of the system is evaluated by classifying 3 types of tea leaves by analyzing their odours. The presented system requires only 4.5 s for the analysis of each analyte [1, 2].

2. Experimental

A generic SnO_2 gas sensor comprises a microheater and a SnO_2 sensing pellet integrated on an insulating ceramic substrate. The structure of a typical chemoresistor is schematically presented in figure 1 [1]. Pellet temperature is elevated by applying different heating voltages to the microheater. The pellet is thermally shocked by applying heating pulses of ~0.5 s duration and different amplitudes to the sensor microheater. The thermal impulse resulting from each pulse is calculated and compared by time integration of its respective V^2/R, where V is the pulse amplitude and R is the resistance of the microheater (57 Ω, for the sensor used). The suitable voltage waveform for thermal shock induction was established by trial and error. In each trial different combination of pulses is applied and its effect on the systems distinguishing power between ethanol and methanol contaminations is estimated. The selected combination of pulses is given in figure 2. The maximum temperature on the pellet is 300°C, which its occurrence coincides t= 3.5 s in figure 2.

At the standby state, the microheater is warmed up by 5V for cleaning the pellet surface from the residues of the previous tests and condensed humidity. After pressing the start button for sampling, for 18 s the voltage level is zero which is followed by a 1.5 s rest at 1 V heating voltage. This initial signal adjusts the microheater temperature and prepares the system for response sampling. Then, different thermal shock pulses are applied to the microheater. Shock pulses are increased in amplitude from 6.5V to 9.3V as shown in figure 2. The duration of each step is 0.5 s.

Numerical information is collected from the transient responses of the sensor at different conditions. The information is used for establishing a database which is analyzed by principle component analysis (PCA) for classification of the data recording conditions. The electronic system designed for performing this process is schematically described in figure 3. The sensor Response voltage is transmitted to the microcontroller, where, it is read by the ADC of the microcontroller and saved. Sampling rate is 100 s^{-1}. Pellet resistance is mathematically calculated. Response signals transmitted to PC by a USB port. The received signal is processed, classified and identified. The system is shown in figure 4.

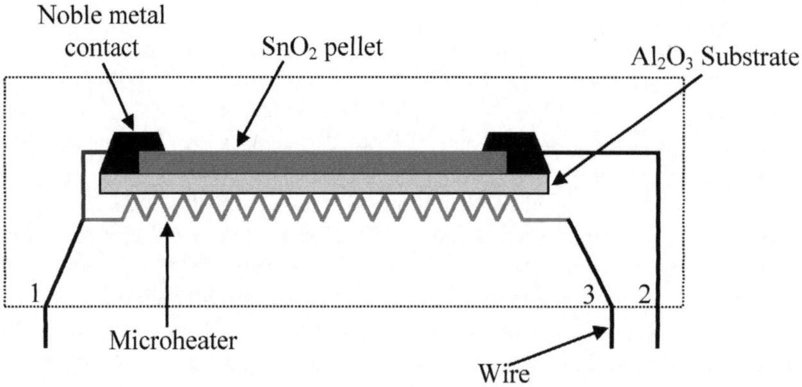

Figure 1. The schematic diagram of the sensor and microheater.

IOP Conf. Series: Materials Science and Engineering **108** (2016) 012019 doi:10.1088/1757-899X/108/1/012019

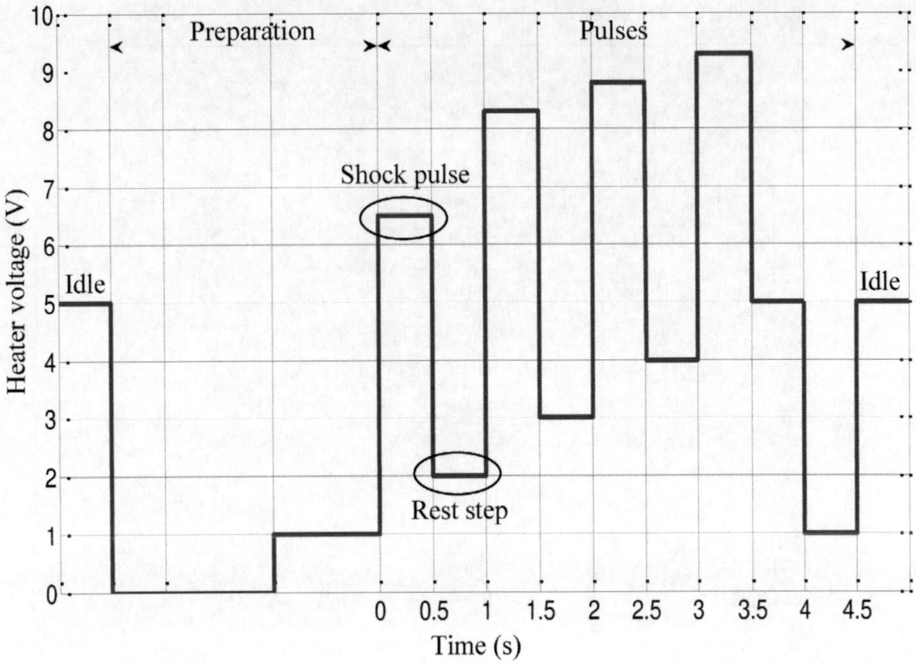

Figure 2. The selected heating voltage waveform applied to induce thermal shock.

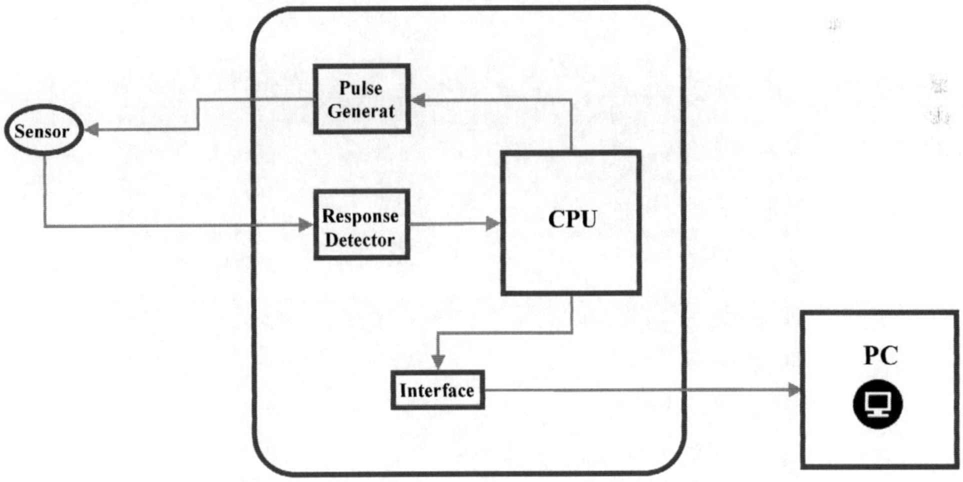

Figure 3. The schematic diagram of the system designed for performing response recording and signal processing for analyte classification.

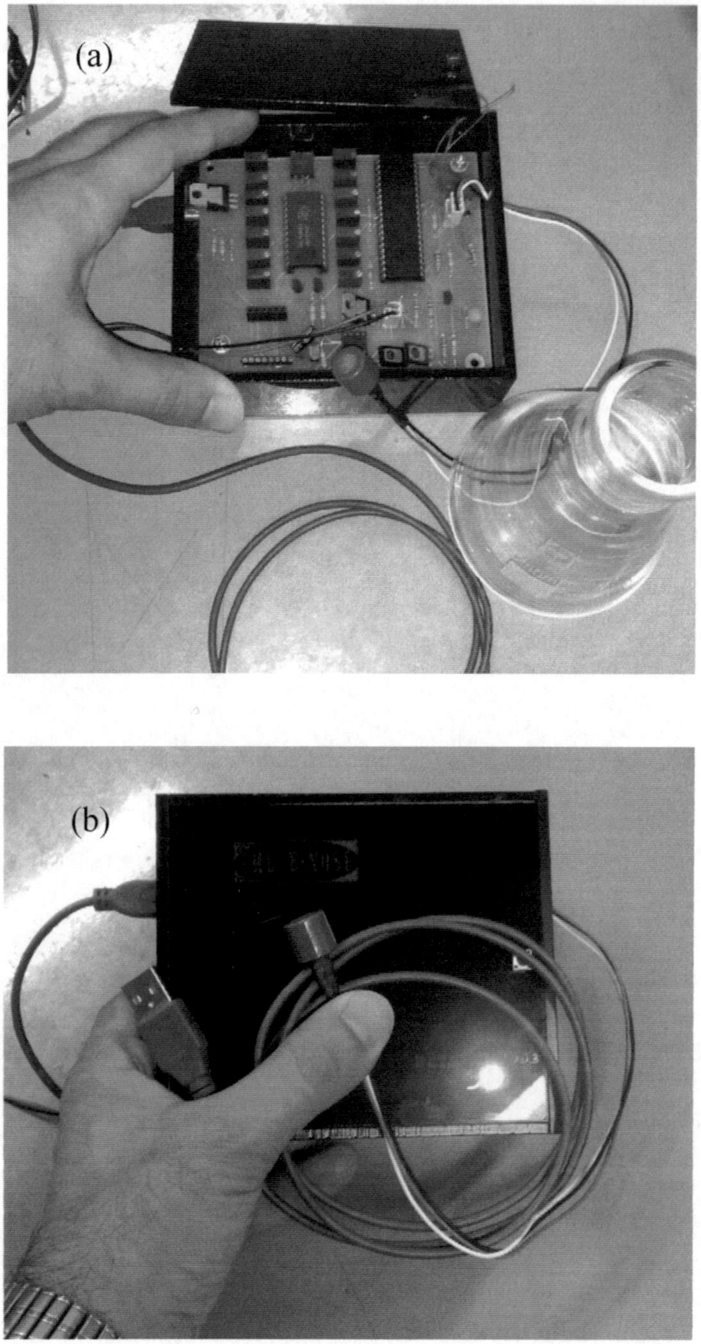

Figure 4. Photo graphs of the fabricated system: (a) the printed board of the CPU; (b) the complete system.

3. Results

The temperature modulated gas sensor was exposed to the odours collected from 3 different types of tea leaves (Earl Gray tea, Ceylon tea, and Iranian tea). The recorded response patterns at different analyte concentrations are shown in Figure 5. The ambient temperature and relative humidity in which the experiments were carried out are 25±2°C and 25±5% respectively. Owing to the negligible environmental fluctuations, there is no need to compensate their effect on the responses [1]. The obtained patterns are normalized between 0 and 1 to diminish the effect of gas concentration level on the shape of the responses. Normalized responses are depicted in Figure 6.

The normalized response patterns obtained from various tea leaves are quite different in shape, so that they can be visually distinguished. However, by the utilization of simple pattern analysis methods analyte classifications can be automated. It is clear in Figure 6 that the normalized responses in the 2.5-3.5 s time range have the most decisive discriminative information. By applying the principal component analysis (PCA) to the feature vector dataset, the number of feature space dimensions is reduces to three. The results, then, can be visualized in a 3-D feature space, as is shown in figure 7. Complete segregation of the different classes is clear in figure 7. Test patterns, obtained in a different work session, are mapped into the same feature space utilizing the same procedure. The results are depicted with filled markers in figure 7. Test data points are all successfully clustered to the correct classes validating the reproducibility of the obtained sensor responses and the efficacy of the fabricated system.

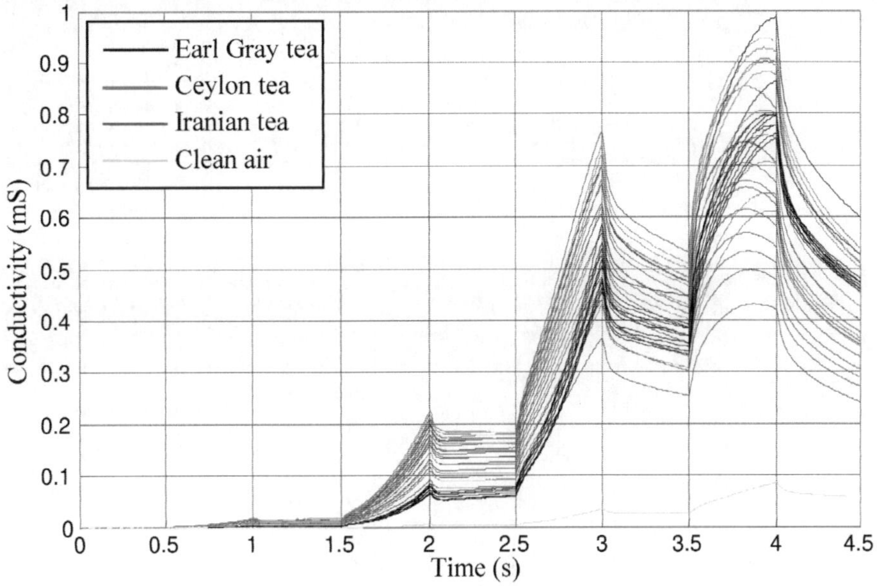

Figure 5. Raw response patterns of the thermal shock-induced gas sensor recorded for different concentration levels of Earl Gray tea (black), Ceylon tea (red), Iranian tea (blue), and clean air (green).

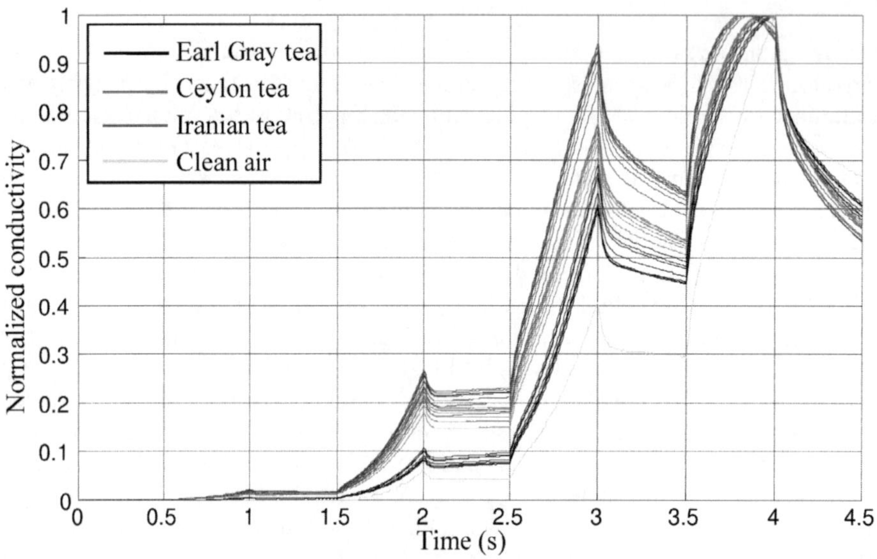

Figure 6. Normalized response patterns related to different concentration levels of Earl Gray tea (black), Ceylon tea (red), Iranian tea (blue), and clean air (green).

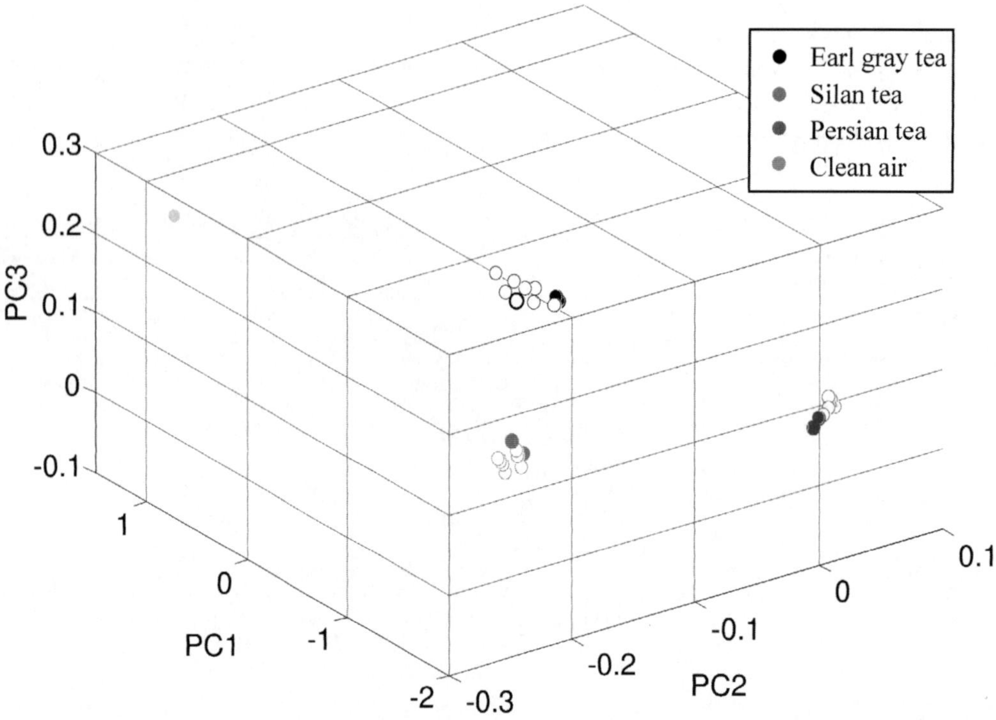

Figure 7. The 3-dimensional feature space obtained by applying PCA mapping to the normalized response patterns of Earl Gray tea (black), Ceylon tea (red), Iranian tea (blue), and clean air (green). Hollow markers are related to system training datasets and filled markers are used for the validation tests.

4. Conclusion

Design, fabrication and test of an electronic nose with a single thermal shock-induced sensor were reported. Accuracy, sensitivity and response time of the system was improved by using a suitable heating voltage waveform. The efficacy of the system in odor classification was demonstrated by utilizing it for the recognition of three different tea leaves. The recognition success rate was 100%. The designed system is cost effective, small and suitable for industrial applications.

References

[1] F. Hossein-Babaei, A. Amini. "A breakthrough in gas diagnosis with a temperature-modulated generic metal oxide gas sensor." *Sensors and Actuators B: Chemical* 166 (2012): 419-425.

[2] F. Hossein-Babaei, A. Amini. "Recognition of complex odors with a single generic tin oxide gas sensor." *Sensors and Actuators B: Chemical* 194 (2014): 156-163.

[3] L. Feng, J. Musto Christopher, W. Kemling Jonathan, H. Lim Sung, S. Suslick Kenneth. "A colorimetric sensor array for identification of toxic gases below permissible exposure limits." *Chem. Commun.* 46.12 (2010): 2037-2039.

[4] M. Ryan, H. Zhou, M. Buehler, K. Manatt, V. Mowrey, S. Jackson, Kisor, K. Adam A. Shevade, M. Homer. "Monitoring space shuttle air quality using the jet propulsion laboratory electronic nose." *Sensors Journal, IEEE* 4.3 (2004): 337-347.

[5] E. Scorsone, A. Maria Pisanelli, C. Krishna. "Development of an electronic nose for fire detection." *Sensors and Actuators B: Chemical* 116.1 (2006): 55-61.

[6] C. Arnold, M. Harms, J. Goschnick. "Air quality monitoring and fire detection with the Karlsruhe electronic micronose KAMINA."*Sensors Journal, IEEE* 2.3 (2002): 179-188.

[7] S. Dragonieri, M. van der Schee, T. Massaro, N. Schiavulli, P. Brinkman, A. Carrat. "An electronic nose distinguishes exhaled breath of patients with Malignant Pleural Mesothelioma from controls." *Lung Cancer* 75.3 (2012): 326-331.

[8] A. PF Turner, M. Naresh. "Electronic noses and disease diagnostics. " *Nature Reviews Microbiology* 2.2 (2004): 161-166.

[9] M. Peris, E. Laura. "A 21st century technique for food control: Electronic noses." *Analytica Chimica Acta* 638.1 (2009): 1-15.

[10] A. Loutfi, S. Coradeschi, G. Mani, P. Shankar, J. Rayappan. "Electronic noses for food quality: A review." *Journal of Food Engineering* 144 (2015): 103-111.

[11] F. Hossein-Babaei, V. Ghafarinia. "Gas analysis by monitoring molecular diffusion in a microfluidic channel." *Analytical chemistry* 82.19 (2010): 8349-8355.

[12] Z. Ouyang, R. Graham Cooks. "Miniature mass spectrometers."*Annual Review of Analytical Chemistry* 2 (2009): 187-214.

[13] M. Peris, E. Laura. "On-line monitoring of food fermentation processes using electronic noses and electronic tongues: a review." *Analytica chimica acta* 804 (2013): 29-36.

[14] A. Vergara, S. Vembu, T. Ayhan, M. Ryan, M. Homer, L. Huerta. "Chemical gas sensor drift compensation using classifier ensembles." *Sensors and Actuators B: Chemical* 166 (2012): 320-329.

[15] F. Hossein-Babaei, M. Paknahad, V. Ghafarinia. "A miniature gas analyzer made by integrating a chemoresistor with a microchannel." *Lab on a Chip* 12.10 (2012): 1874-1880.

[16] E. Martinelli, D. Polese, A. Catini, A. D'Amico, C. Di Natale. "Self-adapted temperature modulation in metal-oxide semiconductor gas sensors." *Sensors and Actuators B: Chemical* 161.1 (2012): 534-541.

[17] F. Hossein-Babaei, A. Hooshyar Zare, V. Ghafarinia, S. Erfantalab, "Identifying volatile organic compounds by determining their diffusion and surface adsorption parameters in microfluidic channels." *Sensors and Actuators B: Chemical* 220 (2015): 607-613.

[18] A. Szczurek, M. Maciejewska. "Virtual sensor array as a tool for classifying air pollution." *International Journal of Environmental Analytical Chemistry* ahead-of-print (2015): 1-14.

[19] F. Hossein-Babaei, M. Hemmati, M. Dehmobed. "Gas diagnosis by a quantitative assessment of the transient response of a capillary-attached gas sensor." *Sensors and Actuators B: Chemical* 107.1 (2005): 461-467.
[20] S. M. Hosseini-Golgoo, F. Hossein-Babaei. "Assessing the diagnostic information in the response patterns of a temperature-modulated tin oxide gas sensor." *Measurement Science and Technology* 22.3 (2011): 035201.
[21] F. Herrero-Carrón, D. J. Yáñez, F. de Borja Rodríguez, P. Varona. "An active, inverse temperature modulation strategy for single sensor odorant classification." *Sensors and Actuators B: Chemical* 206 (2015): 555-563.
[22] R. Gosangi, R. Gutierrez-Osuna. "Active temperature modulation of metal-oxide sensors for quantitative analysis of gas mixtures."*Sensors and Actuators B: Chemical* 185 (2013): 201-210.
[23] M. Roth, R. Hartinger, R. Faul, H-E. Endres. "Drift reduction of organic coated gas-sensors by temperature modulation." *Sensors and Actuators B: Chemical* 36.1 (1996): 358-362.
[24] F. Hossein-Babaei, and V. Ghafarinia. "Compensation for the drift-like terms caused by environmental fluctuations in the responses of chemoresistive gas sensors." *Sensors and Actuators B: Chemical* 143.2 (2010): 641-648.
[25] F. Hossein-Babaei, M. Orvatinia, "Thickness dependence of sensitivity in thin film tin oxide gas sensors deposited by vapor pyrolysis."*International Journal of Engineering Transaction B: Applications* 16.1 (2003): 33-40.

IOP Conf. Series: Materials Science and Engineering **108** (2016) 012020 doi:10.1088/1757-899X/108/1/012020

Refractive index and extinction coefficient of doped polycrystalline silicon films in infrared spectrum

Xia Zhang[1] and Dacheng Zhang [2]
[1] Department of Photo-electronics, Communication University of China, Beijing 100024, China
[2] Institute of Micro-Nano electronics, Peking University, Beijing100871, China

E-mail: zhangxia@cuc.edu.cn

Abstract. The refractive index and extinction coefficient in infrared spectrum of the polycrystalline silicon films with different doped dosages, base on the inverse calculation, are obtained by means of utilizing the measured reflectance and transmittance of a layer of material and multilayer films, and the equations derived from photonics and electromagnetic theory. The calculation results demonstrate that the refractive index of the doped polycrystalline silicon films decreases with the doped dosages increasing and the extinction coefficient increases with the doped dosages increasing for a given wavelength. This method used for determining the refractive index and extinction coefficient of the polycrystalline silicon films is effective and has the advantage of that the measured samples are fabricated simply.

1. Introduction

Polycrystalline silicon (Poly-Si) thin film is commonly used for movable structures in the micro sensor base on microelectronics technology, as its fabrication method is compatible with integrated circuit processes, it exhibits excellent mechanical performance and the heavily doped Poly-Si is a conductor. The physical properties of thin films are different from that of bulk materials. Therefore, the research on the properties of the doped Poly-Si film is helpful for micro sensor design.

Various methods for the determination of thin film parameters was reported due to thin films can be widely applied in modern optical instrument [1]. In the paper, based on the equations derived from photonics and electromagnetic theory, utilizing the value of the measured reflectance and transmittance of a layer of plate and the multi-layer films, the refractive index and extinction coefficient in infrared spectrum of a layer of Poly-Si film with a certain doped dosage are obtained by means of using the inverse calculation.

2. Measurement equipment

A Fourier transform infrared spectrometer, as shown in Fig. 1, is used for measuring the spectral reflectance and transmittance of samples, which has advantages of higher precision, resolution and efficiency. As the surface of the measured samples is not smooth enough, there are specula reflection and diffuse reflection. Therefore, a non-standard auxiliary integrating sphere, as shown in Fig. 2, is employed in the test to collect reflected and transmitted energy [2].

[1] Corresponding author Tel.: +86 010 5779606

Figure 1. Fourier transform infrared spectrometer (EQUINOX 55, Bruker Co.)

Figure 2. Non-standard auxiliary integrating spheres

In order to provide a high single to noise ratio, the detector in Fourier transform infrared spectrometer is cooled by liquid nitrogen.

3. Sample design and parameter determination

3.1. Sample design

A practical measurement setup requires millimetres size membranes for a reasonable signal level, to ensure the reflected and transmitted energy to be measured accurately [3]. However, the fabrication of the sample in which the free-standing suspended poly-Si film with a thickness of 1-2μm or less satisfies this size is more complex

In order to solve the problem brought by the dimensions feature above, a series of samples are designed, fabricated and measured, including a silicon (Si) substrate, silicon oxide (SiO_2) film/Si substrate, and Poly-Si film with different doped dosages, SiO_2 film and Si substrate forming a sandwich structure. In the structure, a layer of SiO_2 film grown on the silicon substrate is used for avoiding doped phosphorus into the silicon substrate, to ensure the accuracy of concentration and uniformity of the impurity doped in the Poly-Si film.

3.2. Parameter determination

3.2.1. A layer of material

The spectral reflectance is the ratio of spectral intensity reflected I_R by a sample to the incident spectral intensity I_0 upon the sample surface. The spectral transmittance is the ratio of the transmitted spectral intensity I_T to the incident spectral intensity I_0.

Using R to denote the reflectance of the sample surface, in accordance with P. Bouguer's law, the intensity of infrared at a depth x below the sample surface is given by [4]

$$I = I_0(1\text{-}R)\exp(-\alpha x) \tag{1}$$

where α is the absorption coefficient.

$$\alpha = \frac{4\pi\kappa}{\lambda} \tag{2}$$

$$R = \frac{(1-n)^2 + \kappa^2}{(1+n)^2 + \kappa^2} \tag{3}$$

where n is the refractive index of the sample, κ is extinction coefficient, λ is the wavelength in vacuum.

For near-normal incidence, according to Equation (1) and considering the multiple reflections in the plate as a sample with the thickness of w, using $R_\Sigma = I_R / I_0$ and $T_\Sigma = I_T / I_0$ to denote the reflectance and the transmittance of the plate, therefore, there are

$$R_\Sigma = R\left[1 + \frac{(1\text{-}R)^2 \exp(\text{-}2\alpha w)}{1\text{-}R^2 \exp(\text{-}2\alpha w)}\right] \tag{4}$$

and

$$T_\Sigma = \frac{(1\text{-}R)^2 \exp(\text{-}\alpha w)}{1\text{-}R^2 \exp(\text{-}2\alpha w)} \tag{5}$$

In accordance with Equations (2)-(5) and the measured dates of R_Σ and T_Σ, the value of the refractive index n and the extinction coefficient κ of the silicon substrate can be obtained by using calculation.

3.2.2. Multilayer material

The scheme of multilayer films is shown as Fig. 3.

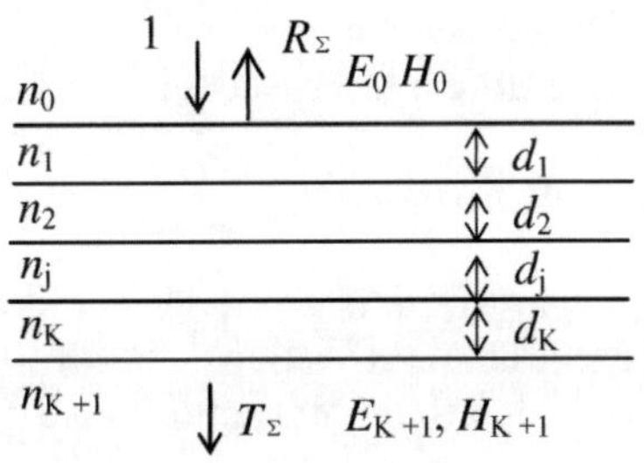

Figure 3. Multilayer films

According to the electromagnetic theory, the equations used for describing the relationship of the reflectance R_Σ and the transmittance T_Σ of multilayer materials is given by [5]

$$R_\Sigma = [(\eta_0 B - C)\cdot(\eta_0 B + C)^{-1}][(\eta_0 B - C)\cdot(\eta_0 B + C)^{-1}]^* \tag{6}$$

$$T_\Sigma = (1 - R_\Sigma)\frac{\eta_{K+1}}{\mathrm{Re}(BC^*)} = \frac{4\eta_0\eta_{K+1}}{(\eta_0 B + C)(\eta_0 B + C)^*} \tag{7}$$

$$\begin{bmatrix} B \\ C \end{bmatrix} = \left\{\prod_{j=1}^{K}\begin{bmatrix} \cos\delta_j & \mathrm{i}\cos\delta_j \cdot \eta_j^{-1} \\ \mathrm{i}\eta_j \sin\delta_j & \cos\delta_j \end{bmatrix}\right\}\begin{bmatrix} 1 \\ \eta_{K+1} \end{bmatrix} \tag{8}$$

$$\delta_j = \frac{2\pi}{\lambda} d_j (n_j^2 - k_j^2 - n_0^2 \sin^2\theta_0 - i2n_j k_j)^{1/2} \tag{9}$$

where j is the number of the film layer (shown as figure 3); n_j and k_j are the refractive index and the extinction coefficient of the film layer respectively; d_j is the thickness of the film layer; θ_0 is the incident angle. For near-normal incidence, $\eta_0 = n_0$ and $\eta_j = n_j$.

Using Equations (6)-(9), and according to the measured value of R_Σ and T_Σ, the value of the refractive index n and the extinction coefficient k of films can be obtained by using Matlab software based on the inverse calculation.

4. Samples and fabrication processes

There are many factors affecting the material parameters of thin films, such as fabrication environment temperate, crystallite dimension and arranged direction, the doped dosages and the uniformity of the film and so on.

In the sample preparation, the n-type silicon wafer of crystal orientation (100), with thickness of 500 μm and resistivity of 2-4 Ω·cm, is selected as the substrate. The size of the samples is 3.5cm×3.5cm. The prepared samples are as follows:

Ⅰ. 1# sample is the substrate with parameters above.

Ⅱ. 2# sample is that a layer of SiO_2 film with 300nm thickness is grown on the substrate at temperature of 1000℃ by thermal oxidation, which is used for the implanting isolation in order to control the doped concentration of within the Poly-Si film.

Ⅲ. 3# series samples are that a layer of Poly-Si film of 1 μm thickness is deposited on 300nm thickness SiO_2 film with the substrate at temperature of 610℃ by low pressure chemical vapour deposition (LPCVD). Then, phosphorus-doped is implanted in a certain dosage, and annealing at temperature of 1000℃. The dosages of sample 3#-1, 3#-2, 3#-3 and 3#-4 are $1.0\times10^{14}cm^{-2}$, $6.0\times10^{14}cm^{-2}$, $1.6\times10^{15}cm^{-2}$ and $5.1\times10^{15}cm^{-2}$, respectively.

5. Test, Data Processing and Result Analysis

5.1. Test and Data Processing

The Fourier transform infrared spectrometer (Section 2) is used for measuring the spectral reflectance and transmittance of each sample at the room temperature in a given infrared spectral window in the air.

Data Processing is as follows:

Ⅰ. Use Equations (2)-(5), in accordance with the value of the measured reflectance R_Σ and the transmittance T_Σ of the Si substrate for a given wavelength, the refractive index and the extinction coefficient of the Si substrate are obtained by using 1st Opt software based on Powell Optimization.

Ⅱ. Appling Equations (6)-(9) to 2# sample, according to the refractive index and the extinction coefficient of the Si substrate, and the value of the measured reflectance and the transmittance of 2# sample, base on inverse calculation, the refractive index and the extinction coefficient of SiO_2 film for a given wavelength is obtained by means of Matlab software.

Ⅲ. Appling Equations (6)-(9) to each of 3# series samples, utilizing the refractive indexes and the extinction coefficients of Si and SiO_2 above, and the value of the measured reflectance R_Σ and the transmittance T_Σ of 3# series samples for a given wavelength, base on the inverse calculation, the refractive index and the extinction coefficient of each of 3# series samples are obtained by using Matlab software.

5.2. Results Analysis

The calculation results and the related parameters of samples are tabulated for a given wavelength of 10μm in Table 1.

Table 1. Sample parameters

Sample number	Implanted dosages cm^{-2}	Refractive index n	Extinction coefficient κ
3#-1	1.0×10^{14}	3.30	3.48×10^{-3}
3#-2	6.0×10^{14}	3.27	1.09×10^{-2}
3#-3	1.6×10^{15}	3.14	2.23×10^{-2}
3#-4	5.1×10^{15}	2.67	2.04×10^{-1}

The calculation results show that the refractive index and extinction coefficient in infrared spectrum of doped Poly-Si films are related to the doped dosages for a given wavelength. The refractive index decreases with the doped dosages increasing, whereas, the extinction coefficient increases with the doped dosages increasing, i.e., the increases in the infrared absorbance of the Poly-Si film depending of the increase in doped dosage in it.

6. Conclusion

The method is effective that the refractive index and extinction coefficient of the Poly-Si films with different doped dosages are obtained by the inverse calculation. This determination of film parameters has the advantage of that samples are fabricated simply. In addition, the calculation results demonstrate that the refractive index and extinction coefficient in infrared spectrum of doped Poly-Si films are related to the doped dosages for a given wavelength.

This paper only explores an approach to obtain the refractive index and extinction coefficient of films. There is still a lot of result analysis and film optimization work to be done.

Acknowledgments

The authors would like to thank National Key Laboratory of Science and Technology on Micro/Nano-Fabrication in Peking University for their help on the sample fabrication. The work is financially supported by National Basic Research Program of China (973) under program No. 2011CB309502.

References

[1] Yongchang L and Weiqiang L 1990 *Principle of Optical filem* (Beijing: National Defense Industry Publish House) p23-53

[2] Guojin F, Yu W, Yu M, Ping L and Chundi Z 2013 A non-standard auxiliary intergrating sphere and correction method for the realization of diffuse reflectance *Optik* **124** 4325-4327

[3] Tuohiniemi M, Blomberg M and Gao F., Infrared spectroscopy study of a poly-silicon film for optimizing the boron-implanting dose 2013 *Journal of Microelectromechanical Systems*. **22** (5), 1207-1212

[4] Sze S M and KWOK K. NG J 2008 *Physics of Semiconductor Devices* (*3rd Edition*) (Xian: Xian Jiao Tong University Press) p 39-41

[5] Jinfa Tang, Peifu G, Xu L, and Haifeng Li 2006 *Modern Optical Thin Film Technology* (Hangzhou: Zhejiang University Press) p 29

Biomaterials and therapeutic applications

Angelo Ferraro

Institute of Physics, Laboratory of Functional programmable materials of quantum electronics for biomedicine and infocommunication applications. Kazan Federal University. 18 Kremlyovskaya Street, Kazan, 420008, Russian Federation

ferraro@eie.gr

Abstract. A number of organic and inorganic, synthetic or natural derived materials have been classified as not harmful for the human body and are appropriate for medical applications. These materials are usually named biomaterials since they are suitable for introduction into living human tissues of prosthesis, as well as for drug delivery, diagnosis, therapies, tissue regeneration and many other clinical applications. Recently, nanomaterials and bioabsorbable polymers have greatly enlarged the fields of application of biomaterials attracting much more the attention of the biomedical community. In this review paper I am going to discuss the most recent advances in the use of magnetic nanoparticles and biodegradable materials as new biomedical tools.

1. Introduction

The family of biomaterials is huge. Nevertheless new materials are under investigation (1; 4). Currently, biomaterials can be subdivided based on the their physicochemical nature. Materials presently used in biomedicine can be: metallic, ceramic, polymeric, composite and biodegradable polymers (3) (Table 1).

Table 1. List of some biomaterials currently used in biomedicine.

Metallic	Vanadium (V), iron (Fe), chromium (Cr), cobalt (Co), nickel (Ni), titanium (Ti), tantalum (Ta), niobium (Nb), molybdenum (Mo), and tungsten (W)
Ceramic	Alumina, zirconia, silicone nitrides, carbons, glass ceramics, dense hydroxyapatites, calciumphosphates and calcium aluminates
Polymeric	Polyvinylchloride, Polyethylene,Polypropylene, Polymethylmetacrylate, Polystyrene, Polyethylenterephthalate, Polytetrafluoroethylene and Polyamide
Composite (4)	Combination of two or more materials e.g.: bisphenol-A-glycidyldimethacrylate (Bis-GMA); triethylenglycoldimetachrylate (TEGDMA)
Biodegradable polymers and materials	Polyhydroxyalkanoates, polyhydroxyvalerate, polyhydroxyhexanoate, Polylactic, Polybutylene succinate , polycaprolactone , Polyanhydrides, Polyvinyl alcohol. Collagen, Elastin, Albumin, Fibrin, Polysaccharides

Modern biomedicine offers a variety of implantable medical devices able to improve the quality of life since they restore or implement body functions. They range from bone prosthesis (5) to devices for continuous glucose monitors (6), brain-stimulating electrodes (7), coagulation sensors (8) and drug delivery systems (9). Biomaterials are also used for diagnosis, therapies (10), tissue regeneration (11) and biomedical research (12; 13). Many of the challenges in producing new medical devices and biocompatible materials are focused on the mitigation of the host foreign-body response, which encapsulates the implant in a fibrotic membrane. In fact, immediately after biomaterial implantation, a normal process of tissue inflammation becomes prolonged and aberrant, producing after several weeks a distinct, avascular, collagenous sheath around the implant (14). The ultimate effect of this body's response is to isolate and neutralize the implanted material with inevitable chronic inflammation. Modifications of material's surface as well as geometry, porosity and mechanical properties are often executed in order to diminish cellular response. These modifications can be easily achieved by coating biomaterials' surface with polysaccharides, polyacrylamide, poly(vinyl alcohol), poly(Nvinyl-2-pyrrolidone) and polyethylene glycol (PEG) (15) or by functionalization (16). Another strategy has been to design implants that release anti-inflammatory drugs. These devices can reduce foreign-body cellular response at the implant sites, but only until the drug is released (17). Furthermore, for many applications such as angioplasty stent implantation are now available bioabsorbable materials that allow to restore proper tissue function and in parallel reduce chronic inflammation thanks to the body's self-absorption (18). This paper will present and discuss clinical purposes and issues related to the use of magnetic nanoparticles and biodegradable materials.

2. Magnetic nanoparticles

2.1. Nanoparticles definition and properties

Nanoparticles are particles with a size between about 1 and 100 nm that show properties that are not found in bulk samples of the same material (19). As other biomaterials, nanoparticles can be fabricated from organic and inorganic matter. Specifically, magnetic nanoparticles are synthesized from a variety of metallic atoms combined eventually with oxygen including iron oxides, such as Fe_3O_4 and γ-Fe_2O_3, spinel-type ferromagnets such as $MgFe_2O_4$, $MnFe_2O_4$, and $CoFe_2O_4$, alloys such as $CoPt_3$ and FePt as well as pure metals such as Fe and Co (20). Successful application of such magnetic nanoparticles in biomedicine is highly dependent on the stability and behaviour of the particles under specific parameters such as those found in the body tissues and fluids. An important and critical feature is that a single magnetic domain must show magnetization only in response to applied magnetic fields. This characteristic, named superparamagnetism, makes magnetic nanoparticles very attractive for biomedical applications because the risk of forming agglomerates is very low in the absence of magnetic fields.

2.2. Body's response to magnetic nanoparticles

As mentioned in the introduction, following the implantation of biomaterials in the human body an inflammatory response is expected. Particles and superparamagnetic nanoparticles are not exempt by this response and since their surface-to-volume ratio is particularly high, for such kind of biomaterials the body's response may be exacerbate with respect to prosthesis or other implants. In fact, independently by the size, naked magnetic nanoparticles will react with biological fluids (blood, plasma, interstitial fluid) immediately. One of the first events is the opsonization of each single particles. Opsonin are specialized circulating proteins belonging to the immune system able to mark, by covering them, foreign materials and microbes in order to facilitate the phagocytosis (a process that eliminates from the human body foreign organisms and tiny parts) (21). A similar process is expected to happen also in vitro when magnetic nanoparticles as well as other nanomaterials are in contact with the culture medium, which is rich of organic molecules (proteins, sugars, vitamins etc.). In this case the phenomenon is known as corana formation (also called crowing formation and is often used as synonymous of opsonization) and is due to the reactivity of nanoparticles surface (e.g. z-potential) (21). Finally, it is important to mention that also the size of the magnetic nanoparticles is critical for a proper biomedical performance. Particles smaller then 10 nm are removed from blood and eliminate with urine because kidneys are not able to hold such small particles; whereas particles bigger than 200 nm are trapped by the liver and also in this case they are removed by the blood pretty soon. In both cases the blood half-life of nanoparticles that are not in that range is a few minutes (21).

2.3 Biomedical application of magnetic nanoparticles
Nowadays plenty biomedical and biotechnological applications have been reported for magnetic nanoparticles. They are suitable for imaging and drug delivery (22), gene therapy and gene delivery (23), isolation and purification of DNA and RNA (24); isolation and purification of proteins (25), hyperthermia and magnetomechanical cell disruption for cancer therapies (26; 27), cell immobilization for bioscience research (28); stem cell therapies for tissue regeneration (29). In spite of these already existing applications, investigation on magnetic nanoparticles is expanding rapidly thanks to the easy manipulation performed using a magnetic field. This characteristic makes these carriers, amongst others, a desirable option for use in biomedical and process engineering applications.

3. Biodegradable materials

3.1. Biodegradable polymers definition and properties
A biodegradable material can be absorbed by the body following hydrolytic or enzymatic degradation that ultimate in its erosion. They can be both synthetically and biologically derived (natural). In table 1 are reported some examples of biodegradable polymers (30). Important characteristics of biodegradable materials are: 1) material should not evoke a sustained inflammatory response; 2) degradation products should be non-toxic; 3) material should have acceptable shelf life; 4) material should be with appropriate mechanical properties.

3.2. Body's response to nanoparticles
Though a biological response is expected to take place after biodegradable materials implantation, it should not evolve in chronic inflammation because the implant undergoes degradation avoiding in this way long-term biocompatibility issues. Furthermore, material bioengineering offers the possibility to design biodegradable polymers embedded with specific anti-inflammatory drugs that reduce the immune response and improve long-term clinical outcomes (31). Hydrolytically degradable polymers present functional groups susceptible to hydrolysis that include esters, orthoesters, anhydrides, carbonates, amides, urethanes, ureas, etc (32). Enzymatic degradation is executed within the body via enzymes and act mainly on protein biodegradable implants. These enzymes can digest specific molecules, such as collagenases for collagen, or act in a more aspecific way such as metalloproteinases. In both cases the biomaterial is split in the corresponding amino acids (30).

3.3. Biomedical application of biodegradable materials
Biodegradable platform materials are required for three-dimensional structures, tissue engineering and remodelling, regenerative medicine, controlled drug delivery and bionanotechnology. Three-dimensional scaffold have been used, for example, to generate in vitro cartilage tissue from stem cell that potentially can be implanted in the human body (33). Zartner and colleagues showed an elegant application of biodegradable material for tissue remodeling by using such kind of martial to restore the functionality of pulmonary artery (18). Moreover, another example of the utility of bioabsorbable fabrics that at the same time can release drug came again from angioplasty. It has been demonstrated that biodegradable polymer drug-eluting stents reduce the risk of stent thrombosis with good clinical outcomes (31). As for nanoparticles, research on biodegradable materials is expanding rapidly and new products are continuously studied. It has been estimated that almost all permanent biostable/biocompatible devices used for therapeutic applications will be replaced by biodegradable devices since offers better clinical performance and quality of life.

References

[1] Papadopoulos N et al (2010). Cyclodextrin inclusion complexes as novel MOCVD precursors for potential cobalt oxide deposition. Applied Organometallic Chemistry 24, 112-121.

[2] Antipas G et al. (2014). Solute–solvent interactions and atomic cohesion in GeSe4 and GeSe4In5 metallic

glasses. Materials Research Express 1, 015202.
[3] Parida P. et al. (2012). Classification of Biomaterials used in Medicine. International Journal of Advances in Applied Sciences 1, 31-35
[4] Salernitano E. et al. (2003). Composite materials for biomedical applications: a review. Journal of Applied Biomaterials & Biomechanics 1, 3-18.
[5] Agarwal R. et al. (2015). Biomaterial strategies for engineering implants for enhanced osseointegration and bone repair. Advanced drug delivery reviews.
[6] Nichols S. P et al. (2013). Biocompatible materials for continuous glucose monitoring devices. Chemical reviews, 113, 2528-2549.
[7] Raoul S. et al. (2003). Staged lesions through implanted deep brain stimulating electrodes: a new surgical procedure for treating tremor or dyskinesias. Movement disorders 18, 933-938.
[8] Maliaritsi E. et al (2006). Coagulation sensors based on magnetostrictive delay lines for biomedical and chemical engineering applications. Journal of magnetism and magnetic materials, 299(1), 41-52.
[9] Helary C. et al. (2015). Recent Advances in Biomaterials for Tissue Engineering and Controlled Drug Delivery. Current pharmaceutical biotechnology, 16(7), 635-645.
[10] Bae K et al. (2011). Nanomaterials for cancer therapy and imaging. Molecules and cells, 31, 295-302.
[11] Tabata Y. (2009). Biomaterial technology for tissue engineering applications. Journal of the Royal Society Interface, 6 (Suppl 3, S311-S324.
[12] Prodělalová J. et al. (2004). Isolation of genomic DNA using magnetic cobalt ferrite and silica particles. Journal of Chromatography A, 1056, 43-48.
[13] Guduru R. et al. (2013). Magneto-electric nanoparticles to enable field-controlled high-specificity drug delivery to eradicate ovarian cancer cells. Scientific reports, 3.
[14] Anderson J. et al. (2008). Foreign body reaction to biomaterials. In Seminars in immunology 20, 86-100
[15] Owens D. et al. (2006). Opsonization, biodistribution, and pharmacokinetics of polymeric nanoparticles. International journal of pharmaceutics, 307, 93-102.
[16] Antipas G. et al. (2013). Experimental and First - Principles Characterization of Functionalized Magnetic Nanoparticles. ChemPhysChem 14, 1934-1942.
[17] Wu P. et al. (2006). Drug/device combinations for local drug therapies and infection prophylaxis. Biomaterials, 27, 2450-2467.
[18] Zartner P. et al. (2005). First successful implantation of a biodegradable metal stent into the left pulmonary artery of a preterm baby. Catheterization and Cardiovascular Interventions, 66, 590-594.
[19] Kreyling W. et al. (2010). A complementary definition of nanomaterial. Nano Today, 5, 165-168.
[20] Lu A. et al. (2007). Magnetic nanoparticles: synthesis, protection, functionalization, and application. Angewandte Chemie International Edition, 46, 1222-1244.
[21] Nel A et al. (2009). Understanding biophysicochemical interactions at the nano–bio interface. Nature materials, 8, 543-557.
[22] Sun C. et al. (2008). Magnetic nanoparticles in MR imaging and drug delivery. Advanced drug delivery reviews, 60, 1252-1265.
[23] Majidi S. et al. (2015). Magnetic nanoparticles: Applications in gene delivery and gene therapy. Artificial cells, nanomedicine, and biotechnology, (0), 1-8.
[24] Berensmeier, S. (2006). Magnetic particles for the separation and purification of nucleic acids. Applied microbiology and biotechnology 73, 495-504.
[25] Safarik I. et al. (2004). Magnetic techniques for the isolation and purification of proteins and peptides. BioMagnetic Research and Technology 2, 7.
[26] Liu x. et al. (2012) Optimization of surface coating on Fe_3O_4 nanoparticles for high performance magnetic hyperthermia agents. Journal of Materials Chemistry 22: 8235-8244..
[27] Kim D. et al. (2010) Biofunctionalized magnetic-vortex microdiscs for targeted cancer-cell destruction. Nature materials 9: 165-171.
[28] Tseng P. et al. (2012). Magnetic nanoparticle-mediated massively parallel mechanical modulation of single-cell behavior. Nature methods 9, 1113-1119.
[29] Cromer B. et al. (2011). Tracking stem cells using magnetic nanoparticles. Wiley Interdisciplinary Reviews: Nanomedicine and Nanobiotechnology, 3, 343-355.
[30] Nair L. et al. (2007). Biodegradable polymers as biomaterials. Progress in polymer science, 32(, 762-798.
[31] Stefanini G. et al. (2012). Biodegradable polymer drug-eluting stents reduce the risk of stent thrombosis at 4 years in patients undergoing percutaneous coronary intervention. European heart journal, ehs086.
[32] Li S. et al. (1999). Hydrolytic degradation characteristics of aliphatic polyesters derived from lactic and glycolic acids. Journal of biomedical materials research, 48(3), 342-353.
[33] Li W et al. (2005). A three-dimensional nanofibrous scaffold for cartilage tissue engineering using human mesenchymal stem cells. Biomaterials, 26, 599-609.

IOP Conf. Series: Materials Science and Engineering **108** (2016) 012022 doi:10.1088/1757-899X/108/1/012022

Improving sensitivity of the polyurethane/CNT laminate strain sensor by controlled mechanical preload

Petr Slobodian, Robert Olejnik, Jiri Matyas, Dipak Gorakh Babar

Centre of Polymer Systems, University Institute, Tomas Bata University, Nad Ovcirnou 3685, 760 01 Zlin, Czech Republic

Corresponding author's e-mail address: slobodian@cps.utb.cz

Abstract. This article describes strain detection potential of polyurethane/CNT layered composite and further possible enhance of its sensitivity to strain, expressed by value of gauge factor, GF, employing its controlled mechanical preload. In course of its fabrication a non-woven polyurethane membrane made by electro spinning was used as filtering membrane for CNT aqueous dispersion. Final CNT polyurethane laminate composite is prepared by compression molding. Produced polyurethane/CNT composite laminate is electrically conductive and high elastic. Its elongation leads to change of its macroscopic electrical resistance. Changes in resistance are further reversible, reproducible and can monitor deformation in real time. Gauge factor reaches very high values around 8 for strain reaching 3.5 % comparing with conventional metallic strain gauges. Finally, controlled mechanical preload significantly increases value of GF. For example for value of 8.1 % of preload value of GF reaches 23.3 for strain 3.5 %.

1. Introduction

The carbon nanotube networks (CNT-Ns) can proportionally transfer their unique properties into composites and bring about substantial improvements in their properties such as the strength, the electrical and thermal conductivity. For instance, CNT-Ns as a strain sensor has potential advantage in that the networks have high sensitivity to local distributions of stress and strain as well as the ability to sense stresses and strains in different directions in the host materials [1]. In this respect the strain sensitivities of epoxy [1], poly(methylmethacrylate) [2] and thermoplastic polyurethane composites [3] with MWCNT-Ns or SWCNT-Ns have already been studied. Moreover, there are also some studies on the stimulation effect of the functionalization of CNT-Ns with plasma [4,5] as well as the carboxylic acid and amine groups [3] on the electrical properties [4,6] and mechanical properties [4,5] of polyurethane [3,6], polyimide [4] and polycarbonate [5] composites with CNT-Ns.

In this study, the stimulation effect of preloading on the electromechanical properties of thermoplastic polyester-based polyurethane (TPU) laminate with MWCNT-Ns is investigated under elongation and elongation/relaxation cycles and compared with corresponding properties of CNT-N/TPU laminates without preloading.

2. Experimental

MWCNT/Polyurethane laminate tested as a strain sensitive element is a thermoplastics polyurethane (TPU- Desmopan 385S) based composite with layered network made of entangled multi-walled carbon nanotubes of Buckypaper. Sensory layer was made by entangled multi-walled carbon nanotubes (MWCNT-Sun Nanotech Co. Ltd.). PU non-woven filtering membranes were prepared by electrospinning process and aqueous dispersion of MWCNT was vacuum filtration through it. Polyurethane filter with MWCNT network was melt welded (at 175 °C) onto the surface of TPU plate 2 mm thick. MWCNTs were analysed via transmission electron microscopy (TEM) using microscope JEOL JEM 2010 at the accelerating voltage of 160 kV. TPU filtering membrane and MWCNT/PU composites were analysed by scanning electron microscope (SEM) NOVA NanoSEM 450 (FEI). The change of electrical resistance of carbon nanotubes network/polyurethane laminates in extension/relaxation cycles were measured lengthwise by the two-point technique using multimeter Sefram 7338. For attachment of two copper electrodes to the MWCNT network of the composite samples a screw mechanism was used.

3. Results and discussion

The scanning electron microscope (SEM) analysis of mentioned PU non-woven filtering membrane is presented in Fig. 1 a). The transmission electron microscopy (TEM) micrograph in Fig. 1 b) gives a detail view of used nanotubes. Individual nanotubes are entrapped by filtering membrane in the course of MWCNT aqueous dispersion vacuum filtration. The nanotubes infiltrate into membrane pores which are finally blocked by a filtering cake made of pure network of entangled MWCNT [7]. SEM micrograph in Fig. 1 c) shows the upper surface of MWCNT network. MWCNT network is a porous and electrically conductive structure created by entangled nanotubes with electrical conductive junctions between them. The electrical conductivity of the MWCNT network is predominantly determined by the contact resistance in junctions of crossing nanotubes, rather than by resistance of MWCNTs itself. The nanotubes are much shorter than the dimension of straining sample and inter-tube contacts act as parallel resistors between highly conductive MWCNT segments. The final step of the laminate preparation is the melt welding of PU filter and MWCNT resulting about 2 mm thick composite plates. Electromechanical sensitivity of the prepared laminate was tested using dog-bone shape specimens cut out from two-layer laminate, see Fig. 1 d).

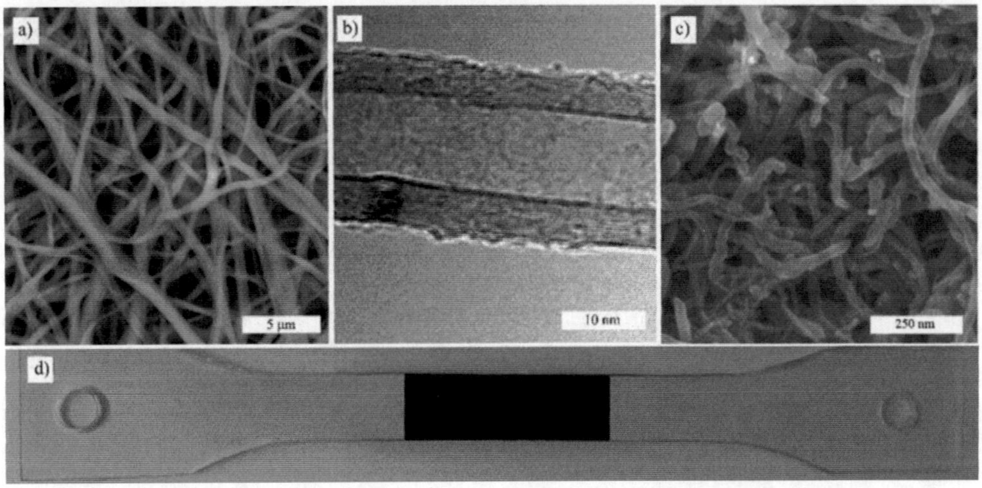

Fig. 1 a) SEM micrograph of polyurethane non-woven filtering membrane, b) TEM micrograph of used nanotubes, c) SEM image of the surface of entangled MWCNT network, d) Photograph of PU dog-bone shaped specimen, for the tensile test, with the fixed stripe of MWCNT/PU laminate (black).

The electromechanical behavior of carbon nanotubes network/polyurethane laminates and their preload stimulated form are presented in the following Fig. 2 a-b. The data represents five extension/relaxation cycles (the cycle deformation gradually increases in each 2 minutes cycle up to maximal strain value 3.5 %). The extension/relaxation cycle was five times repeated. The stimulated laminate was prepared by preload elongation 8.1 % lasting 10 minutes. The electromechanical testing was then performed after 20 hours of specimen relaxation. The extension of the each test specimen denoted by black lines, initiates resistance increase (illustrated by red lines). The strain is defined $\varepsilon = \Delta L/L_0$, where ΔL represents the change in specimen length and L_0 is the initial length before the first elongation. Resistance changes are defined as the relative resistance change $\Delta R/R_0 = (R-R_0)/R_0$, where R0 is the electrical resistance of the measured sample before the first elongation, and R is the resistance while elongating. After specimen unloading, deformation of polyurethane specimen decreases what is monitored by the resistance decrease. In the course of strain cycles, some irreversible changes in strain and the relative resistance are observed due to possible residual plastic deformation of TPU and irreversible damage of the electrically conductive MWCNT network. In the second and following cycle series the difference between responses is not such significant comparing with the first cycle series. Similar strain stabilization effect is observed for preloaded specimen when in consecutive cycles the relative resistance change or irreversible part are nearly equal. Moreover, the data in Fig. 2 demonstrates preload stimulation of electro-mechanical sensitivity. For example, the final value at the end of the fifth cycle of the fifth series reaches value of relative resistance change for no preload laminates 25.9 %. On the other hand, the corresponding value at preload 8.1 % achieves the resistance change 72 %.

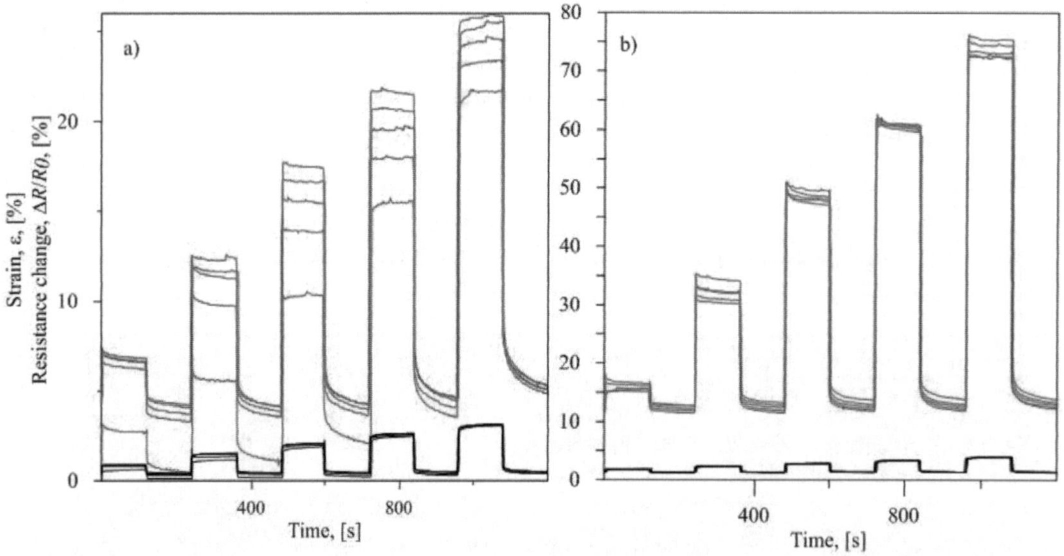

Fig. 2 Response of the relative resistance change, $\Delta R/R_0$, to applied strain, ε, for MWCNT/PU specimen at different values of preload strain. A) No preload specimen, B) preload 8.1 %. The tensile strain step increases in a five consecutive cycles which are five times repeated. The value of strain steps is denoted by black lines and the relative resistance change by red lines.

The data in Fig. 2 reveals also a significant effect of applied preload on the increase of relative resistance change. The change can be successfully quantified by means of a gauge factor (GF), which defines sensitivity of strain gauge as the relative resistance change divided by the applied strain, GF = $(\Delta R/R_0)/\varepsilon$. According to calculated GF values presented in Fig. 3, GF for the non-preload laminate is nearly independent on applied strain in the first cycle series since only slowly increases from values of 4.4 to 5.4 is observed. In the following series of cycles GF gradually increases up to values of about 8.2. At the same time, the values are stabilized by the applied strain cycles.

At the same time, the values are stabilized by the applied strain cycles. It is partially because of the irreversible resistance increase but this increase can be probably also explained as a mechanical stimulation to increase the resistance in the following series of imposed deformation cycles. A similar phenomenon takes place also when the stimulation by preload deformation 8.1 % was applied and GF reaches values around 23.3, respectively. It was found that preload of laminates to higher values of strain, than is its range in the measurement (up to 3.5 %), stabilizes GF values having only small variation in subsequent deformation cycles.

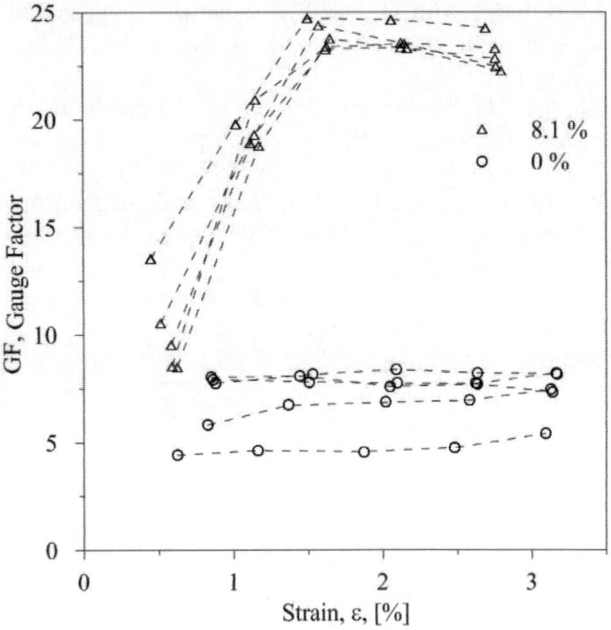

Fig. 4 Dependence of the gauge factor, GF, determined for MWCNT/PU laminate on applied strain for specimen with different values of preload indicated. Each experiment was performed as the tensile strain step increases measurement five times repeated.

4. Conclusions

Multi-walled carbon nanotubes were used in their pure form to prepare entangled networks as a part of MWCNT/TPU laminates. The laminate is made by taking a non-woven polyurethane filtering membrane and enmeshing it with carbon nanotubes. This innovative procedure eliminates the laborious process that is usually used, i.e. peeling off the nanotube network from the common micro-porous (polycarbonate, nylon) filter followed by the network polymeric impregnation to increase its compactness. Prior to deformation, the laminate was subjected to elongation 8.1 %. The subsequent testing consisted of the measurement of laminate deformation by its electrical resistance change in the course of elongation and elongation/relaxation cycles. The results show that the laminate after preloading has an enhanced electrical resistance and increased gauge factor in comparison with the non-preloaded samples.

Acknowledgement
This project was supported by the National Budget of the Czech Republic within the framework of the Centre of Polymer Systems project (reg. no.: CZ.1.05/2.1.00/03.0111).

IOP Conf. Series: Materials Science and Engineering **108** (2016) 012022 doi:10.1088/1757-899X/108/1/012022

References

[1] Rein MD, Breuer O, Wagner HD. Sensors and sensitivity: Carbon nanotube buckypaper films as strain sensing devices. Compos Sci Technol 2011;71:373-381.

[2] Kang I, Schulz MJ, Kim JH, Shanov V, Shi D. A carbon nanotube strain sensor for structural health monitoring. Smart Mater Struct 2006;15:737-748.

[3] Lima AMF, de Castro VG, Borges RS, Silva GG. Electrical conductivity and thermal properties of functionalized carbon nanotubes/polyurethane composites. Polímeros: Ciencia e Tecnologia 2012;22:117-124.

[4] Jiang Q, Li Y, Xie J, Sun J, Hui D, Qiu Y. Plasma functionalization of bucky paper and its composite with phenylethynyl-terminated polyimide. Compos Part B-Eng, 2013;45:1275-1281.

[5] Pötschke P, Zschoerper NP, Moller BP, Vohrer U. Plasma functionalization of multiwalled carbon nanotube bucky papers and the effect on properties of melt-mixed composites with polycarbonate. Macromol Rapid Comm 2009;30:1828-1833.

[6] R. Benlikaya, P. Slobodian, P. Riha. Enhanced strain-dependent electrical resistance of polyurethane composites with embedded oxidized multiwalled carbon nanotube networks. J Nanomat, 2013; Article ID 327597:10 pp.

[7] Slobodian P, Riha P, Saha P. A highly-deformable composite composed of an entangled network of electrically-conductive carbon-nanotubes embedded in elastic polyurethane. Carbon 2012;50:3446-3453.

IOP Conf. Series: Materials Science and Engineering **108** (2016) 012023 doi:10.1088/1757-899X/108/1/012023

Influence of losses on the output voltage of ferrite transformer in case of strong magnetic field in the core

A. Gulbis[1] **D. Malinovska**[2] **D. Stepins**[3]
[1,2,3]Faculty of Electronics and Telecommunications, Riga Technical University, Azenes str. 12, Riga, LV 1048, Latvia

E-mail: [1]arnis.gulbis@rtu.lv, [2]diana.malinovska@rtu.lv, [3]deniss.stepins@rtu.lv,

Taking into account the nonlinear properties of ferrites the output voltage of toroidal transformer with magnetically soft ferrite core is calculated. Calculations are based on methods analogous to those used in nonlinear optics. Initially demagnetized ferrite is assumed to be isotropic. At this condition apart from the harmonic on fundamental frequency the output voltage of transformer contains also higher harmonics of odd order. Nonlinear susceptibilities of third and fifth order are used in calculations Magnetic losses are taken into account considering linear and nonlinear susceptibilities as being complex. These losses affect amplitudes of harmonics in the output voltage of transformer and create phase shifts between components of magnetization on different frequencies in the core and the intensity of exciting harmonic magnetic field. Results obtained in calculations are confirmed by experimental measurement of output voltage of toroidal transformer with MnZn ferrite core. In order to calculate amplitudes of harmonics in the output voltage of transformer numerical values of susceptibilities in question should be known.

1. Introduction

In last decades due to their magnetic and other properties ferrites have found applications in different areas of electronics. In general the magnetization of ferrites is a nonlinear function of the strength of the magnetic field. If the magnetic field applied is sufficiently weak ferrites can be considered as being linear substances. However, there are applications were nonlinear properties of ferrites play an important role, e.g., in [1] an application is considered where intermodulation distortions of ferrites should be taken into account. In [2], we considered the influence of nonlinear properties of ferrite on the voltage of toroidal transformer with a soft ferrite (narrow hysteresis loop) core. The output voltage of transformer was calculated in a simplified case when losses in the core were not taken into account. The goal of this paper is to generalise the phenomenological description developed in [2] by taking into account also losses in the core.

From the experimental point of view it was very well known that the sufficiently strong magnetic field in the core leads to the distortion of the signal shape and the appearance of harmonics. Results obtained in [2] are based on the following statements. Magnetization $\vec{M}$ of ferrite can be expanded in power series of the intensity of the magnetic field $\vec{H}$ [3,4]. In symbolic form one can write

$$\vec{M} = \chi^{(1)}\vec{H} + \chi^{(2)}\vec{H}\vec{H} + \chi^{(3)}\vec{H}\vec{H}\vec{H} + \chi^{(4)}\vec{H}\vec{H}\vec{H}\vec{H} + \chi^{(5)}\vec{H}\vec{H}\vec{H}\vec{H}\vec{H} + \cdots, \qquad (1)$$

where $\chi^{(1)}$ is the linear but $\chi^{(2)}$ - $\chi^{(5)}$ are nonlinear magnetic susceptibilities. In fact susceptibilities in (1) are tensors of corresponding rank.

Initially demagnetised ferrite core as polycrystalline substance is isotropic due to the arbitrary orientation of crystallites. This means that properties of ferrite du not depend from the direction in space, respectively, should stay invariant under the transform $\vec{H} \to -\vec{H}$. The consequence of this is that susceptibilities of even order in (1) are zero. Indeed, the product of even number of $\vec{H}$ doesn't change the sign at transform $\vec{H} \to -\vec{H}$ while the magnetization $\vec{M}$ does. From this it follows that at transform $\vec{H} \to -\vec{H}$ $\chi^{(2k)} = -\chi^{(2k)}$ (*k* is a natural number). This is possible only if $\chi^{(2k)} = 0$. In contrast, susceptibilities of odd number don't change the sign at transform $\vec{H} \to -\vec{H}$ and, therefore, in general are nonzero. In this paper we will confine ourselves to the consideration of susceptibilities up to the fifth order including.

2. Magnetization of the core

For toroidal cores the cylindrical system of reference is appropriate. In what follows it is supposed that only the component H_φ of the magnetic field is nonzero. Due to isotropy the magnetization has the same direction. Besides it is also supposed that the harmonic magnetic field acts on the core and the wavelength of the field is much more grater then the linear dimensions of the core. Then

$$\mathrm{H}_\varphi(\omega) = H_\varphi \cos\omega t = \left[H_\varphi(\omega)\exp(-i\omega t) + H_\varphi(-\omega)\exp(i\omega t)\right], H_\varphi(-\omega) = H_\varphi^*(\omega) = \frac{1}{2}H_\varphi . \quad (2)$$

In (2) *i* is the imaginary unit and * means the complex conjugated expression.
Let us writer the magnetization of the core as a sum of three terms

$$\mathrm{M}_\varphi = \mathrm{M}_\varphi^{(1)} + \mathrm{M}_\varphi^{(3)} + \mathrm{M}_\varphi^{(5)}, \quad (3)$$

where the first one is determined by the linear susceptibility, the second one by the cubic susceptibility and the last one by the susceptibility of the fifth order. Let us first consider $\vec{\mathrm{M}}_\varphi^{(3)} = \vec{M}_\varphi^{(3)} + \vec{M}_\varphi^{(3)*}$. Due to analogy between the electrical polarisation and the magnetization methods used in nonlinear optics [3,4] can be applied here. For magnetization one can write

$$M_\varphi^{(3)} = \sum_{klm} \chi_{4\varphi}^{(3)}(\omega_{k+l+m}, \omega_k, \omega_l, \omega_m) H_\varphi(\omega_k) H_\varphi(\omega_l) H_\varphi(\omega_m) \exp[-i(\omega_k + \omega_l + \omega_m)t]. \quad (4)$$

In (4) $\chi_{4\varphi}^{(3)}$ means $\chi_{\varphi\varphi\varphi\varphi}^{(3)}$ and ω_{k+l+m} stands for $\omega_k + \omega_l + \omega_m$. At perturbation of the core with a harmonic field only two frequencies *ω* and *–ω* are present. Then $\vec{M}_\varphi^{(3)}$ has two time dependent terms, one with the fundamental frequency *ω* and another with the triple frequency *3ω*. The first one is

$$\begin{aligned}\vec{M}_\varphi^{(3)}(\omega) = &\left[\chi_{4\varphi}^{(3)}(\omega,\omega,\omega,-\omega) + \chi_{4\varphi}^{(3)}(\omega,\omega,-\omega,\omega) + \chi_{4\varphi}^{(3)}(\omega,-\omega,\omega,\omega)\right] \\ &\times \left|H_\varphi(\omega)\right|^2 H_\varphi(\omega)\exp[-i\omega t]\end{aligned} \quad (5)$$

From the symmetry properties of nonlinear susceptibilities [3,4] it follows that all susceptibilities in (5) are equal and, therefore

$$\vec{M}_\varphi^{(3)}(\omega) = 3\chi_{4\varphi}^{(3)}(\omega,\omega,\omega,-\omega)\left|H_\varphi(\omega)\right|^2 H_\varphi(\omega)\exp[-i\omega t]. \quad (6)$$

The second one is the following

$$\vec{M}_\varphi^{(3)}(3\omega) = \chi_{4\varphi}^{(3)}(3\omega,\omega,\omega,\omega) H_\varphi^{\ 3}(\omega)\exp[-i3\omega t]. \quad (7)$$

Completely analogously one can consider terms caused by susceptibilities of the first and the fifth order. Magnetization caused by the susceptibility of the first order is

$$M_\varphi^{(1)}(\omega) = \chi_{2\varphi}^{(1)}(\omega) H_\varphi(\omega)\exp[-i\omega t] \quad (8)$$

Correspondingly, susceptibility of the fifth order determine 3 terms each on different frequencies. They are as follows

$$\vec{M}_{\varphi}^{(5)}(\omega)=10\chi_{6\varphi}^{(5)}(\omega,\omega,\omega,\omega,-\omega,-\omega)|H_{\varphi}(\omega)|^{4}H_{\varphi}(\omega)\exp[-i\omega t], \tag{9}$$

$$\vec{M}_{\varphi}^{(5)}(3\omega)=5\chi_{6\varphi}^{(5)}(3\omega,\omega,\omega,\omega,\omega,-\omega)|H_{\varphi}(\omega)|^{2}H_{\varphi}^{3}(\omega)\exp[-i3\omega t], \tag{10}$$

$$\vec{M}_{\varphi}^{(5)}(5\omega)=\chi_{6\varphi}^{(5)}(5\omega,\omega,\omega,\omega,\omega,\omega)H_{\varphi}^{5}(\omega)\exp[-i5\omega t]. \tag{11}$$

In (9) and (10) symmetry properties of the susceptibility of the fifth order analogous to those used in (5) are taken into account.

3. Magnetic flux and the output voltage of transformer

The magnetic flux in the core is given by $\Phi=\mathrm{B}_{\varphi}S$ where the induction of the magnetic field $\mathrm{B}_{\varphi}=\mu_0(\mathrm{H}_{\varphi}+\mathrm{M}_{\varphi})$ on the basis of (6) to (11) can be expressed as a sum of three terms

$$\mathrm{B}_{\varphi}=\mathrm{B}_{\varphi}(\omega)+\mathrm{B}_{\varphi}(3\omega)+\mathrm{B}_{\varphi}(5\omega), \tag{12}$$

where

$$\mathrm{B}_{\varphi}(\omega)=\mu_0\left(1+\chi_{2\varphi}^{(1)}(\omega)+\frac{3}{4}\chi_{4\varphi}^{(3)}(\omega)H_{\varphi}^{2}+\frac{5}{8}\chi_{6\varphi}^{(5)}(\omega)H_{\varphi}^{4}\right)\frac{H_{\varphi}}{2}\exp(-i\omega t)+c.c., \tag{13}$$

$$\mathrm{B}_{\varphi}(3\omega)=\mu_0\left(\chi_{4\varphi}^{(3)}(3\omega)+\frac{5}{4}\chi_{6\varphi}^{(5)}(3\omega)H_{\varphi}^{2}\right)\frac{H_{\varphi}^{3}}{8}\exp(-i3\omega t)+c.c, \tag{14}$$

$$\mathrm{B}_{\varphi}(5\omega)=\mu_0\chi_{6\varphi}^{(5)}(5\omega)\frac{H_{\varphi}^{5}}{32}\exp(-i5\omega t)+c.c. \tag{15}$$

In (13) to (15) *c.c.* stands for the complex conjugated expression and for the sake of simplicity the following designations are introduced
$\chi_{4\varphi}^{(3)}(\omega)=\chi_{4\varphi}^{(3)}(\omega,\omega,\omega,-\omega)$, $\chi_{6\varphi}^{(5)}(\omega)=\chi_{6\varphi}^{(5)}(\omega,\omega,\omega,\omega,-\omega,-\omega)$, $\chi_{4\varphi}^{(3)}(3\omega)=\chi_{4\varphi}^{(3)}(3\omega,\omega,\omega,\omega)$, $\chi_{6\varphi}^{(5)}(3\omega)=\chi_{6\varphi}^{(5)}(3\omega,\omega,\omega,\omega,\omega,-\omega)$ and $\chi_{6\varphi}^{(5)}(5\omega)=\chi_{6\varphi}^{(5)}(5\omega,\omega,\omega,\omega,\omega,\omega)$.

Due to magnetic losses in the core all susceptibilities in (13) to (15) become complex, respectively have the form

$$\chi^{(2k+1)}=\mathrm{Re}\,\chi^{(2k+1)}+\mathrm{Im}\,\chi^{(2k+1)},\ k=0,1,2. \tag{16}$$

Inserting (16) into (13) to (15) one obtains

$$\mathrm{B}_{\varphi}=A\cos(\omega t-\delta_1)+C\cos(3\omega t-\delta_3)+D\cos(5\omega t-\delta_5), \tag{17}$$

where

$$A=\mu_0\left\{\left[1+\operatorname{Re}\chi_{2\varphi}^{(1)}(\omega)+\frac{3}{4}\operatorname{Re}\chi_{4\varphi}^{(3)}(\omega)H_\varphi^{\ 2}+\frac{5}{8}\operatorname{Re}\chi_{6\varphi}^{(5)}(\omega)H_\varphi^{\ 4}\right]^2+\left[\operatorname{Im}\chi_{2\varphi}^{(1)}(\omega)+\frac{3}{4}\operatorname{Im}\chi_{4\varphi}^{(3)}(\omega)H_\varphi^{\ 2}+\frac{5}{8}\operatorname{Im}\chi_{6\varphi}^{(5)}(\omega)H_\varphi^{\ 4}\right]^2\right\}^{\frac{1}{2}}H_\varphi, \tag{18}$$

$$C=\frac{\mu_0}{4}\left\{\left[\operatorname{Re}\chi_{4\varphi}^{(3)}(3\omega)+\frac{5}{4}\operatorname{Re}\chi_{6\varphi}^{(5)}(3\omega)H_\varphi^{\ 2}\right]^2+\left[\operatorname{Im}\chi_{4\varphi}^{(3)}(3\omega)+\frac{5}{4}\operatorname{Im}\chi_{6\varphi}^{(5)}(3\omega)H_\varphi^{\ 2}\right]^2\right\}^{\frac{1}{2}}H_\varphi^3 \tag{19}$$

$$D=\frac{\mu_0}{16}\left\{\left[\operatorname{Re}\chi_{6\varphi}^{(5)}(5\omega)\right]^2+\left[\operatorname{Im}\chi_{6\varphi}^{(5)}(5\omega)\right]^2\right\}^{\frac{1}{2}}H_\varphi^5 \tag{20}$$

and

$$\delta_1=\arctan\frac{\operatorname{Im}\chi_{2\varphi}^{(1)}(\omega)+\frac{3}{4}\operatorname{Im}\chi_{4\varphi}^{(3)}(\omega)H_\varphi^{\ 2}+\frac{5}{8}\operatorname{Im}\chi_{6\varphi}^{(5)}(\omega)H_\varphi^{\ 4}}{1+\operatorname{Re}\chi_{2\varphi}^{(1)}(\omega)+\frac{3}{4}\operatorname{Re}\chi_{4\varphi}^{(3)}(\omega)H_\varphi^{\ 2}+\frac{5}{8}\operatorname{Re}\chi_{6\varphi}^{(5)}(\omega)H_\varphi^{\ 4}}, \tag{21}$$

$$\delta_3=\arctan\frac{\operatorname{Im}\chi_{4\varphi}^{(3)}(3\omega)+\frac{5}{4}\operatorname{Im}\chi_{6\varphi}^{(5)}(3\omega)H_\varphi^{\ 2}}{\operatorname{Re}\chi_{4\varphi}^{(3)}(3\omega)+\frac{5}{4}\operatorname{Re}\chi_{6\varphi}^{(5)}(3\omega)H_\varphi^{\ 2}}, \tag{22}$$

$$\delta_5=\arctan\frac{\operatorname{Im}\chi_{6\varphi}^{(5)}(5\omega)}{\operatorname{Re}\chi_{6\varphi}^{(5)}(5\omega)}. \tag{23}$$

E.m.f. per turn E of toroidal transformer with ferrite core equals to the voltage per turn and is given by formula

$$E=-\frac{d\Phi}{dt}=A\omega\sin(\omega t-\delta_1)+3C\omega\sin(3\omega t-\delta_3)+5D\omega\sin(5\omega t-\delta_5). \tag{24}$$

In order to obtain the output voltage of transformer E given by (24) should be multiplied by the number of turns N in the output winding. If losses are absent, respectively, imaginary parts of all susceptibilities are zero then from (17) to (24) follow corresponding formulae obtained in [2]. Besides, the phase shifts between all magnetization components on different frequencies and the exciting magnetic field disappear.

For illustrative purposes we present here experimental measurements of the output spectrum of toroidal transformer with MnZn ferrite core (Fig. 1). It was checked that the spectrum of generator used in the input of transformer contained only the harmonic of fundamental frequency. Measurements

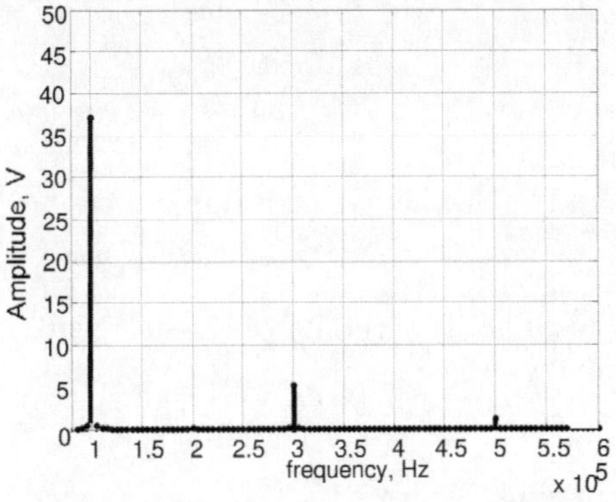

Figure 1. Output spectrum of transformer with MnZn ferrite core at fundamental frequency 100kHz.

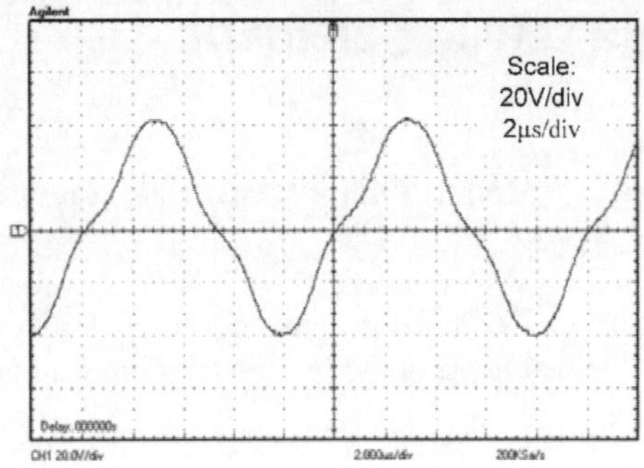

Figure 2. Shape of the output voltage of transformer at harmonic input.

were performed at the saturation of the magnetic induction in the core (see Fig.2). From Fig. 1 the presence of third and fifth harmonics in the output voltage of transformer is clearly seen. Measured output quantities at fundamental frequencies 1 and 10 kHz are similar to those shown on Fig. 1 and 2.

4. Conclusions

Methods used in nonlinear optics successfully can be applied to description of nonlinear properties of soft ferrites. An initially demagnetized ferrite can be considered as being isotropic. Due to this apart from fundamental harmonics of odd order appear in the output of transformer with soft ferrite core. Magnetic losses present in the core create the phase shifts between components of magnetization on different frequencies and the intensity of exciting magnetic field. They affect also numerical values of amplitudes of harmonics. In order to calculate these amplitudes numerical values of susceptibilities in question should be known.

Acknowledgements

This investigation was partly financed by Latvian Council of Science (grant No 467/2012). Authors are grateful to Professor Janis Jankovskis for valuable discussions.

References

[1] Edridge T 2006 Intermodulation distortion in medium-power drop-in ferrite isolators and circulators *Microwave Journal* **49** 90
[2] Gulbis A, Stepins D and Malinovska D 2015 Nonlinear Properties of Soft Ferrites *Advanced Materials Research* **1117** 56-59
[3] Landau L D and Lifshitz E M 1982 *Electrodynamics of continuous media* **8** 1236
[4] Boyd R W 2008 *Nonlinear Optics* 640

IOP Conf. Series: Materials Science and Engineering **108** (2016) 012024 doi:10.1088/1757-899X/108/1/012024

High elastic polyurethane/carbon nanotube composite laminate for structure health monitoring by gain shifting of antenna sensing element

Robert Olejnik, Petr Slobodian, Jiri Matyas, Dipak Gorakh Babar

Centre of Polymer Systems, University Institute, Tomas Bata University, Nad Ovcirnou 3685, 760 01 Zlin, Czech Republic

corresponding author's e-mail address: rolejnik@volny.cz

Abstract. The composite of carbon nanotubes and polyurethane (PU) was prepared by simple filtration technique. The PU nonwoven filtration membrane was prepared by electrospinning. A layer of carbon nanotubes was prepared by vacuum filtration on the surface of PU membrane. The resulting composite was subsequently placed on highly elastic polyurethane substrate. The contribution shows an efficient method of preparing the sensing element for monitoring the state of strain of loaded structures by using highly elastic polyurethane / carbon nanotubes composite. This sensor has been involved as passive antenna with stable resonance frequency of 650 MHz. When it is get deformed in the range from 0 to 3.5% the sensor gain was changing from -39 dB to - 19.45 dB. But if it is get deformed by 15% and again measured strain from 0 to 3.5%, sensor gain was changing from -33 dB to -12.3 dB, which clearly indicates the damage of structure.

1. Introduction

Carbon nanotubes network in the form of very thin layer was prepared by simple filtration method. For filtration was used non-woven structured polyurethane membrane prepared by electrospinnig method. The network was placed on polymer substrate by hot pressing method and the composite laminate was made [1]. The carbon nanotubes composite laminate has the new additional properties such as the strength, the electrical and thermal conductivity.
PU/CNT network as a strain sensor for structure health monitoring has potential advantage in that the networks have high sensitivity to local distributions of stress and strain as well as the ability to sense stresses and strains in different directions [2]. PU/CNT network as a strain sensor can be used for monitoring of crack formation or propagation in the monitored material or construction [3]. There are a lot of technique how monitored and analyzed the signal from sensing element. The sensing element signal can be measured as a change of resistivity during applying strain or change of voltage when we use Wheatstone bridge and many others.

The idea of this article is make a fusion between very elastic polymer material in the form of layered composite and carbon nanotubes as an electrical conductive material in the form of network [4]. The composite was used in the form of micro-strip antenna sensing element. The sensing element combined very elastic nonconductive polymer on one side, with stress/strain sensitive conductive layer on other side.

In this study, the stimulation effect of preloading on the gain changing properties of thermoplastic polyester-based polyurethane (TPU) laminate with MWCNT-Ns is investigated under elongation and elongation/relaxation cycles and compared with corresponding properties of CNT-N/TPU laminates without preloading. CNT-N/TPU laminates strain sensing element is able to sensitive react a clearly shows the change of gain after preloading which simulate the damage of construction.

2. Experimental

Purified MWCNT produced by chemical vapor deposition (CVD) of acetylene were supplied by Sun Nanotech Co. Ltd., China. According to the supplier, the nanotube diameter is 10-30 nm, length 1-10 µm, purity >90% and (volume) resistivity 0.12 cm-1. Further details on the nanotubes and the results of TEM analysis can be found in our paper [8] 2.65 g of MWCNT, sodium dodecyl sulfate (SDS) and 1-pentanol. Consequently, NaOH aqueous solution was added to adjust pH to the value of 10. The final nanotube concentration in the suspension was 0.49 wt.%, concentration of SDS and 1-pentanol 0.1M and 0.14M, respectively. The suspension was sonicated in Dr. Hielscher GmbH apparatus (ultrasonic horn S7, amplitude 88 µm, power density 300 W/cm^2, frequency 24 kHz) for 15 minutes and the temperature of ca 40°C.

The polyurethane granules were dissolved in a mixture of dimethyl formamide/methyl isobutyl ketone (Penta Chemikalie, Czech Republic) with volume ratio 3:1. The polymer weight concentration was adjusted to 16 % (w/v) and the mixture electric conductivity to 30 □s/cm by adding NaCl in order to optimize the process. The total number of nozzles was 18, the length of nozzles 30 mm, the distance between nozzles 20 mm, the nozzle internal diameter 1.2 mm and the outer diameter 2.2 mm. The electric voltage was set to 75 kV (Matsusada DC power supply), the temperature 21 ± 2 °C, the relative humidity 35 ± 2.5%.

Trough above prepared membrane was filtrated carbon nanotubes dispersion. The dispersion was filtrated by using standard water vacuum filtration apparatus. Prepared filtration cake was rinsed by 65 °C hot water and then by methanol. Filtration membrane with carbon nanotubes layer was placed between two filtration paper and leave without disturbing for 24 hours to reach dry PU/CNT composite structure. From PU/CNT layer was cut sample with dimension 10 x 50 mm. The simple was place on PU base material and put together by molding machine at 175 °C. The PU/CNT composite laminate in the form of strip was mounted in the creep test apparatus. Creep test apparatus was used for extension/relaxation cyclic testing.

The speciment was equipped with elastic wire. This wire was connected with ground plane. The ground plane was equipped with coaxial cable which is connected with spectral analyzer. The sample is used as an antenna. The measurements of the antenna were performed using the R&S FSH3 spectrum analyzer (the model with a tracking generator and without a preamplifier) and FSH-Z2 bridge capable of measuring within a bandwidth of 100 kHz to 3 GHz. The frequency range was set from 100 kHz to 1.5 GHz.

3. Results and discussion

The transmission electron microscopy (TEM) micrograph in Fig. 1 a) gives a detail view of used nanotubes. There is clearly visible multi wall structure of carbon nanotubes. Individual nanotubes are entrapped by non-woven filtration membrane during vacuum filtration of MWCNT aqueous dispersion. The nanotubes are infiltrated into membrane pores and the filtering cake is made. [7]. The scanning electron microscope (SEM) analysis of mentioned PU non-woven filtering membrane is presented in Fig. 1 b). SEM micrograph in Fig. 1 c) shows the upper surface of MWCNT network. MWCNT network is a porous and electrically conductive structure created by entangled nanotubes with electrical conductive junctions between them. The electrical conductivity of the MWCNT network is predominantly determined by the contact resistance in junctions of crossing nanotubes, rather than by resistance of MWCNTs itself. The nanotubes are much shorter than the dimension of straining sample and inter-tube contacts act as parallel resistors between highly conductive MWCNT segments. Fig. 1 d) shows detail view of created carbon nanotubes entangled network. The final step of the laminate preparation is the melt welding of PU filter and MWCNT resulting about 2 mm thick composite plates. Electromechanical sensitivity of the prepared laminate was tested using dog-bone shape specimens cut out from two-layer laminate, see Fig. 1 e).

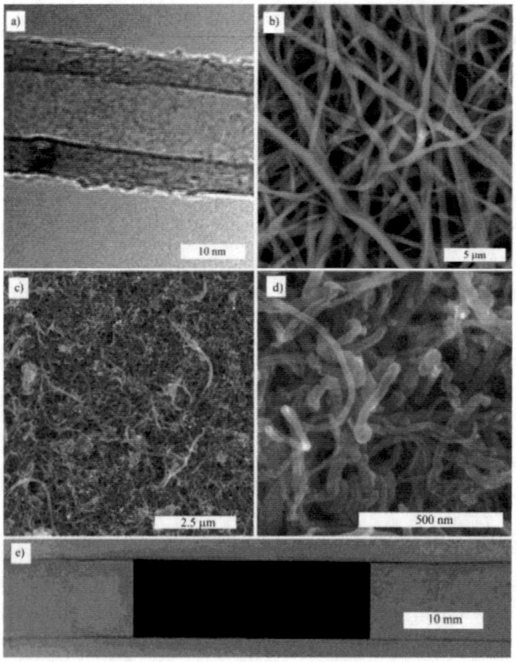

Fig. 1 a) TEM micrograph of multi wall carbon nanotube, b) SEM micrograph of polyurethane non-woven filtering membrane, c) SEM image of the surface of entangled MWCNT network, d) Detail of entangled MWCNT network e) Image of PU strip shaped specimen, for the stress/stain test for gain shift measurement with the fixed stripe of MWCNT/PU laminate (black).

The mechanical/electromagnetical behavior of carbon nanotubes network/polyurethane laminates in the form without preloading is presented in Fig. 2a. The data represents ten extension/relaxation cycles. The cycle deformation gradually increases in each 5 minutes cycle up to maximal strain value approximately 3.5 %. The preload stimulated form are presented in the following Fig. 2 b. The data represents ten extension/relaxation cycles. The cycle deformation gradually increases in each 5 minutes cycle up to maximal strain value approximately 3.25 %. The stimulated laminate was prepared by preload elongation 15%. The mechanical/electromagnetical testing was then performed after 20 hours of specimen relaxation. Both speciments form (without and preloaded) was simultaneously tested as an antenna and also extension/relaxation cycles. The gain of this antenna was measured. The strain is defined $\varepsilon = \Delta L/L_0$, where ΔL represents the change in specimen length and L_0 is the initial length before the first elongation.

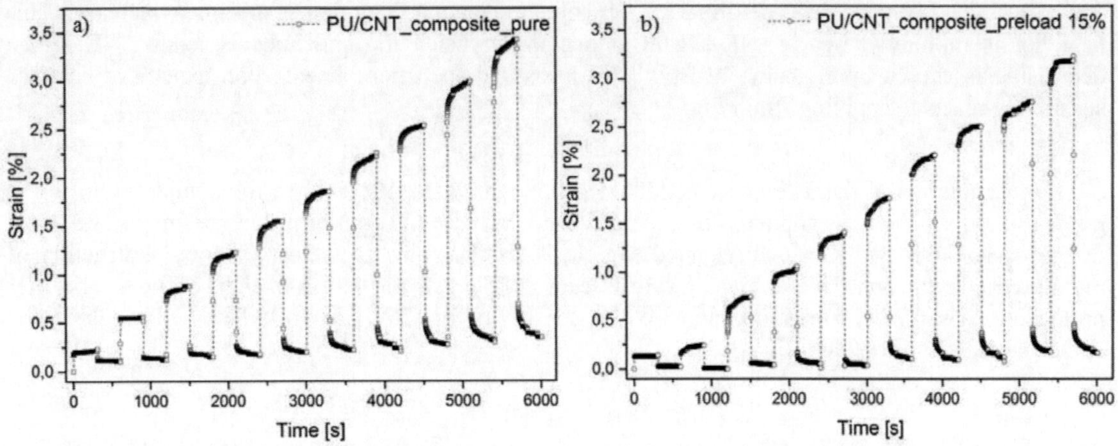

Fig. 2 Mechanical response to applied strain, ε, for MWCNT/PU laminate composite sample. a) No preload specimen, b) preload 15 %. The tensile strain step increases in a ten consecutive extension/relaxation cycles.

In the course of strain cycles, some irreversible changes in strain and the relative resistance are observed due to possible residual plastic deformation of TPU and irreversible damage of the electrically conductive MWCNT network. In the second and following cycle series the difference between responses is not such significant comparing with the first cycle series. Similar strain stabilization effect is observed for preloaded specimen when in consecutive cycles the relative resistance change or irreversible part are nearly equal

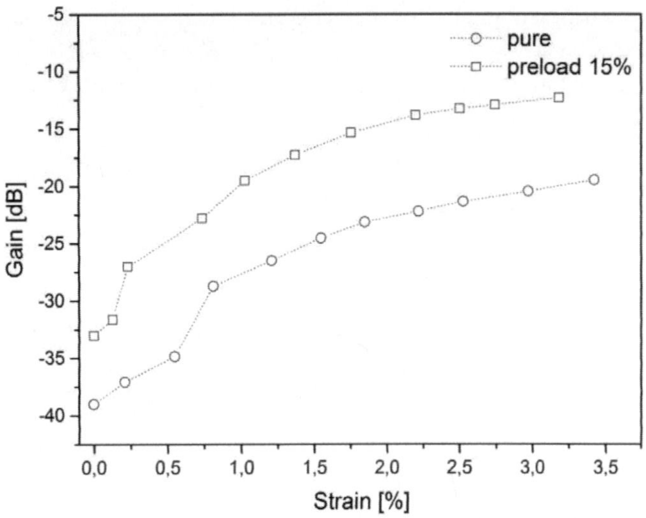

Fig. 3 Dependence of the gain change determined for MWCNT/PU laminate on applied strain. Pure MWCNT/PU laminate (cycle). Applied strain in the range from 0 to 3.5 %. 15 % Preload sample (square). Applied strain in the range from 0 to 3.25 %.

After specimen loading, deformation of polyurethane/CNT laminate specimen increase. This behavior is monitored by gain. The gain is decreasing when the specimen is loading. The gain decreasing is caused by changing resistivity of measured specimen. The carbon nanotubes network sensitively reacts to applying strain Fig. 3.

For simulation of damage construction 15% preload was applied. Then the similar strain in the range fom 0 to 3.25% was applied. Preloading caused decrease the gain of tested specimen. The shape of the curve strain vs. gain of preloaded sample is similar. This fact clearly shows destruction of monitored construction. Decreasing of gain is caused of crack formation in the CNT network after preloading. The crack formation increase the sensitivity of CNT network due to increasing the contacts point in the network Fig. 3.

Conclusions

High elastic polyurethane/carbon nanotubes laminate can be used as a sensing element for deformation detection. Sensing element works as an antenna with stable resonant frequency 650 MHz. For the testing sensing element was applied extension/relaxation cycles. The cycle deformation gradually increases in each 5 minutes cycle up to maximal strain value approximately 3.5 %. Ten consecutive extension/relaxation cycles was applied. During the applying each cycle also the gain response was measured. The same sample was 15% preload to simulate destruction of monitored construction. . Than the strain in the range from 0 – 3.25% was applied. The change of gain is visibly different. This fact clearly shows damage of monitor construction. The principle of function can be described as a change of resistivity of carbon nanotubes network during applying strain. The presented sensor can be used for structure health monitoring(SHM). This type of nondestructive inspection of monitored construction is very important of permanently used construction for instance bridges, towers or blades of wind power plant. This sensing element is made from polymer, very sensitive to applied strain, small in size.

Acknowledgement
This project was supported by the National Budget of the Czech Republic within the framework of the Centre of Polymer Systems project (reg. no.: CZ.1.05/2.1.00/03.0111).

References

[1] I. Kang, Y.Y. Heung, J.H. Kim, J. Won Lee, R. Gollapudi, S. Subramaniam, et al. Composites: Part B, 37 (2006), pp. 382–394
[2] Kang I., Schulz M.J., Kim J.H., Shanov V., Shi D. A carbon nanotube strain sensor for Structural health monitoring. Smart Mater. Struct. 2006;15:737–748.
[3] Yamada T., Hayamizu Y., Yamamoto Y., Yomogida Y., Izadi-Najafabadi A., Futaba D.N., Hata K. A stretchable carbon nanotube strain sensor for human-motion detection. Nat. Nanotechnol. 2011;6:296–301.
[4] C.L. Li, E.T. Thostenson, T.W. Chou Compos Sci Technol, 68 (2008), pp. 1227–1249

IOP Conf. Series: Materials Science and Engineering **108** (2016) 012025 doi:10.1088/1757-899X/108/1/012025

Simple estimation of dipole source z-distance with compact magnetic gradiometer

M Janošek, A Platil[1] and J Vyhnánek
Czech Technical University in Prague, Technicka 2, 166 27 Prague, Czech Republic

E-mail: platil@fel.cvut.cz

Abstract. A compact magnetometer/gradiometer with combined homogeneous and gradient outputs facilitates precise measurement of both H and G values with good spatial and temporal coherence. By evaluating combination of both signals, it is possible to estimate distance to a dipole source with relatively small error and largely independent from precise knowledge of source strength, orientation and lateral displacement. The performance is limited primarily by ambient noise. With an AC-driven source, tool navigation or distance sensing is also possible.

1. Introduction

We have previously presented a compact (10-cm head size, 3-cm base) axial magnetic gradiometer with two ring-core fluxgate sensors and combined homogeneous feedback and gradient feedback coils [1]. This arrangement provides simultaneously information about the homogeneous field strength B_z as well as the gradient dB_z/d_z in one axis with minimal geometric error. Thanks to that we can utilize simple method of dipole source distance estimation [2, 3]. Such an information is needed e.g. for unexploded-ordnance-detection or mine-hunting [4]. In our method, we do not obtain the radial distance, but the z-, or the vertical distance to the detected object, which is one of the most important informations to distinguish between deep and strong sources and shallow and weak ones. Assuming constant background conditions (both homogeneous field H and gradient G), the vertical distance (z-component) to a source positioned along the gradiometer axis can be estimated with relatively low error simply from the ratio of H_z and G_{zz} signals, largely independent from radial distance to the dipole, and its strength. If we calculate $3\times H_z/G_{zz}$, we get a figure characterizing the distance to source directly from one reading - not depending on its actual amplitude – this approach can be used also with AC-current driven coils for distance measurements or tool navigation tasks.

2. Experiment

The sensor was fixed in space after H and G readings in "infinity" (i.e. far from dipole source) were recorded as offsets for later subtraction from data. The dipole source in our case is small ($13\times4\times5 mm^3$) NdFeB permanent magnet. In the first experiment, the sensor and dipole were aligned on the same axis (z), with the dipole axis oriented either coaxially or perpendicularly - see figure 1 (left). The vertical distance (z) was modified and the corresponding distance estimate from H and G

[1] To whom any correspondence should be addressed.

data was calculated. In the following experiment, the vertical (z) distance was fixed at 15 cm and lateral displacement (y) was modified - again, see figure 1 (left) for the geometry of the setup.

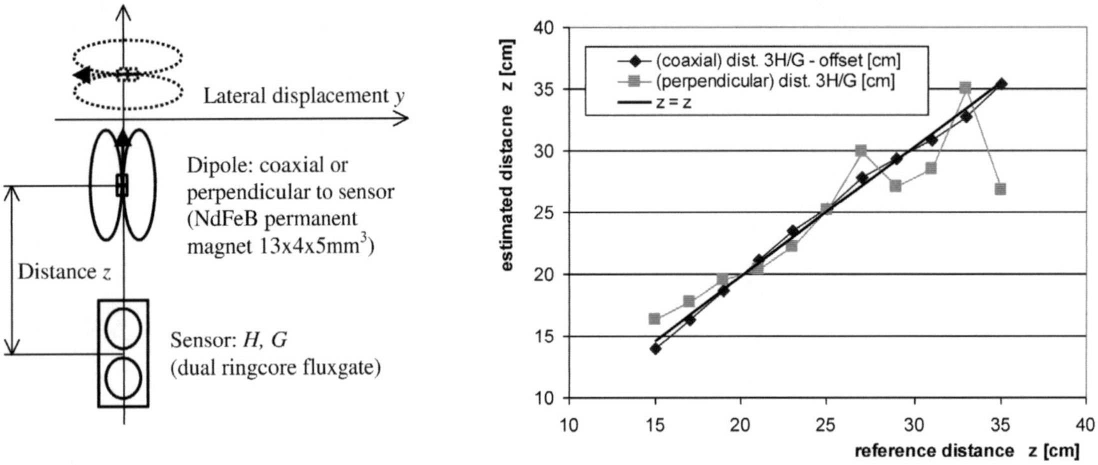

Figure 1. Left: the geometrical situation of dipole and sensor. Right: the estimated distance vs. real (reference) distance to a dipole in coaxial or perpendicular orientation to sensor. For coaxial orientation, an offset must be subtracted from the estimate. Here, the lateral displacement y is 0 cm.

3. Theory

As was shown in [2] and several related papers, measurement of full tensor gradient and field vector provides information for dipole source position estimate, irrespective of dipole orientation relative to sensor. In [2] the source is loop antenna transmitting AC signal. In our approach, we simplified the setup to just combined single-axis magnetometer and gradiometer. Also the dipole in our case was permanent magnet within Earth's field. In spite of these limitations, we achieved reasonably good position estimates, albeit in limited range of distances.

If we assume z- orientation of the magnetic dipole (magnetic moment m_z), we can write for the z-component of the magnetic field B_z:

$$B_z = -\nabla\left(\frac{\mu_0 \cdot m_z \cdot z}{4\pi \cdot z^3}\right) = -\nabla\left(\frac{\mu_0 \cdot m_z}{4\pi \cdot z^2}\right) = \frac{2\mu_0 \cdot m_z}{4\pi \cdot z^3} \tag{1}$$

The G_{zz} gradient can be then written as:

$$G_{zz} = \frac{\partial B_z}{\partial z} = -\nabla\left(\frac{\mu_0 \cdot m_z}{2\pi \cdot z^3}\right) = \frac{-3\mu_0 \cdot m_z}{2\pi \cdot z^4} \tag{2}$$

By dividing B_z and G_{zz}, we get finally: $z = -3 \times \dfrac{B_z}{G_{zz}}$ (3)

This result is consistent with the tensor representation presented in [2].

4. Results and discussion

As shown in figure 1 (right), the estimated distance from $3 \times H/G$ matches with reference distance quite well in 15-30 cm range. The estimate for coaxial case contained an unwelcome position offset (about 7cm), but after subtraction thereof, the error is 1cm or less. The data from perpendicular case are more influenced by noise due to lower dipole signal strength in 2nd Gauss position.

IOP Conf. Series: Materials Science and Engineering **108** (2016) 012025 doi:10.1088/1757-899X/108/1/012025

In the second measurement, the influence of lateral displacement (along y-axis) on distance (z) estimate was examined. The dependence of homogeneous field H_z and G_{zz} gradient on y-axis position of dipole is shown in figure 2, again for coaxial and perpendicular orientation of dipole. The position y = 0 corresponds to alignment of dipole with sensor from first experiment. From the same data, position estimates are again calculated. The quality of estimated z-distance quickly deteriorates by more than 100 % - see figure 3 and figure 4, but for small lateral displacements (< 5 cm) the error is quite acceptably low. Dashed line represents true z-distance 15 cm. Note that in figure 3, the estimated distance z is not corrected for offset.

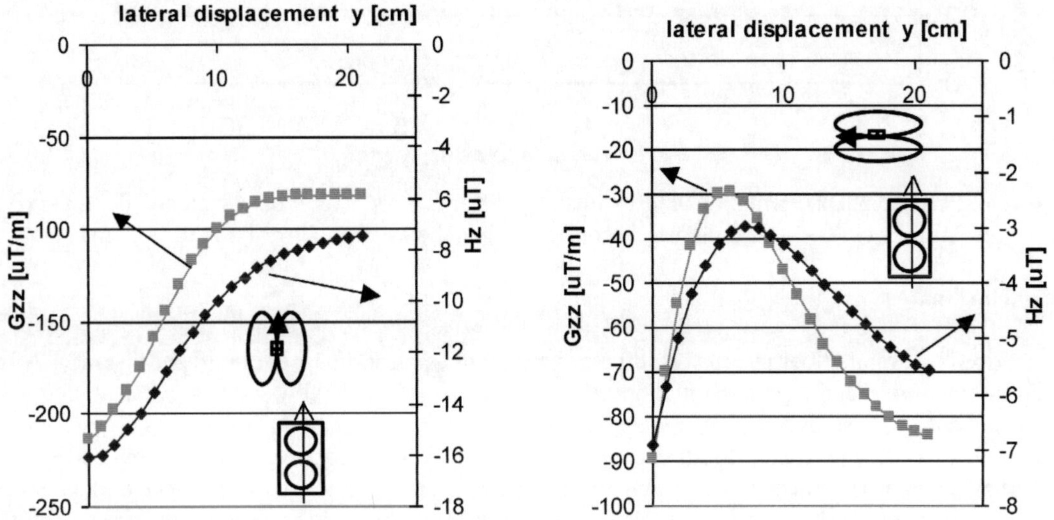

Figure 2. The measured homogeneous magnetic field and gradient in coaxial (left) and perpendicular (right) orientation of dipole to sensor vs. lateral displacement along y-axis.

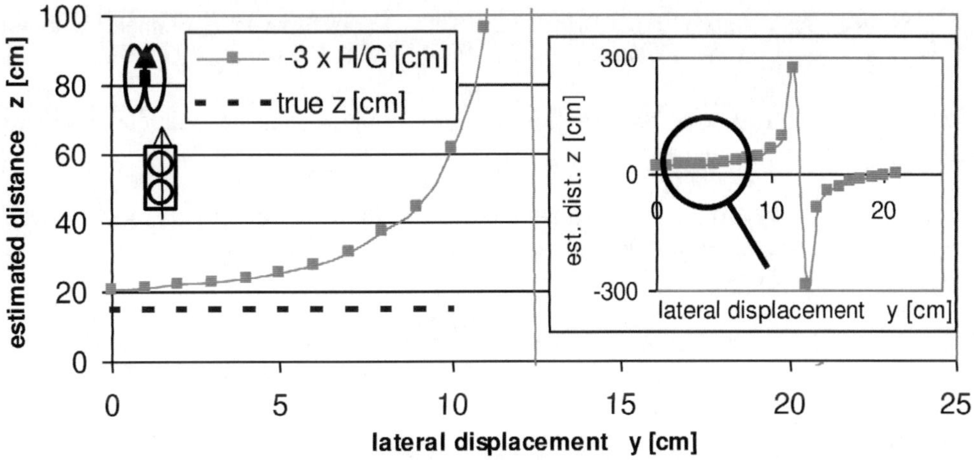

Figure 3. Coaxial orientation. The estimate of distance z vs. lateral displacement along y-axis. In limited range of lateral displacements (< 5cm), the estimate error is small (< 5cm after offset removal). The inset shows overall response for large lateral displacements (detail in main graph).

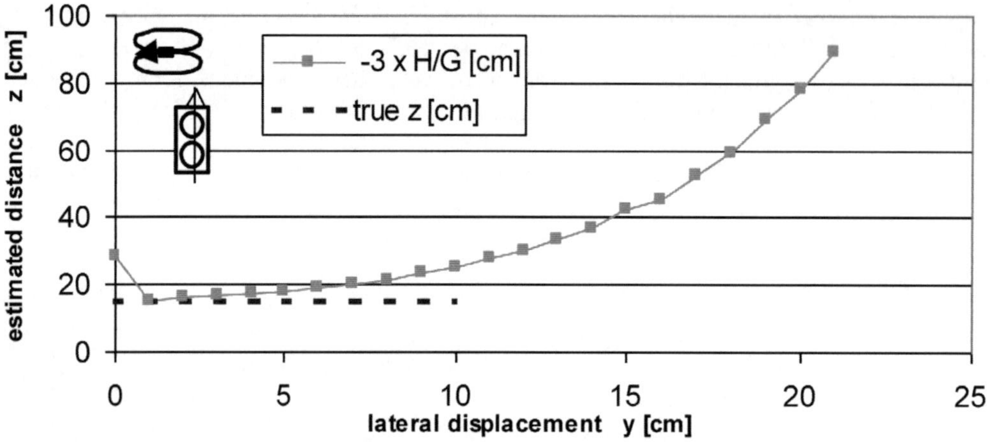

Figure 4. Perpendicular orientation. The estimate of distance z vs. lateral displacement along y-axis. In limited range of lateral displacements (1 - 7cm), the estimate error is small (< 3cm).

5. Conclusions

We conducted several experiments with dipole distance estimation in 15 cm to 30 cm distance. After offset subtraction, the distance estimation error in our experiment with small NdFeB source was < 5 cm for coaxial case and < 3 cm for perpendicular case in 15 cm to 30 cm dipole-to-gradiometer distance. Moreover, the z-axis distance estimation was largely independent (within some limits) to axial misalignment (lateral shift) between the source and gradiometer axes. The performance was mainly limited by ambient noise, which was about 100× larger that our gradiometer noise (1 nT/m/√Hz). AC-driven coils experiment was also conducted showing feasibility of the proposed algorithm e.g. for tool navigation purposes or for distance sensors.

References

[1] Janosek M, Platil A, Vyhnanek J and Brinek J 2013, Dual-core fluxgate gradiometer with gradient feedback *IEEE Sensors 2013 Proceedings,* DOI: 10.1109/ICSENS.2013.6688287

[2] Nara T, Suzuki S and Ando S 2006, A closed-form formula for magnetic dipole localization by measurement of its magnetic field and spatial gradients *IEEE Transactions on Magnetics* **42** 3291-93.

[3] Young, JA, Clark DA 2010, Magnetic tensor gradiometry in the marine environment *International Conference on Electromagnetics in Advanced Applications (ICEAA)* pp 701-4, DOI: 10.1109/ICEAA.2010.5650898

[4] Wiegert R, Oeschger J and Tuovila E 2007, Demonstration of a novel man-portable magnetic STAR technology for real time localization of unexploded ordnance *OCEANS 2007, IEEE conference*, 2007, DOI: 10.1109/OCEANS.2007.4449229

IOP Conf. Series: Materials Science and Engineering **108** (2016) 012026 doi:10.1088/1757-899X/108/1/012026

Low-noise magnetic observatory variometer with race-track sensors

M Janošek[1], V Petrucha[1] and M Vlk[2]
[1]Czech Technical University in Prague, Technicka 2, 166 27 Prague, Czech Republic
[2]Institute of Geophysics, Bocni II/140, 141 31 Prague, Czech Republic

E-mail: janosem@fel.cvut.cz

Abstract. We present a low-noise, high-stability observatory magnetometer with race-track sensors, as developed by the Czech Technical University in Prague for National Observatory of Athens. As opposed to the standard instruments, we used our novel race-track fluxgate sensors with planar oval core which were cut by state-of-the art pico-second UV-laser. The noise performance of the complete electronics and sensor chain is below 6 pT/√Hz @ 1 Hz. The electronics uses 24-bit 200-Hz A/D converter with simultaneous sampling and all digital processing is done in FPGA. The variometer with the sensors mounted on a MACOR cube has been successfully calibrated by scalar method.

1. Introduction

The requirements on observatory variometers, as they are in service either officially in IAGA network, or for other purposes, are very demanding. It is necessary to achieve very high stability and low noise to record truly the diurnal Earth's field variations and possible magnetic storms (the IAGA dynamic range requirement is ± 3000 nT). Standard full-field magnetometers can be used for this purpose, if their limited dynamic range (about 130 dB for 24-bit converters) is not a problem. However, if there is a requirement of achieving very low noise, i.e. «20 pT/√Hz @ 1 Hz, compensation of the main Earth's field components (horizontal and vertical in the northern hemisphere) is inevitable [1]. Such low noise variometer, if having suitable bandwidth, can be used for advanced ionospheric or geomagnetic studies (i.e. observing Schumann resonances and other effects). We have implemented a low noise compensation of the main field components, which allowed us to use our low-noise race-track sensors (< 6 pT/√Hz) in the variometer.

2. Magnetometer construction

For the variometer, we used in-house race-track fluxgate sensors with laser-cut cores, slightly modified and downscaled of the heritage CTU sensor [2]. Two of the sensors (N-S and vertical) have an additional coil wound, which is used for main field component offset in the respective direction. For that purpose, we have implemented an ultra-low-noise (< 0.5 pT/√Hz) current source using LTC6655 reference, which feeds the additional coil wound directly on top of the feedback coil of the sensor. In that manner, the possible mutual angular imperfections and mainly their temperature instabilities are minimized. The sensor triplet (each sensor dimension are approx. $30x8x1mm^3$) is mounted on a solid MACOR holder maintaining large thermal conductivity and geometric stability, nevertheless, also the temperature of the holder is monitored – see Fig. 1. The sensor holder is attached to marble base plate, which is to be leveled. The "standard part" of the magnetometer electronics relies on a "standard low-noise magnetometer" manufactured by the CTU and CSRC (Czech Space Research Centre) company, it uses FPGA for signal clock generation and ADC driving and the power supply for the analog part is galvanically isolated. The electronics returns uncalibrated ADC data on RS232 for the three magnetometer axis and also the head temperature measurement. With the help of the low-noise current-source for the NS and vertical field offsets, the baseline noise due to electronics/ADC resolution is less than 1 pT/√Hz@ 1 Hz in the final variometer range of ±3750 nT.

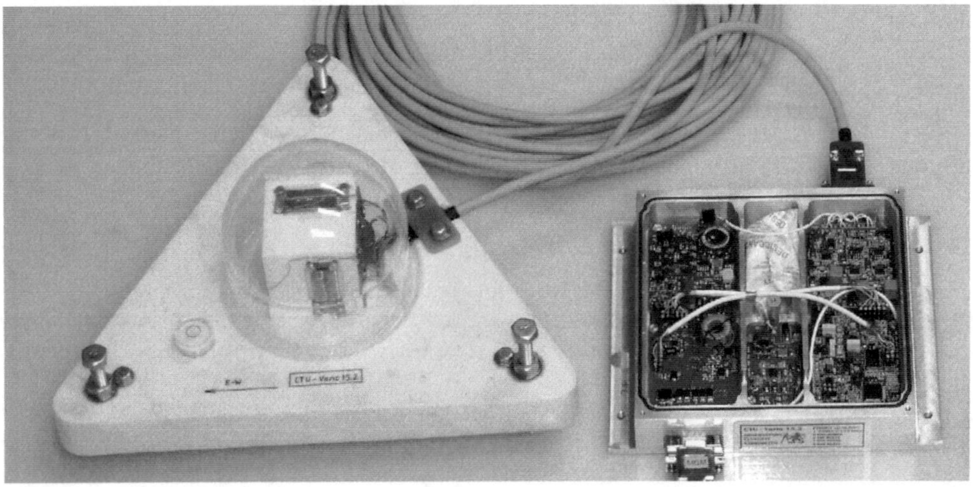

Figure 1. The presented variometer – left: the triaxial race-track fluxgate triplet mounted on MACOR holder on marble base plate, right: the electronics (cover removed).

3. Variometer calibration

A non-trivial task is variometer calibration. We used following approach using scalar calibration technique [3, 4]:

- First, the offset fields have been disabled and the magnetometer has been operated in ±75000 nT range, and a scalar calibration was done. The calibration RMS error was < 2nT.
- After the calibration, the sense resistor (the magnetometer is feedback operated) was measured .
- The resistor has been replaced with 20x larger value (again precisely measured after soldering), and the gain coefficients have been recalculated.
- In this manner, we can use the scalar calibration results, which are comparable or superior to standard techniques utilizing coils and flux density standards [5].

Valuable information has been obtained in trial tests of the magnetometer (before range expansion) during 1-month testing at the Budkov observatory (IAGA BDV). As seen from table 1, we could see ageing of the sense resistors (Vishay PLT thick film series) – the values changed by 250-300 ppm after one-month burn-in. T_c of the gain channels could be also computed: it was 8, 16 and 9 ppm/K, respectively which is a combination of thermal expansion of the compensating coil and sense resistors. From the table, it can be also seen that the mutual angular position was stable with temperature.

From the comparison of the total field computed from the variometer and values from standard instruments at the observatory, we can see that the gains have finally settled after 14 days (Figure 3).

Table 1 – The calibration results – after one month of running and with changed temperature.

Calibration date	temp [°C]	S1 [normalized]	S2 [normalized]	S3 [normalized]	O1 [nT]	O2 [nT]	O3 [nT]	Φ1 [°]	Φ2 [°]	Φ3 [°]
14.12.2014	12	1.2996	1.3068	1.2935	-18.77	-189.59	8.46	0.71	-0.09	0.11
23.1.2015	19	1.2991	1.3064	1.2933	-17.30	-190.25	7.95	0.72	-0.08	0.08
23.1.2015	6	1.2989	1.3062	1.2931	-18.22	-197.94	9.58	0.71	-0.09	0.08

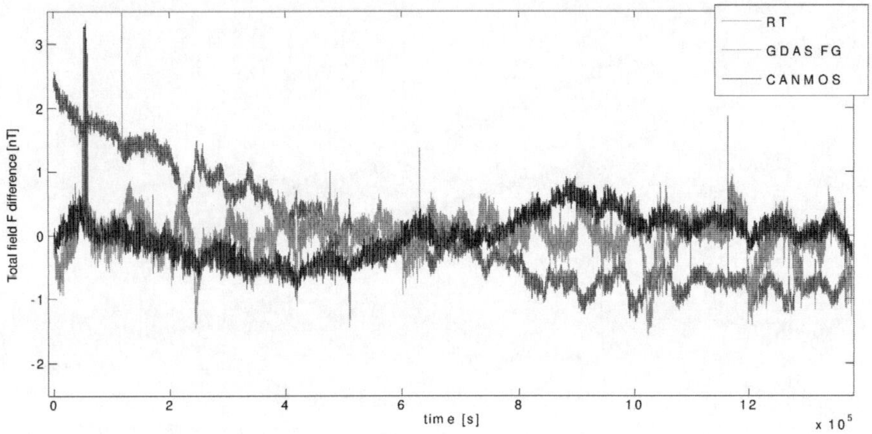

Fig. 3 – Instrument stability (before range expansion and with offsets off) – 14 days of total field (F) are displayed. Blue – CTU variometer, red and black – instruments of BDV observatory

4. Noise performance

Since the sensor head is too large to be tested in our in-house magnetic shield and also because of the need of creating a low-noise counter-acting magnetic field (simulating the NS and vertical Earth's field component), we have decided to do this test in the calm magnetic field at the BDV observatory. Typical record is depicted in the spectrogram on Figure 4: it can be seen that even at the quiet locality, man-made AC noise is present in the low-frequency spectra. AC traction noise at 16 2/3 Hz from Austria / Germany railways is also visible as burst. The source of the 3 Hz noise is still unknown.

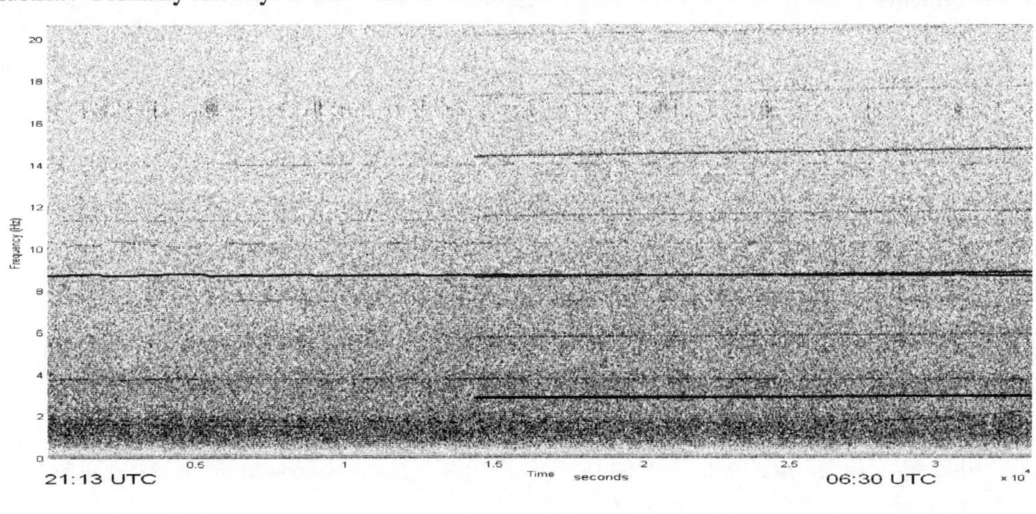

Fig. 4 – The 0.01-20 Hz spectrogram during 8-hours of logging (vertical axis)

If we have chosen a quiet part of the day, we were able to compute noise spectra as depicted in Fig.5: it can be seen, that in the vertical axis, the measured magnetic noise PSD was better than 6 pT/√Hz @ 1 Hz; however the EW and NS axes were noisier which is presumably by the magnetic field noise at the locality since the sensors in the triplet perform equally well.

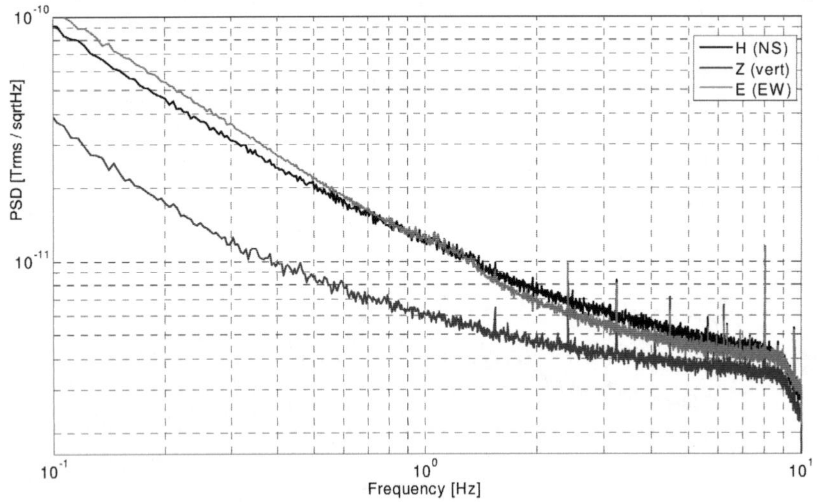

Fig. 5 – The magnetic field noise at the BDV observatory as logged with the variometer.

5. Conclusion

We have successfully built and calibrated a closed-loop operated, observatory variometer with race-track sensors. Its measured noise performance in real conditions of < 6 pT/√Hz @ 1 Hz is up to our knowledge on the state of the art in the field. We have used a simple yet effective calibrating method to obtain the instrument parameters. Further improvements are sought in terms of fluxgate sensors performance with a target of < 3 pT/√Hz. In this case however, from our experience, a large shielded room and low-noise artificial magnetic field generator would be necessary to confirm the instrument performance.

Acknowledgment
This research has been on-contract supported by the National Observatory of Athens. We would like to thank also to Dr. Tomas Bayer from the Budkov observatory for his help during calibrations at the Budkov observatory and Jan Vyhnanek of CTU for participating on the instrument development.

References

[1] Vlk M 2013, Modernisation of the Narod fluxgate electronics at Budkov Geomagnetic Observatory, *EGU General Assembly 2013, Vienna, Austria*, EGU2013-7721

[2] Ripka P 1993, Race-track fluxgate sensors, *Sensors and Actuators A: Physical*, **37** 417-421.

[3] Petrucha V et al 2009, Automated system for the calibration of magnetometers, *Journal of Applied Physics*, **105(7)** 07E704.

[4] Olsen N et al 2003, Calibration of the Ørsted vector magnetometer. *Earth, planets and space*, **55**(1) 11-18.

[4] Zikmund A and Janosek M 2014, Calibration procedure for triaxial magnetometers without a compensating system or moving parts, *IEE International Instrumentation and Measurement Technology Conference (I2MTC) Proceedings*, 473-476

IOP Conf. Series: Materials Science and Engineering **108** (2016) 012027 doi:10.1088/1757-899X/108/1/012027

Engineering of composite metallic microfibers towards development of plasmonic devices for sensing applications

A Petropoulou[1,3], G Antonopoulos[1], P Bastock[2], C Craig[2], G Kakarantzas[1], D W Hewak[2], M N Zervas[2], C Riziotis[1]

[1] Theoretical and Physical Chemistry Institute, Photonics for Nanoapplications Laboratory, National Hellenic Research Foundation, Athens 116 35, Greece

[2] Optoelectronics Research Centre, University of Southampton, Southampton SO17 1BJ, United Kingdom

[3] Department of Informatics and Telecommunications, University of Peloponnese, Tripolis 22100, Greece

E-mail: apetropoulou@eie.gr, Riziotis@eie.gr, M.N.Zervas@soton.ac.uk

Abstract. The paper discusses the analysis of tapered hybrid composite microfibers based on a metal-core and dielectric-cladding composite material system. Its advantages over the pure metal tips conventionally used, are the inherent enhanced environmental robustness due to inert borosilicate cladding and the capability of multiple excitation of the tapered nanowire through the length of the fiber due to the enabled total internal reflection at the borosilicate/air interface. Simulations through finite element method (FEM) have demonstrated an improved field enhancement at the tapered region of such microfibers. Furthermore, experimental results on tapering in copper based microfibers together with light coupling and propagation studies will be demonstrated revealing the potential for the development of plasmonic devices for sensing applications.

1. Introduction

Surface plasmon polaritons (SPPs) are surface electromagnetic waves travelling along a metal/dielectric interface, whose field decays exponentially in both media. SPPs have attracted great interest and have found many applications in biosensing, surface-enhanced Raman spectroscopy (SERS) and also microscopy, since they offer high field enhancement and confinement beyond the diffraction limit [1].

The field enhancement at the apex of metallic tips have been extensively studied theoretically [2-4], however it has been clearly identified lately the need for further research towards the engineering of physically realizable and robust plasmonic structures with enhanced performance. In this study the performance of tapered hybrid copper-core/borosilicate-cladding microfiber is analyzed. Their advantages over the metal tips are the absence of impurities and the multiple excitation of the nanowire through the length of the fiber due to the repeated total internal reflection at the silicate/air interface [5]. High field enhancement is expected at the apex of such tapers.

2. Modelling

For metallic core nano-wires of radius R_1 surrounded by a dielectric layer of radius R_2, the amplitudes of the electric and magnetic fields for TM_{01} mode [6] are:

$$\boldsymbol{E}(\rho) = \begin{cases} A_1\left[-i\left(\frac{\beta}{\alpha_1}\right)I_1(\alpha_1\rho)\hat{e}_\rho + I_0(\alpha_1\rho)\hat{e}_z\right], & \rho \le R_1 \\ \left(\frac{\beta}{\alpha_2}\right)[-iA_2I_1(\alpha_2\rho) + B_2K_1(\alpha_2\rho)]\hat{e}_\rho + [A_2I_0(\alpha_2\rho) - iB_2K_0(\alpha_2\rho)]\hat{e}_z, & R_1 \le \rho \le R_2 \\ A_3\left[\left(\frac{\beta}{\alpha_3}\right)K_1(\alpha_3\rho)\hat{e}_\rho - iK_0(\alpha_3\rho)\hat{e}_z\right], & \rho \ge R_2 \end{cases} \quad (1)$$

$$H_\varphi(\rho) = k \times \begin{cases} A_1\left[-i\left(\frac{\varepsilon_1}{\alpha_1}\right)I_1(\alpha_1\rho)\right], & \rho \le R_1 \\ \left(\frac{\varepsilon_2}{\alpha_2}\right)[-iA_2I_1(\alpha_2\rho) + B_2K_1(\alpha_2\rho)], & R_1 \le \rho \le R_2 \\ A_3\left[\left(\frac{\varepsilon_3}{\alpha_3}\right)K_1(\alpha_3\rho)\right], & \rho \ge R_2 \end{cases} \quad (2)$$

where β is the propagation constant, ε_1, ε_2 and ε_3 are the relative permittivities of metal core, dielectric cladding and air respectively, $\alpha_j = (\beta^2 - \varepsilon_j k^2)^{1/2}$ for j=1,2,3, I and K are the modified Bessel functions of the first and second kinds and A_1, A_2 and A_3 are the complex constants.

The dispersion equation can be derived by the continuity of the tangential components of the fields at the interfaces $\rho = R_1$ and $\rho = R_2$:

$$\varepsilon_1\alpha_2I_1(\alpha_1R_1)[\alpha_2\varepsilon_3K_1(\alpha_3R_2)M_{00} + \alpha_3\varepsilon_2K_0(\alpha_3R_2)M_{10}] = -\varepsilon_2\alpha_1I_0(\alpha_1R_1)[\alpha_2\varepsilon_3K_1(\alpha_3R_2)M_{01} + \alpha_3\varepsilon_2K_0(\alpha_3R_2)M_{11}] \quad (3)$$

where $M_{ab} = I_a(\alpha_2R_2)K_b(\alpha_2R_1) - (-1)^{a+b}I_b(\alpha_2R_1)K_a(\alpha_2R_2)$.

Figures 1 and 2 show the real and imaginary part of the effective refractive index (n_{eff}) as a function of the wavelength for different radii of the copper core. As the radius of the copper tip decreases, both the real and imaginary parts of n_{eff} increase. This means that, toward the apex of the metal tip, the effective wavelength of the plasmon wave decreases and the losses increase. For radii of the metal tip larger than 50nm the losses are relatively low. As the radius decreases the losses become significant and they are minimum for wavelength around 800nm.

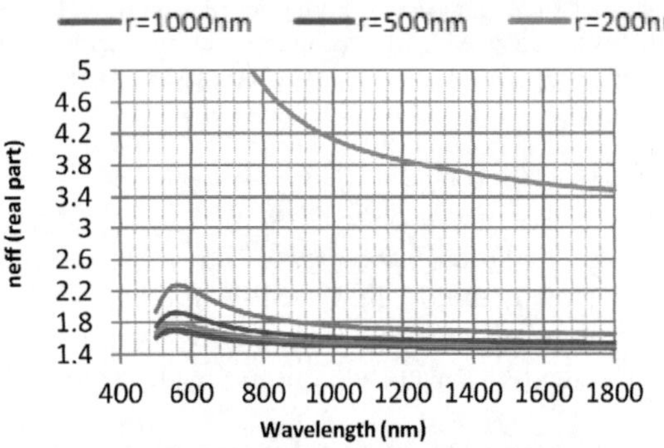

Figure 1. Real part of the effective refractive index (n_{eff}) as a function of the wavelength for different radii of the copper core.

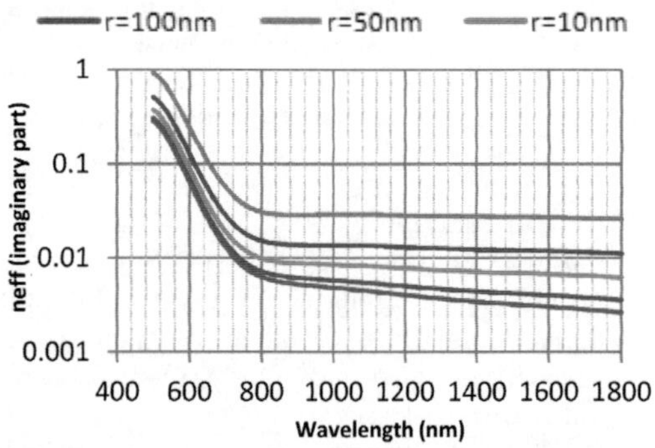

Figure 2. Imaginary part of the effective refractive index (n_{eff}) as a function of the wavelength for different radii of the copper core.

2.1. FEM simulations of tapered composite microwire

The simulated structure is a copper cone with semi-angle, a=3.43°, input radius of r_{input}=150nm and output radius r_{output}=3nm to avoid singularities. The cone is surrounded by a borosilicate layer with a thickness that decreases so that the ratio of outer radius/core radius remains constant (~ 12.5). The launched mode is TM_{01} (P_{input}=1W) and the transmission is along the z direction. Figure 3 shows the maximum normalized electric field at the apex of the tip as a function of the wavelength. We observe that periodic peaks appear in the field enhancement spectrum. These peaks appear when the tip length is more than 1 μm due to SPPs that propagate along the tip. When the SPPs meet the end of the tip, they are reflected and form cavity modes [7].

The highest enhancement factor, defined as the maximum intensity I at the apex of the tip divided by the intensity I_o at the copper/borosilicate surface at the input of the fiber tip, is ~1.1x10^4 for λ=837nm (Fig. 4).

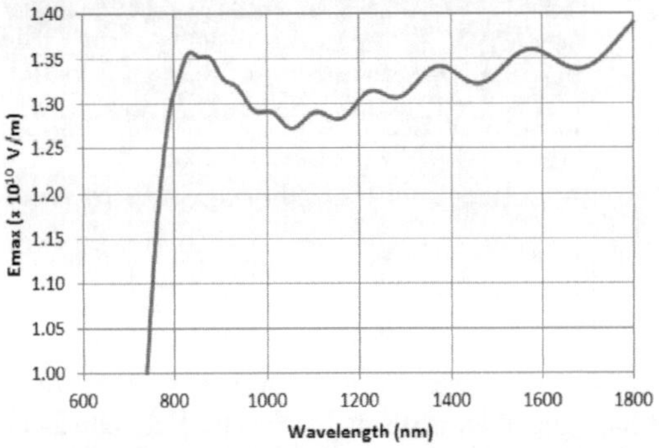

Figure 3. The maximum normalized electric field at the apex of the tip as a function of the wavelength.

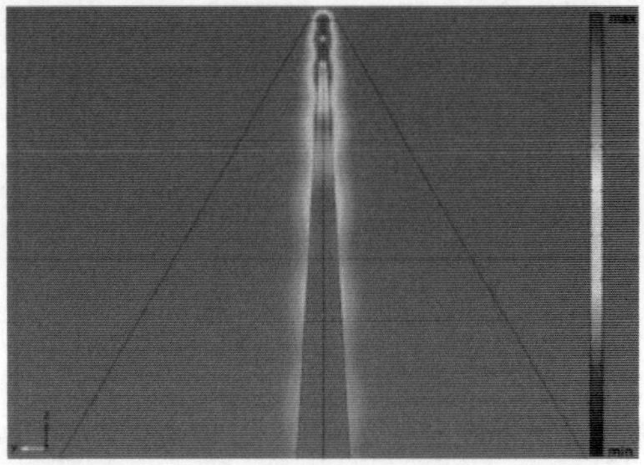

Figure 4. Normalized electric field for λ=837nm.

3. Experimental

The hybrid metal-core/dielectric-cladding microfibers that we used (Fig. 5), manufactured within the ORC [8], have copper core and borosilicate cladding with diameters of 2μm and 25μm, respectively. The nanowires are excited by coupling light of 671nm into the silicate enclosure. As expected we observe two peaks in the 2D beam profile at the copper/silica interface for the fundamental mode (Fig.6). The two observed peaks correspond in good agreement to the illustrated modelled mode profile. In order to obtain the beam profile we used a 60x objective lens to image the near-field intensity distribution at the fiber output and a scanning slit optical beam profiler.

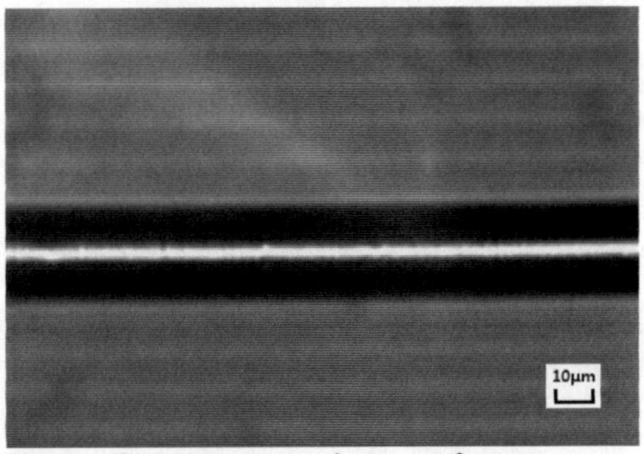

Figure 5. Microscope image of copper core micro-wire with borosilicate cladding.

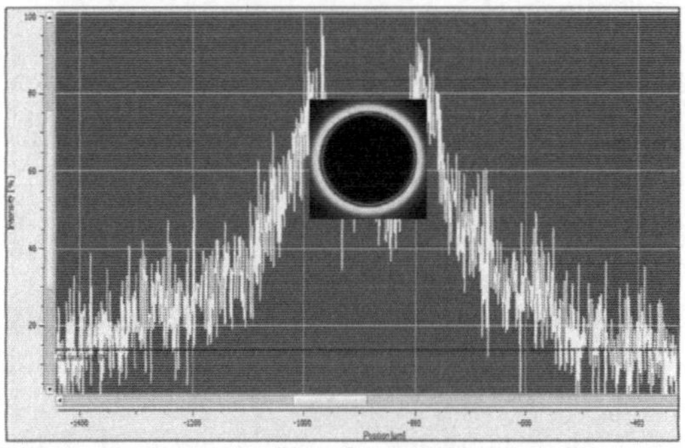

Figure 6. Beam profile of a hybrid microfiber and the corresponding modelled TM_{01} mode profile.

3.1. Tapering process

The microfiber was tapered with the method of heating and stretching [9]. Two linear stages stretch the microfiber while a third one moves the flame along the fiber axis (Fig. 7). Due to the low melting point of copper the fast and cold method was used in order to avoid the breakdown of the microfiber. The final waist radius of the copper core and the final waist length was 200nm and 13mm, respectively (Fig. 8).

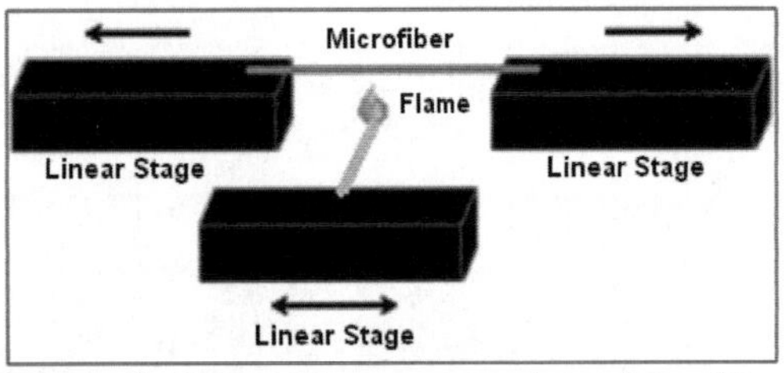

Figure 7. Schematic representation of the fiber tapering rig.

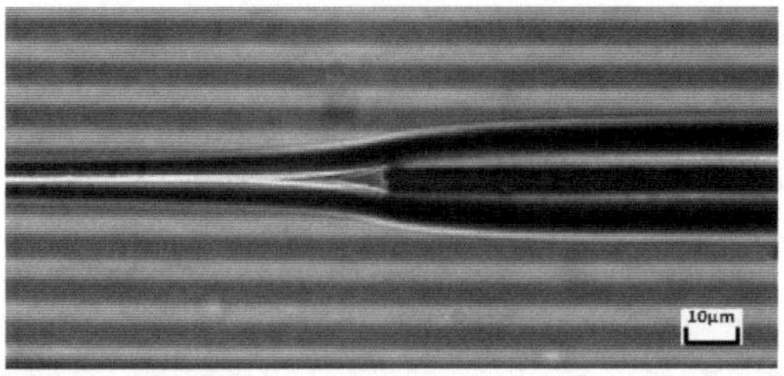

Figure 8. Microscope image of a tapered microfiber after the end of the tapering process.

4. Conclusions

The demonstrated results suggest that the hybrid copper-core/dielectric cladding microfibers are a very promising platform that can provide high field enhancement and confinement. The field enhancement expected at the apex of these tips is $E_f \sim 1.1 \times 10^4$ at λ=837nm making their use for sensing applications feasible. Their advantage over bare metal tips used is their low loss due to multiple reflections at the borosilicate/air interface that can provide higher intensities at their apex. We were able to taper them down to a core radius of 200nm using the heating and stretching method. Further research is ongoing for the optimization of the tapering conditions in a variety of metal-silicate composite systems, in order to achieve uniform and extremely thin tapered microfibers with optimised field enhancement.

Acknowledgements

This work was funded by the General Secretariat for Research and Technology, Greece (Project Polynano-Kripis 447963).

References

[1] Stockmann M I 2011 *Opt. Express* **19** 22029
[2] Stockmann M I 2004 *Phys. Rev. Lett.* **93** 137404
[3] Uebel P, Bauerschmidt S T, Schmidt M A and Russell P St J 2013 *Appl. Phys. Lett.* **103** 021101
[4] Zaccaria R P, Alabastri A, De Angelis F, Das G, Liberale C, Toma A, Giugni A, Razzari L, Malerba M, Sun H B and Di Fabrizio E 2012 *Phys. Rev.* B **86** 035410
[5] Gholipour B, Nalla V, Bastock P, Khan K, Craig C, Hewak D W, Zheludev N I and Soci C 2014 *CLEO '14*
[6] Handapangoda D, Premaratne M, Rukhlenko I D and Jagadish C 2011 *Opt. Express* **19** 16058
[7] Zhang W, Cui X and Martin O J F 2009 *J. Raman Spectrosc.* **40** 1338
[8] Bastock P. 2015 "Manufacturing Novel Fibre", *Doctor of Philosophy.* Optoelectronics Research Centre. University of Southampton.
[9] Birks T A, Wadsworth W J and Russell P St J 2000 *Opt. Lett.* **25** 1415

Construction of an AMR magnetometer for car detection experiments

V Fúra, V Petrucha and A Platil[1]
Czech Technical University in Prague, Technicka 2, 166 27 Prague, Czech Republic

E-mail: platil@fel.cvut.cz

Abstract. A new construction of magnetometer with commercially available AMR (anisotropic magnetoresistive) sensors intended for vehicle detection experiments is presented. Initial experiments with simple AMR gradiometer indicated viability of the approach in a real-world setup. For further experiments and acquisition of representative data, a new design of precise multi-channel magnetometer was developed. The design supports two models of commercial AMR sensors: the proven and reliable, but obsolete Honeywell HMC1021-series sensors and newly available Sensitec AFF755B sensors. In the comparison the two types are similar in most achieved parameters, except offset stability in flipped operation regime. Unfortunately, the new AFF755B sensors seem to have perhaps inferior coupling of the flipping (set/reset) coil to the ferromagnetic core that causes insufficient saturation of the AMR material. The issue is being solved by Sensitec, current deliverables of the AFF755B have "product sample" status (September 2015).

1. Introduction

In the framework of an industrial cooperation project, we develop car detection system with AMR sensors as the preferred solution. Initial experiments with simple AMR gradiometer made of pair of Phillips KMZ51 (with fluxgate magnetometer Billingsley TFM100G2 as a reference) indicate viability of the scheme in accordance with literature [1, 2, 3]. The typical car signature is well in excess of 1 μT deviation from background in 4 m distance. In order to conduct further experiments and accumulate larger amount of representative real-world data in the field of car detection, a new construction of an AMR magnetometer is developed, offering flexible operating parameters (sample rate, flipping, feedback compensation).

Well established and proven AMR sensors (Honeywell HMC1001, HMC1021 and Phillips KMZ51) have unfortunately recently became obsolete and stock shortages are imminent. For modern constructions, new types of sensors with (at least nominally) comparable parameters are available, namely AFF755B from Sensitec. In order to compare the new sensors with their obsolete counterparts, we decided to make overall testing of their parameters when used in new AMR magnetometer. The design supports two possible types of 3-axis sensor probe construction, with slightly different magnetic field feedback compensation coils. Internal (on chip) compensation/test coils are not used

[1] To whom any correspondence should be addressed.

due to poor coil constant and thus high current needed. The flipping is frequently used in AMR sensors to periodically re-magnetize the ferromagnetic sensor core using another built-in (on chip) coil and short intense unipolar current pulses. It is also possible to alternate the core magnetization between two polarities (N-S or S-N) using positive and negative current pulses, effectively alternating the sensor response polarity which may help in suppressing offsets and/or hysteresis [4, 5, 6]. The flip current RMS value (i.e. thermal effects) must be carefully considered, which dictates short maximum pulse duration.

2. Experiment

The analog output signals from sensors are processed and sampled by a quad (or octal), max. 144 kSa/s, simultaneous sampling 24-bit delta sigma ADC (ADS1278 by Texas Instruments). In the basic configuration, three magnetic channels (x, y, z) and temperature is measured at 10.3 kSa/s rate. Lower than maximum sample rate is used in order to save power. Optionally, another 3-axis probe can be added, thus forming gradiometric configuration. Full scale range of the instrument is ±350 μT. The device is controlled by a 32-bit microcontroller Microchip PIC32MX795F512L (an FPGA is considered as an alternative solution), permitting change of many parameters on the fly.

In the current test implementation the feedback compensation of measured field is permanently active. The flipping pulses (set/reset in Honeywell parlance) can be optionally activated at 10 kHz (at the expense of theoretically limited signal bandwidth), deactivated, or single-fired at request. With flipping active, the signal is preprocessed with switching synchronous detector. The digital output rate can be configured in wide range (1 to 10 300 Sa/s) in order to suit target application. The output data are averages of raw inputs sampled at 10.3 kSa/s. Sleep mode with ADC and flipping deactivation is available. The block diagram of one channel is shown in figure 1. The data are transmitted via USB or RS232 interface to the host computer, processed and graphically displayed in custom application written in National Instruments LabView.

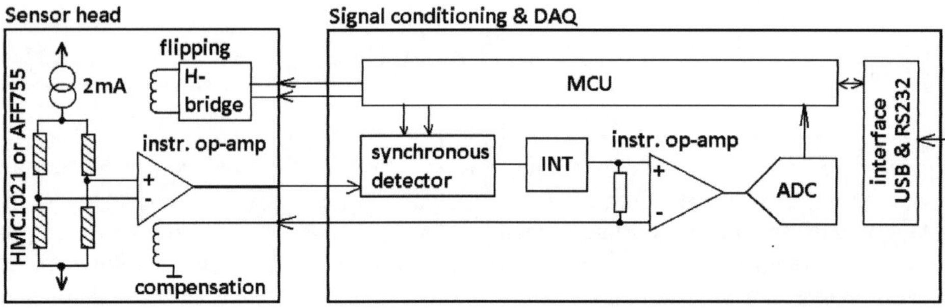

Figure 1. Block diagram of the magnetometer (one channel of analog signal is shown).

The magnetic calibration (determining offsets, scale factors and angular deviations from x-y-z orthogonality) was performed by scalar calibration method in homogenous field using nonmagnetic positioning device [7]. The noise data were acquired with sensors in a 6-layer permalloy tubular shield in lab environment. In all cases, the noise spectra are calculated from output digital data incoming at 206 Sa/s rate (each output value is an average of 50 raw ADC readings sampled at 10.3 kSa/s).

3. Results and discussion

The main focus of the current development was comparison of classical but obsolete sensors of Honeywell HMC1021-series with newly available Sensitec AFF755B-series. In most aspects, these two models of sensors are quite comparable, e.g. open loop sensitivity is about 1 mV/V / 100uT in both models. However, we have observed strong differences in offset stability and noise behavior of these two sensors in flipped and non-flipped mode.

Firstly, the noise of electronics measured without the sensors connected was only some 46 pT/sqrtHz@1Hz, so it is safely below either sensor noise level. The output noise PSD (power spectral density) with HMC1021 sensors in operating mode *with* flipping (current pulse 454 mApk) is about 240 pT/sqrtHz@1Hz as shown in figure 2 (the best case was about 120 pT/sqrtHz@1Hz). Interestingly, *without* flipping the noise was somewhat worse, typically 800 pT/sqrtHz@1Hz.

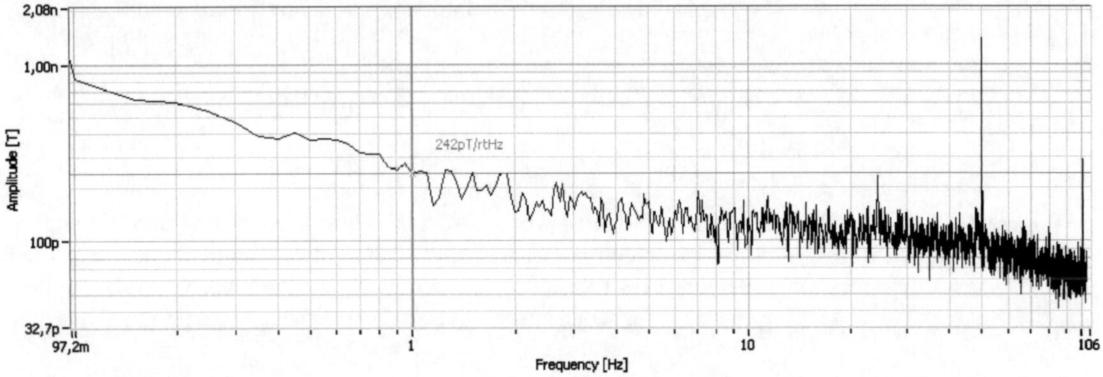

Figure 2. The noise PSD of the electronics with HMC1021 sensors, with flipping.

For the AFF755B sensors *without* flipping the PSD was similar (300 pT/sqrtHz@1Hz) to that of HMC1021 *with* flipping. However, *with* flipping, the AFF755B noise did not improve as expected, but was much worse: in the order of 10 nT/sqrtHz@1Hz - see figure 3.

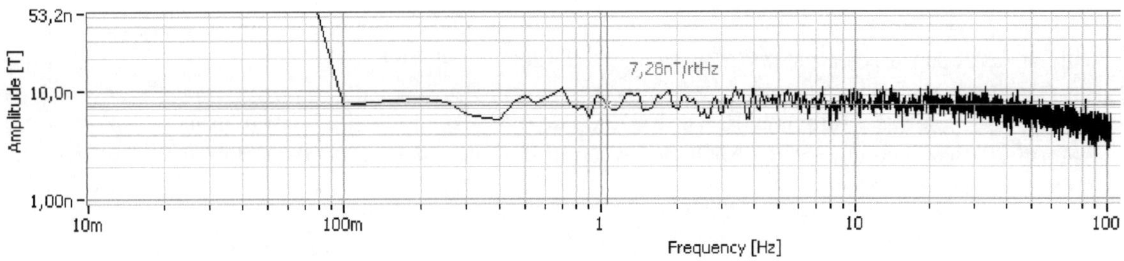

Figure 3. The noise PSD of the electronics with AFF755B sensors, with flipping.

Increasing the flipping current amplitude did not improve the noise either. The figure 4 (left) provides comparison of PSD measured values for various flipping currents. It was also observable that output offset in AFF755B is quite unstable with flipping - see figure 4 right.

4. Conclusions

We have constructed multiple-axis AMR magnetometer with feedback compensation and compared two types of commercially available AMR sensors. The only significant difference observed between the two models of AMR sensors was in flipped mode. The flipping (set/reset) in classical Honeywell HMC1021 works very well, suppressing offsets without compromising the noise quality. Also the current peaks needed for effective flipping is reasonably small: 450 mApk. In contrast, in Sensitec AFF755B sensors (sample deliverables), similar or even higher level of flipping current peaks perhaps

IOP Conf. Series: Materials Science and Engineering **108** (2016) 012028 doi:10.1088/1757-899X/108/1/012028

does not guarantee good saturation of the core and results in unacceptable offset instability after each flip pulse (the noise PSD is in the order of 10 nT/sqrtHz@1Hz, i.e. 50x worse than HMC1021).

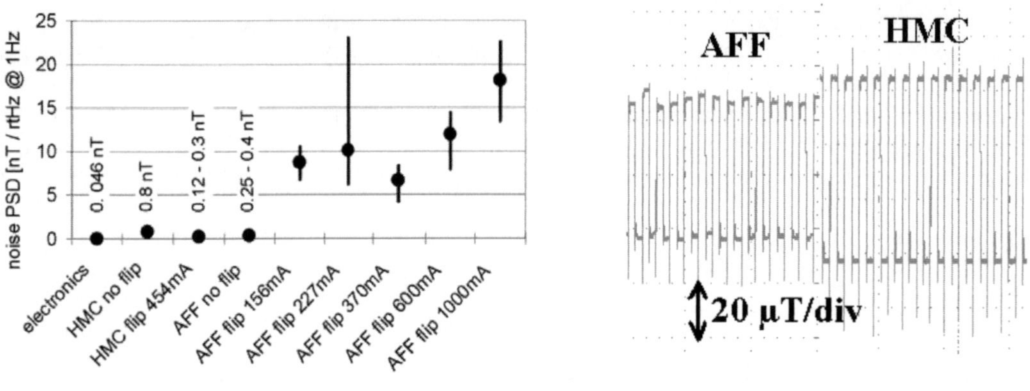

Figure 4. Left: noise PSD values for HMC and AFF sensors for various conditions and flipping current peak values. Right: oscilloscope screenshot of AFF and HMC sensors output signal (measured at the output of synchronous detector), offset instability caused by flipping is clearly visible for the AFF sensor.

Acknowledgments

The project TE02000202 is being realized with financial support of TACR, the Technology Agency of the Czech Republic.

References

[1] Kang M H, Choi B W, Koh K C, Lee J H and Park G T,, Experimental study of a vehicle detector with an AMR sensor, *Sensors and Actuators A* **118** (2005) 278–284

[2] Taghvaeeyan S and Rajamani R, The Development of Vehicle Position Estimation Algorithms Based on the Use of AMR Sensors, *IEEE Trans. Intel. Transportation Syst.*, Vol. **13**, No. 4, Dec 2012, 1845–1854

[3] Yang B, Lei Y, Vehicle Detection and Classification for Low-Speed Congested Traffic With Anisotropic Magnetoresistive Sensor, *IEEE Sensors Journal*, Vol. **15**, No. 2 (2015), 1132–1138

[4] Hauser H, Fulmek P L, Haumer P, Vopalensky M and Ripka P, Flipping field and stability in anisotropic magnetoresistive sensors, *Sensors and Actuators A: Physical* Volume **106**, Issues 1–3, 15 September 2003, 121–125

[5] Ripka P, Vopalensky M Platil A, Döscherd M, Lenssen K-M H, Hauser H, AMR magnetometer, *Journal of Magnetism and Magnetic Materials*, Volumes **254–255**, Jan. 2003, 639–641

[6] Vopalensky M, Ripka P, Platil A, Precise magnetic sensors, *Sensors and Actuators A: Physical*, Volume **106**, Issues 1–3, 15 September 2003, 38–42

[7] Petrucha V, Kaspar P, Calibration of a Triaxial Fluxgate Magnetometer and Accelerometer with an Automated Non-magnetic Calibration System, *IEEE SENSORS 2009 Conference*, 1510–1513, DOI: 10.1109/ICSENS.2009.5398466

IOP Conf. Series: Materials Science and Engineering **108** (2016) 012029 doi:10.1088/1757-899X/108/1/012029

Investigate the magnetic behaviour of thermal treated carbon steel

N P Kanelopoulos[1]
Laboratory of Physical Metallurgy, NTUA, Zografos, 15780, Greece

E-mail: philopimin@hotmail.com

Abstract. The present paper investigates the utilization of the magnetic hysteresis loops and Barkhausen Noise for the non-destructive characterization of annealed and quenched carbon steels samples. The resulting magnetic properties were further evaluated by examining the microstructure of the samples by using scanning electron microscopy.

1. Introduction

Steel is a composite ferrous metal alloy with varying mechanical properties with a wide range of applications. Industry's needs for the enhancement of its mechanical properties has led to the development of new alloys of steel with composite microstructures. It is known that exposing steels to high heat affects their mechanical properties (e.g. tensile strength, yield point), as well as their physical and chemical properties (e.g. change of crystal structure) [1-2].

Several methods have been developed in the resent years for the non-destructive testing of various grades of ferromagnetic steels [3-14]. It is well known that the magnetic non-destructive methods are influenced by the microstructural features of the examined materials [15-40].

The scope of the present paper is the documentation of changing parameters in the magnetic properties of micro-alloyed, high strength steel samples, which have been submitted to a predefined temperature annealing above the austenitic temperature, followed by quenching in oil, and the correlation of those with the respective changes in mechanical properties.

2. Experimental procedure

Micro-alloyed carbon steel were investigated with the chemical composition shown in Table 1. Before the annealing the as-received sample was cut, according to the ASTM E8 Standard (figure 1). Before the annealing process, the surface of these specimens was cleaned, degreased and dried. Then, the as-received sample was heated up to 891 °C for 5 hours and subjected to quenching in oil.

Table 1. Chemical composition in % wt of the examined steel

C	Si	Mn	Al	Cu	P	S	Ni	Cr	Mo	Fe
0.31	0.28	0.57	-	-	0.020	0.020	-	0.93	0.19	Balanced

Figure 1. As received sample.

[1] Phone number : +30 210 772 2183, E-mail : philopimin@hotmail.com

Metallographic specimens were prepared according to the standard procedure from the as-received and heat-treated samples and etched with 2% nital solution. Microscopic examinations were carried out by JEOL JSM 6100 Scanning Electron Microscopy (SEM), which is equipped with a Noran TS 5500 Microanalyzer for EDS analysis. Additionally, both Vickers micro-hardness and uniaxial tensile tests were performed, in order to investigate the variations in the mechanical properties of the samples.

For the Magnetic Barkhausen noise (MBN) measurements, a triangular shaped magnetic field produced by a generator was applied on the surface of the samples by an electromagnet. The discrete changes in the resulting local magnetization (MBN) were sensed at the ends of a receiving coil as voltage pulses. The MBN profile measurements were recorded.

3. Results

The microstructure of the as-received sample consisted of non-equiaxed polygonal ferrite and perlite grains (figure 1a). After annealing and quenching, a fully martensitic microstructure was revealed (figure 1b). Some acicular ferrite was still present in the microstructure. The high annealing temperature increased the austenite volume fraction, which then transformed to martensite upon quenching in the oil.

(a) (b)

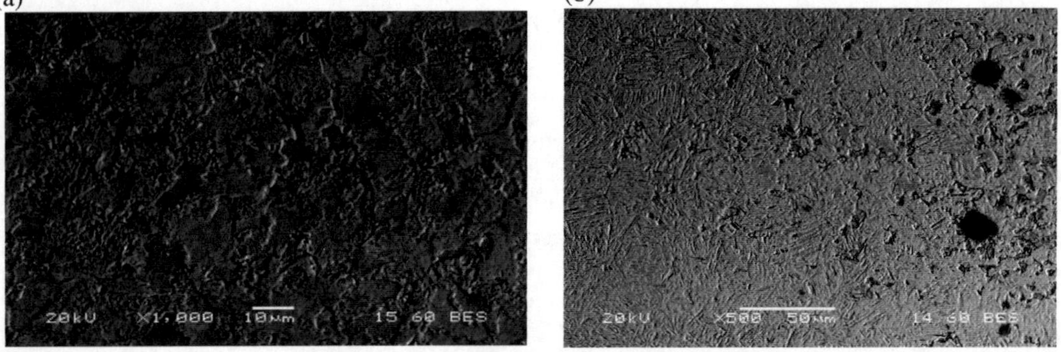

Figure 1. Microstructure of the (a) as received and (b) the annealed samples.

Microstructural changes influence the mechanical properties. The strength values of the annealed sample were higher than the strength of the as-received steel. The higher strengths of the quenched steel are known to be due to the presence of the harder martensite phase (figure 2). The increment of the yield strength and the diminishment of both the elongation and the ductility were also pronounced.

Microstructural changes and variations in the mechanical properties were also affect the magnetic properties of the materials and corresponding MBN profiles. A double-peak MBN profile was observed in the as-received sample. The narrower and higher peak in the MBN profile of the as-received sample indicated that an increased amount of reverse magnetic domain walls were moved in a limited range of input voltage, over shorter displacements, within the ferrite-perlite matrix. A very small peak size it was noticed at a position close to the lower excitation voltages in the MBN profile of the as-received sample. This was associated with the existence of higher amounts of magnetically soft ferrite in the untreated sample.

However, the peak's height in the MBN profile of the quenched samples were reduced and its position was also shifted to excitation voltage, as compared to the as-received sample. This was associated with the restricted displacement of domain walls in the hard martensite structure, which act as strong pinning during the magnetization procedure. The peak's position increased with the increase in the yield and ultimate tensile strengths of the examined steels. Thus, when the microstructure is dominated by high dislocation density, due to the presence of martensite phase, a single-peak MBN profile was observed.

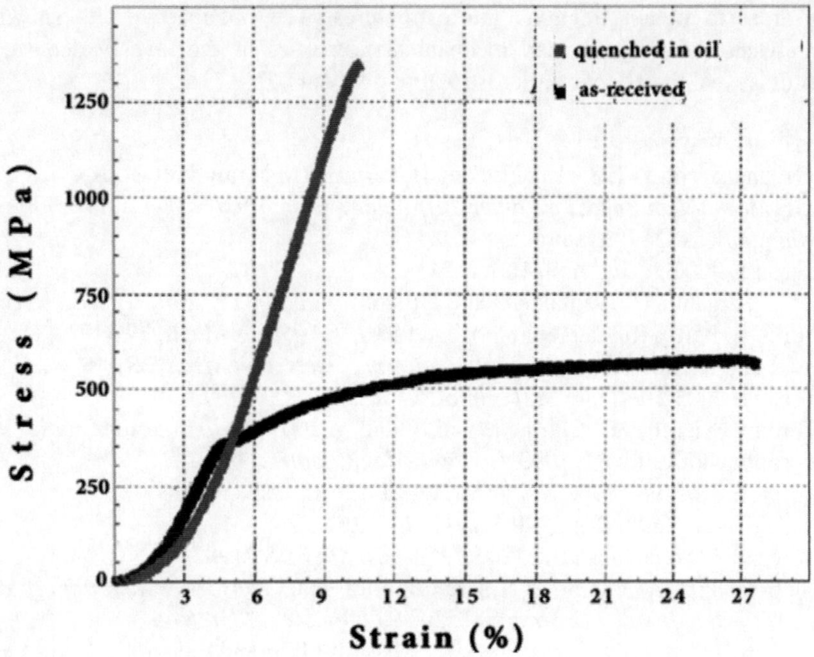

Figure 2. Stress – Strain curves of the samples.

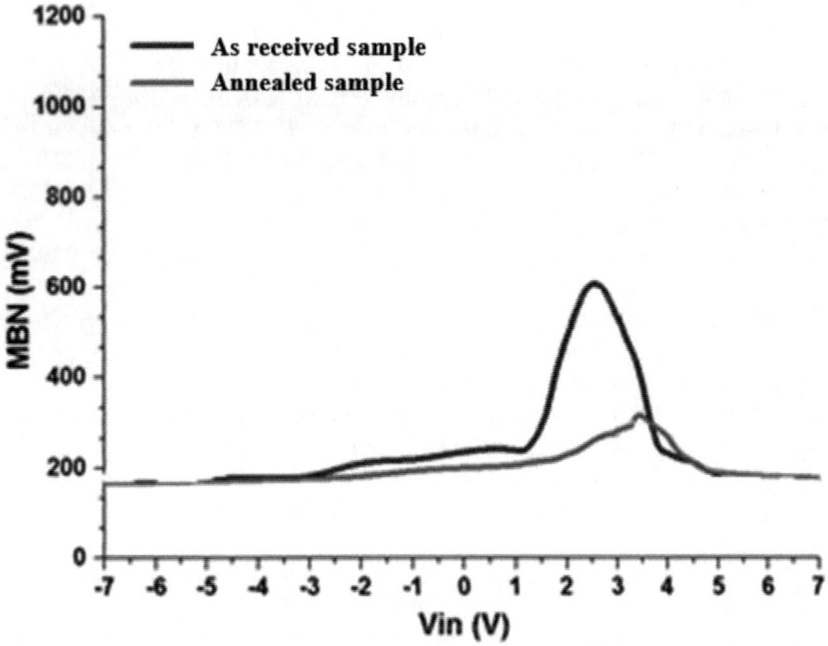

Figure 3. MBN profile of the samples.

4. Conclusions

A micro-alloyed carbon steel sample was intercritically heat treated on a constant temperature and oil quenched in order to obtain fractions microstructures. The obtained results shown the indirect relationship between the magnetic and mechanical properties of the investigated steel samples, in which this results reveal the effect of microstructure on them.

5. References

[1] Dieter GE and Bacon D 1986 *Mechanical Metallurgy* (McGraw-Hill Book Co)
[2] Dieter GE *Mechanical Behavior Under Tensile and Compressive Loads, Mechanical Testing and Evaluations* (ASM International)
[3] Hristoforou E 2002 *J. Opt. Adv. Mat.* **4** 245
[4] Kosmas K, Sargentis C, Tsamakis D and Hristoforou E 2005 *J. Mat.Proc.Tech.* **161** 359
[5] Hristoforou E, Reilly R E and Niarchos D 1993 *IEEE Trans. Magn.* **29** 3171
[6] Kosmas K and Hristoforou E 2007 *Int. J. of App. Electr. and Mech.* **25** 319
[7] Hristoforou E 2003 *Meas. Sci. & Technol.* **14** R15
[8] Hristoforou E, Niarchos D, Chiriac H and Neagu M 2001 *Sensors & Actuators A* **92** 132
[9] Hristoforou E and Reilly RE 1993 *J. Magn. Magn. Mater.*, **119** 247
[10] Hristoforou E, Kosmas K and Kollar M 2008 J. Electr. Eng. 59 90
[11] Hristoforou E and Reilly R E 1990 *Trans. Mag.* **26** 1563
[12] Hristoforou E and Niarchos D 1992 *IEEE Trans. Mag.* **28** 2190
[13] Dimitropoulos P D, Avaritsiotis J N, Hristoforou E 2003 *Sensors & Actuators A* **107** 238
[14] Giouroudi I, Ktena A and Hristoforou E 2004 *J. Opt. Adv. Mat.* **6** 45
[15] Ktena A, Hristoforou E, Gerhardt GJL, Missell FP, Landgraf FJG, Rodrigues Jr DL and Alberteris-Campos M 2014 *Physica B: Coned Matter* **435** 109
[16] Fisher O and Schneider J 2003 *J. Magn.Magn. Mater.* **302** 254
[17] Oshihiko O, Masaaki Y K and Atsuhito H 2008 *J. Magn. Magn. Mater.* **320** 2430
[18] Kumar PR, Samajdar I, Viswanathan NN, Singal V and Seshadri V 2003 *J. Magn. Magn. Mater.* **264** 75
[19] Bertotti G 2008 *J. Magn. Magn. Mater.* **320** 2436
[20] Cunha MA and Paolinelli SC 2008 *J. Magn. Magn. Mater.* **320** 2485
[21] Vourna P, Ktena A, Tsakiridis P E and Hristoforou E 2015 *Measurements* **71** 31
[22] Vourna P, Ktena A, Tsakiridis P E and Hristoforou E 2015 *NDT & E Int.* **71** 33
[23] Vourna P, Ktena A and Hristoforou E 2014 *IEEE Trans. Mag.* **50** 1
[24] Moses AJ, 1990 *J. Magn. Magn. Mater.* **112** 150
[25] Sidor Y, Kovac F and Kvackaj T 2007 *Acta Mater.* **55** 1711
[26] Sha YH, Zhang F, Zhou SC, Pei W and Zuo L, 2008 *J. Magn. Magn. Mater.* **320** 393
[27] Augustyniak B, Piotrowski L, Chmielewski M, Kosmas K and Hristoforou E 2010 *IEEE Trans. Magn.* **46** 544
[28] Vértesy G, Mészáros I and Tomáš I 2013 *NDT & E Int.* **54** 107
[29] Vourna P and Ktena A 2011 *Key Eng. Mat.* **495** 257
[30] Franco FA, González MFR, De Campos MF and Padovese LR 2013 *J. Nondestruct. Eval.* **32** 93
[31] White S, Krause T and Clapham L 2007 *Meas. Sci. Technol.* **18** 3501
[32] Capó Sánchez J, Campos MF and Padovese L R 2012 *J. Magn. Magn. Mater.* **324** 11-14
[33] Piotrowski L, Augustyniak A, Chmielewski M, Hristoforou E and Kosmas K 2010 *IEEE Trans. Magn.* **46** 239
[34] Kasai N, Koshino H, Sekine K, Kihira H and Takahashi M 2013 *J. Nondestruct. Eval.* **32** 277
[35] Lindgren M and Lepistö T 2004 *NDT & E Int.* **37** 403-410
[36] Vincent A, Pasco L, Morin M, Kleber X and Delnondedieu M 2005 *Acta Mat.* **53** 4579-4591
[37] Moorthy V, Shaw B A and Hopkins P 2006 *J. Magn. Magn. Mater.* **299** 362-375
[38] Blaow M, Evans JT and Shaw BA 2005 *Acta Mat.* **53** 279-287
[39] Vourna P, Hervoches C, Vrána M, Ktena A and Hristoforou E 2015 *IEEE Trans.* **51** 6200104
[40] Stefanita CG, Clapham L and Atherton DL 2000 *J. Mater. Sci.* **35** 2675

Oxygen adsorption at noble metal/TiO_2 junctions

F. Hossein-Babaei[1,a], Navid Alaei-Sheini[1,b], Mehdi M. Lajvardi[1,c]
[1]Electronic Materials Laboratory, Electrical Engineering Department, K. N. Toosi University of Technology, Tehran 16317-14191, Iran

[a]fhbabaei@kntu.ac.ir, [b]navid_alaei@ee.kntu.ac.ir, [c]mm.lajvardi@gmail.com

Abstract. Electric conduction in titanium dioxide is known to be oxygen sensitive and the conductivity of a TiO_2 ceramic body is determined mainly by the concentration of its naturally occurring oxygen vacancy. Recently, fabrications and electronic features of a number of noble metal/TiO_2-based electronic devices, such as solar cells, UV detectors, gas sensors and memristive devices have been demonstrated. Here, we investigate the effect of oxygen adsorption at the noble metal/TiO_2 junction in such devices, and show the potentials of these junctions in chemical sensor fabrication. The polycrystalline, poly-phase TiO_2 layers are grown by the selective and controlled oxidation of titanium thin films vacuum deposited on silica substrates. Noble metal thin films are deposited on the oxide layers by physical vapor deposition. Current-voltage (I-V) diagrams of the fabricated devices are studied for Ag/, Au/, and Pt/TiO_2 samples. The raw samples show no junction energy barrier. After a thermal annealing in air at 250°C, I-V diagrams change drastically. The annealed samples demonstrate highly non-linear I-V indicating the formation of high Schottky energy barriers at the noble metal/TiO_2 junctions. The phenomenon is described based on the effect of the oxygen atoms adsorbed at the junction.

1. Introduction

We have recently reported that the electron energy barrier formed at the junction between a noble metal and titanium dioxide depends on the oxygen content of the interface [1], which strongly affects the main electronic features of such contacts. Nobel metal/TiO_2 contacts are utilized as active components in TiO_2-based electronic devices or for connecting such devices to the outside circuitry. Many TiO_2-based electronic devices, such as those used or considered as solar cells [2, 3], varistors [4], ultraviolet detectors [5, 6], memristors [7, 8], and gas sensors [9, 10] are exposed to corrosive environments and/or operate at elevated temperatures and their connections to the external circuitry require utilizing noble metals. In some of these devices, the contact between the noble metal and TiO_2 plays an active role in the operation of the device, as well [11, 12].

Here, we report fabrication and characterization of Schottky diodes with platinum-, gold-, and silver-TiO2 structures on silica substrates and show that the reverse currents in these devices are sensitive to environmental parameters. The reverse current of these diodes can be drastically reduced by thermal annealing in air or pure oxygen; in the case of silver-TiO_2 diodes, the reverse current at -0.5 V reduces down to 4×10^{-11} Amm^{-2}. It is shown that these metal/metal oxide junctions are suitable platforms for the fabrication of a number of low power consuming sensors.

2. Experimental

The metal/TiO_2 contacts examined are of the structure schematically shown in Figure 1. The oxide layers are produced on silica substrates by the direct oxidation of 400 nm-thick titanium thin films previously deposited by thermal evaporation in vacuum. The oxide layers are 700 nm thick. The oxidation process is controlled and stopped when a 30 nm-thick layer of metallic titanium still remains unoxidized. This degree of accuracy in oxidation process stoppage is achieved by monitoring the electrical conductance and, hence, thickness, of the remaining titanium layer. As shown in Figure 1, the titanium conductance is continuously read via two aluminum test points deposited on the titanium. It is shown that a thin layer of aluminum can protect titanium from oxidation [13]. The grown oxide layer is 700 nm thick and contains both rutile and anatase phases as described in reference [1].

Noble metal layer (anode) is deposited by thermal evaporation of the 5N purity metal in vacuum; selective area (1.5 mm × 5 mm) deposition is achieved by photoresist liftoff. A fine gold wire is cemented to the anode using silver paste; the position of the connection is on the extended tail of the noble metal layer outside the TiO_2 layer area. Wire connection to the titanium thin film (cathode) underneath the TiO_2 layer is made in a similar way on both of the aluminum pads (Figure. 1), which are connected together outside the device to form the cathode connection wire. After preliminary electrical examinations, the fabricated diode is annealed in different atmospheric conditions at 250 °C for 120 min while it is DC-biased at -0.5 V. Samples are reexamined after completion of the annealing process.

The electrical examination of the fabricated noble metal/TiO_2/Ti structures is carried out by plotting their current vs. voltage (I-V) characteristics. The I-V diagrams of each sample are plotted at different conditions, before and after annealing process. The experimental setup is shown in Figure. 2. A source measure unit (Keithley-238) is utilized for the point-by-point measurements. Device currents, as low as 10^{-10} A, are measured in a Faraday cage. All of the diodes examined are light sensitive [14- 16]; all I-V measurements are carried out in a dark box. All reported observations are reproducible and the described device features have been re-checked on five samples.

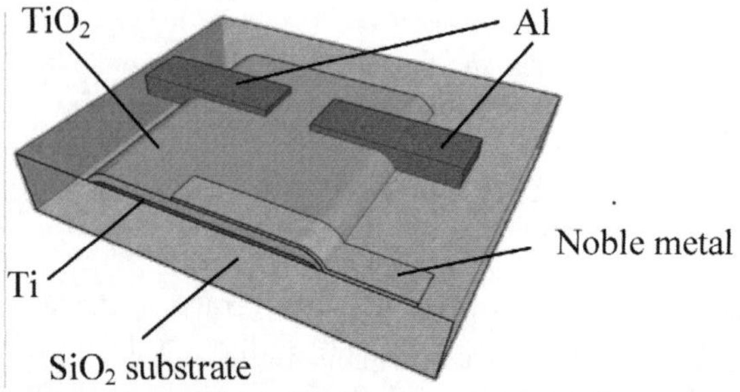

Figure 1. The schematic diagram of a noble metal/TiO_2 diode.

3. Results and Discussion

Each noble metal/TiO_2/Ti diode consists of a noble metal/TiO_2 junction and a TiO_2/Ti contact connected in series. The TiO_2/Ti contact is ohmic [10, 12] and, hence, the nonlinear I-V graphs obtained for the fabricated noble metal/TiO_2/Ti structure is expected to stem from the nonlinearity of the noble metal/TiO_2 junction.

According to Figure. 3a, all samples have almost symmetric I-V characteristics and indicate almost zero junction barriers prior to thermal annealing. The symmetric nonlinearity observed in their respective I-V characteristics around V=0 (the vertical axes in all diagrams presented in Figure. 3) stems from the combination of different conduction mechanism, such as Pool-Frenkel [17], through the trap-populated titanium oxide adjacent to the interface. After annealing at 250 °C in air, all samples turned into Schottky diodes of high energy barriers. The observation of instability between ohmic and diode-like behaviors for silver/SnO_2 contacts has also been reported [18]. The after annealing I-V characteristics of all samples are presented in Fig. 3b demonstrating drastic change of characteristics

IOP Conf. Series: Materials Science and Engineering **108** (2016) 012030 doi:10.1088/1757-899X/108/1/012030

(compare Figure. 3a with Figure. 3b). The obtained Arrhenius diagrams are given in Fig. 4a-c resulting close activation energies for the leakage currents of all annealed diode samples regardless of the nature of the metal/ TiO_2 junction involved. This is attributed to the electron leakage through the traps in the carrier depleted region of the titanium dioxide layer adjacent to the junction. Domination of carrier leakage mechanisms such as Pool-Frenkel makes the reverse current independent from the barrier height established at the junction.

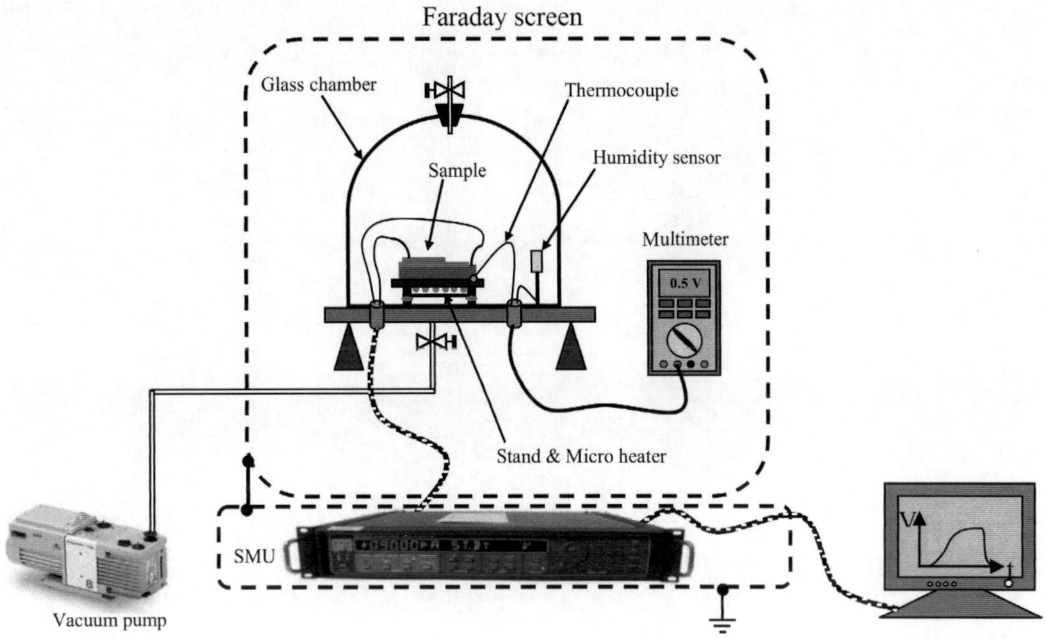

Figure 2. The experimental setup for electrical examinations.

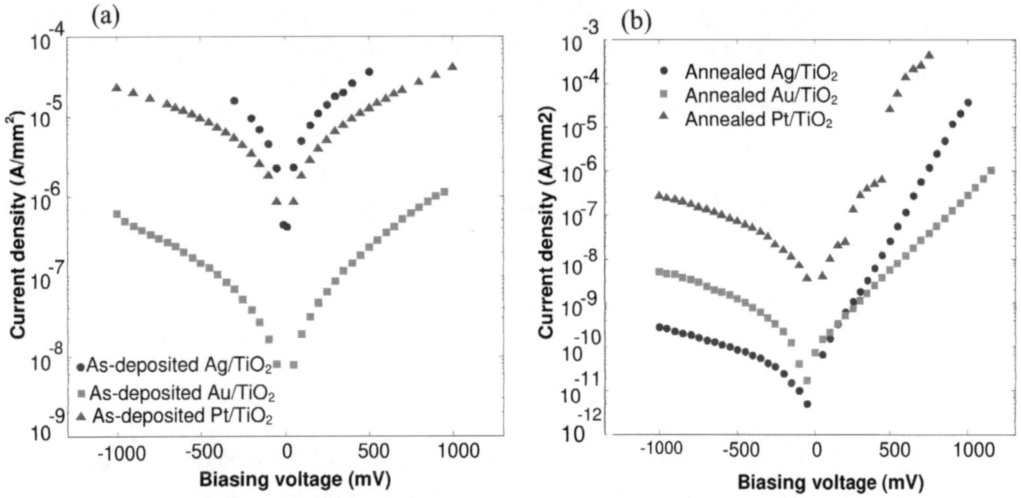

Figure 3. The I-V characteristics obtained for all types of fabricated noble metal/TiO_2/Ti diodes before (a) and after (b) the thermal annealing process.

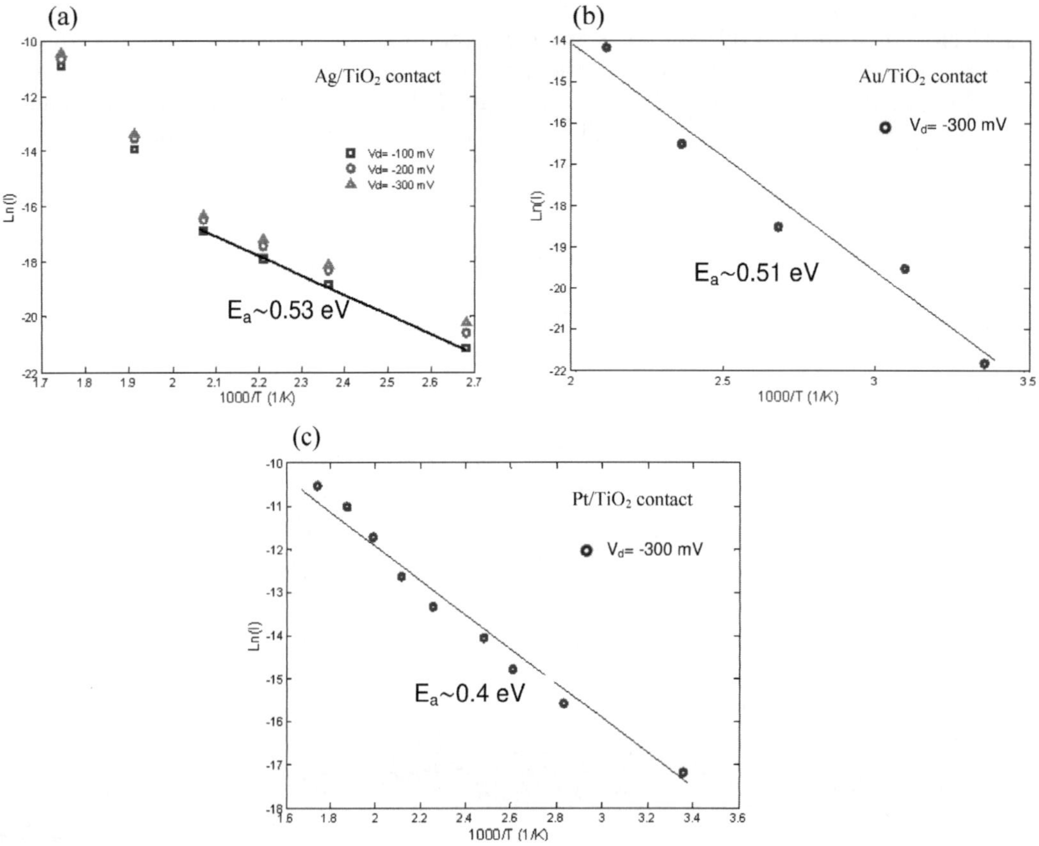

Figure 4. Arrhenius diagrams plotted for the reverse currents of (a) Ag/TiO_2, (b) Au/TiO_2, and (c) Pt/TiO_2 contacts.

To describe the zero junction barriers observed in all samples prior to the annealing regardless of the nature of the anode, Ag, Pt or Au, we note that in the electrical evaluations of parallel mechanisms, the one with the lowest barrier has the most profound effect. The quantitative information regarding the work functions of the metallic anodes and the electron affinity of rutile and anatase are given in reference [1]. However, in most of the junction cases examined, the observed zero barriers cannot be described based on commonly used concept that the junction energy barrier height is equivalent to difference between the metal work function and semiconductor electron affinity. Only in the case of silver/TiO_2 junction, this concept predicts a close to zero barrier height at the junction. For this prediction, the work function of silver, W_{Ag}, is assumed to be 4.3 eV [19, 20] for the polycrystalline silver layers, and the electron affinity of TiO_2 is taken to be 4.8 eV for rutile and 5.1 for anatase [21]. The latter parameter is different for rutile and anatase phases [22]; the value introduced is that of rutile as it results in a lower barrier height. This is because, in complex junctions, charge carriers choose the lowest barrier/resistance path and, hence, I-V characteristics of the junction is expected to approximately indicate the minimum barrier level taken place at the junction. On the other hand, none of the contacts possible in the cases of Au and Pt, according to this concept, can explain the observed zero barrier. We attribute this to the abundance of vacuum-generated vacancies at the surface of the TiO_2 layer during the metal deposition process, as it has been shown that high oxygen vacancy concentration can collapse the interface barrier in M/TiO_2 junctions [7]. Thermal annealing in air, compensates the oxygen vacancies and increases the oxygen concentration at the interface, which allows energy barrier buildup. Similar annealing processes, carried out in vacuum or Ar atmospheres has no effect on the reverse currents because they fail to provide oxygen required for barrier build up, and the I-V characteristics remain symmetrical. The I-V characteristics plotted for all samples after thermal annealing in air are profoundly asymmetric and rectifying, indicating high interface energy barriers formed. The barrier heights calculated based on the first quarter of the I-V diagrams are reported in reference [1] indicating barrier heights of about 0.9 eV for all samples. This is consistent with the reported independence of the junction barrier height formed on the TiO_2 from the nature of the metallic anode [23].

The electron affinity for rutile has been reported to be in the 3.9-4.8 eV range [21, 24, 25] while that of anatase is 5.1 eV [21]. Hence, any metallic anode is expected to form a smaller interface barrier with anatase than with rutile. Regarding the anode work function, however, the smallest is expected to form the lower barrier. Work functions of Pt, Au and Ag at various forms are given in reference [1] indicating that the smallest values are related to their polycrystalline form. W_{Pt} for polycrystalline platinum is reported to be 5.6 eV [26] and it has been reported to grow as much as 1.0 eV upon oxygen adsorption to its surface. Hence, oxygen adsorption on Pt at its junction with anatase can increase the junction barrier from 0.5 to 1.5 eV.

For Au/TiO_2 contact, however, these numerical values lead to a barrier height approximately around 0.4 eV. This is in contrast with the observed 0.9 eV junction barrier. It appears that higher levels of oxygen coverage occurs on the gold surface at the junction than that reported for the (110) surface of a god crystal layer [27]. The results indicate that, indeed, gold surface at the junction with TiO_2 can accommodate oxygen atoms stably up to high coverage levels. This would result in a larger work function for the gold electrode and further increase of the junction barrier.

W_{Ag} reported for polycrystalline silver [19] is smaller than the electron affinity of anatase [21]. Hence, zero energy barrier is predicted for the Ag/TiO_2 junction from this stance. However, W_{Ag} has been reported to grow up to 7.0 eV when silver surface is covered with an oxygen monolayer. Silver differs from gold in that, at room temperature, oxygen is stable up to 25% coverage [28, 29], which according to theoretical predictions [28], translates to a W_{Ag} of 6.2 eV. This considerable change in W_{Ag}, results in a 1.0 eV barrier.

4. Conclusions

Major electronic features of noble metal/TiO_2 junctions were studied. While all vacuum-deposited noble metal anodes formed ohmic contacts with TiO_2, an hour of thermal annealing at 250 °C in air converted them all to Schottky junctions of ~0.9 eV energy barrier heights. The barrier height formed after annealing in air is almost independent from the nature of the noble metal anode. Thermal annealing in inert or reducing atmospheres failed to assist barrier formation. The dependence of the barrier height on the oxygen partial pressure of the annealing atmosphere indicates possibility of sensing oxidizing or reducing atmospheres with different noble metal/TiO_2 diodes.

References

[1] Hossein-Babaei F, Lajvardi M M and Alaei-Sheini N 2015 The energy barrier at noble metal/TiO2 junctions *Appl. Phys. Lett.* **106** 083503
[2] McFarland E W and Jing T 2003 A photovoltaic device structure based on internal electron emission *Nature* **421** 616-8
[3] Kim H S *et al.* 2013 High efficiency solid-state sensitized solar cell-based on submicrometer rutile TiO_2 nanorod and $CH_3NH_3PbI_3$ perovskite sensitizer *Nano Lett.* **13** 2412-7
[4] Bueno P R *et al.* 1996 Effect of Cr_2O_3 in the varistor behaviour of TiO_2 *J. Materials Sci. Lett.* **15** 2048-50
[5] Hossein-Babaei F, Lajvardi M M and Boroumand F A 2012 Large area Ag–TiO_2 UV radiation sensor fabricated on a thermally oxidized titanium chip *Sens and Actuators A: Physical* **173** 116-21
[6] Kong X *et al.* 2009 Metal-semiconductor-metal TiO_2 ultraviolet detectors with Ni electrodes *Appl. Phys. Lett.* **94** 123502
[7] Yang J J *et al.* 2008 Memristive switching mechanism for metal/oxide/metal nanodevices

Nature nanotechnology **3** 429-33.
[8] Kolka Z, Biolek D, Biolkova V 2015 Improved Model of TiO_2 Memristor *Radioengineering* **24**
[9] Kim I D *et al.* 2006 Ultrasensitive chemiresistors based on electrospun TiO_2 nanofibers *Nano Lett.* **6** 2009-13.
[10] Hossein-Babaei F and Rahbarpour S 2011 Titanium and silver contacts on thermally oxidized titanium chip: electrical and gas sensing properties *Solid-State Electronics* **56** 185-90.
[11] Yang J J *et al.* 2011 Metal/TiO_2 interfaces for memristive switches *Appl. Phys. A: Materials Science & Processing* **102** 785-9
[12] Hossein-Babaei F and Rahbarpour S 2011 Separate assessment of chemoresistivity and Schottky-type gas sensitivity in M–metal oxide–M′ structures *Sens and Actuators B: Chemical* **160** 174-80
[13] Raybould D, Chipko P, Fischer W E 2006 *US Patent* 7,135,238
[14] Xue H *et al.* 2007 TiO_2 based metal-semiconductor-metal ultraviolet photodetectors *Appl. Phys. Lett.* **90** 201118
[15] Xing J *et al.* 2011 Highly sensitive fast-response UV photodetectors based on epitaxial TiO_2 films *J. of Physics D: Applied Physics* **44** 375104
[16] Huang H *et al.* 2011 Low dark current MSM UV photodetectors with Pt schottky contacts *Electron Device Lett.* **32** 530-2
[17] Ohring M 2001 Materials science of thin films. *Academic press*
[18] Hossein-Babaei F, Moghadam S and Masoumi S 2015 Forming ohmic Ag/SnO_2 contacts *Materials Lett.* **141** 141-4
[19] Dweydari A W and Mee C H B 1975 Work function measurements on (100) and (110) surfaces of silver *physica status solidi* **27** 223-230
[20] Michaelson H B 1977 The work function of the elements and its periodicity *J. of Appl. Phys.* **48** 4729-33
[21] Scanlon D O *et al.* 2013 Band alignment of rutile and anatase TiO_2 *Nature materials* **12** 798-801
[22] Tang H *et al* 1994 Electrical and optical properties of TiO_2 anatase thin films *J. of Appl. Phys.* **75** 2042-7
[23] Zhong N, Shima H, Akinaga H 2010 Rectifying characteristic of Pt/TiO_x/metal/Pt controlled by electronegativity *Appl. Phys. Lett.* **96** 2107
[24] Könenkamp R and Rieck I 2000 Electrical properties of schottky diodes on nano-porous TiO_2 films *Materials Science and Engineering: B* **69** 519-21
[25] Schierbaum K D *et al.* 1991 Schottky-barrier and conductivity gas sensors based upon Pd/SnO_2 and Pt/TiO_2 *Sens and Actuators B: Chemical* **4**.1 87-94
[26] Jakobi K, 3.1.2.4 Work Function Data, Ed. by G. Chiarotti, *Springer Materials, The Landolt–Börn stein Database* http://www.springermaterials.com DOI:10.1007/10086058_16
[27] Xu Y, Mavrikakis M 2003 Adsorption and dissociation of O_2 on gold surfaces: effect of steps and strain . *J. of Phys. Chemistry B* **107** 9298-307.
[28] Li W X, Stampfl C and Scheffler M 2002 Oxygen adsorption on Ag (111): A density-functional theory investigation *Physical Review B* **65** 075407
[29] Allen M W, Alkaisi M M and Durbin S M 2006 Metal Schottky diodes on Zn-polar and O-polar bulk ZnO *Appl. Phys. Lett.* **89** 3520

Ni/TiO_2 Ultraviolet Detector

Mehdi Mohamadzade Lajvardi[1, a], Mojtaba Jahangiri[2, b]

[1, 2]Electronic Materials Laboratory, Electrical Engineering Department, K. N. Toosi University of Technology, Tehran 16315-1355, Iran

[a]mm.lajvardi@gmail.com; [b]mojtaba.jahangiri@ee.kntu.ac.ir;

Abstract. The fabrication technology of solid-state photon detectors based on semiconductors other than silicon is yet to mature, but their recent progress opens new possibilities. Such devices are especially attractive for ultraviolet radiation level measurements because semiconductor materials with band gaps larger than 3.0 eV can be used as "visible-blind" detectors, the operation of which do not require using visible light filters. Here, fabrication and characterization of a UV detector based on nickel/titanium dioxide Schottky junction is reported. The operation of the device is described based on the photoelectric mechanism taking place in the carrier-depleted oxide adjacent to the Ni layer. Simplicity of fabrication, cost-effectiveness and fast response are the positive features of the device. These features of the device are compared with those of the previously reported Ag/TiO_2 UV detectors.

1. Introduction

Recent civil and military applications of ultraviolet (UV) detectors [1], such as solar UV level monitoring [2], biological and chemical process measurement [3], flame sensors [4] and secure non-line-of-sight communication [5] have made researchers to develop varieties of UV detectors. Wide band-gap semiconductors such as ZnO, SiC, GaN and TiO_2 are becoming dominant in UV photodetector fabrication since they are intrinsically visible-blind, working properly in hostile environments and enabling room temperature operation. Among these semiconductors, TiO_2 has been considered for many optoelectronic applications including photocatalysis and solar energy conversion [6, 7] owing to its facile preparation and low cost fabrication process. TiO_2 band gap is 3.2 eV for its anatase phase and 3.0 eV for the rutile. Such a wide band gap makes TiO_2 suitable for UV detection as it is insensitive to the background visible and infrared lights.

Different structures are used for photodetector fabrication, among which Schottky-type metal-semiconductor junctions, due to their simple manufacture, absence of high temperature diffusion process and its high speed of response are common [8].The most prominent parameter of a Schottky photodiode is its Schottky barrier height. For a metal connected to an n-type semiconductor of smaller work function, the Schottky barrier (Φ_b) is defined by [9-11]:

$$\phi_b = \phi_m - \chi_s \quad (1)$$

Where, Φ_m is the metal work function and χ_s is the semiconductor electron affinity. As TiO_2 work function (φ_{TiO2}=4.6 eV for rutile phase [12]) is smaller than nickel work function (φ_{Ni}=5.1 eV [13]), Ni/TiO_2 contact acts as a Schottky diode. Assuming that the electron affinity of TiO_2 is 3.9 eV [14], according to (1), the barrier height formed at the Ni/TiO_2 interface is 1.2 eV.

Taking the nickel contact layer thin enough (50 nm), it allows UV photons to reach the carrier depletion region at the Ni/TiO_2 contact and generate carriers to participate in the device current. Photons with energies more than 3 eV, if absorbed in the depletion region of the oxide layer, would generate electron-hole-pairs (EHP) because of the band-to-band excitation in the semiconductor. The generated carriers are swept in different directions by the internal electric field and participate in the device current. The light photons with photon energy less than 3 eV cannot be absorbed in the device and create zero current.

In this work, the current-voltage characteristics and the UV photoelectric properties, such as photo response, response time and spectral response of Ni/TiO_2 junctions fabricated on thermally oxidized titanium chips are reported. The results demonstrate that the Ni/TiO_2 UV sensor responds much faster than other similar devices.

2. Experimental

Pure titanium foil (99.7% purity, 0.9 mm thick) is cut into 8 mm×8 mm chips. The chips are polished and then cleaned by the means of ultrasonification in acetone for five minutes followed by rinsing with deionized water and drying with nitrogen. The rutile layer is grown thermally on titanium substrate in an electric furnace at 700°C in clean air for 30 minutes. SEM micrograph of the oxidized surface is shown in Figure 1. The electrical contact between the native oxide layer and the titanium substrate is supposed to be an ohmic contact since the work function of titanium and titanium dioxide are reported to be 4.3 [13] and 4.6 eV [12], respectively. The Ni/TiO_2 contact is formed by thermal evaporation of 45 nm thick nickel thin film on to the surface of titanium dioxide layer. The contact area is 2 mm^2. Two wire segments are attached by silver paste to the Ti and Ni electrodes. The Ni-TiO_2-Ti structure of the device is schematically given in Figure 2.

Figure 1. SEM image of the TiO_2 layer grown by the oxidation of the titanium chip at 700°C for 30 min in air.

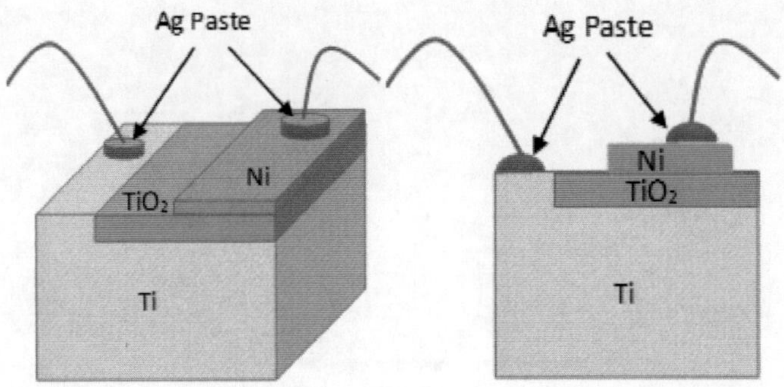

Figure 2. The schematic diagram of the Ni/TiO_2 sample.

Source measure unit (SMU), Keithly 238, and lambda EZ 201 monochromator are utilized for measuring the current vs. voltage (I-V) characteristic and spectral response of the device, respectively. To measure the relationship between the photo current and UV light intensity, the device is exposed to a UV LED. The transient photo response is measured by a sampling oscilloscope, Arma D-1040. All measurement is carried out at room temperature.

3. Results and discussion

3.1. Electrical Properties of Ni-TiO_2 contact

Current-voltage characteristics measured by SMU for Ni/TiO_2 Schottky contact before and after 15 minutes annealing at 190°C under -300 mV biasing voltage are shown in Figure 3 and Figure 4, respectively. For better visualization, Figure 5 shows the I-V characteristics in a semi-logarithmic scale. As it is clear in the Figures 3-5, rectification increases 4 times after annealing. The role of thermal annealing in air in junction barrier formation is described by Hossein-Babaei et al. based on the physicochemical parameters of the constituting materials [9]. The contact behavior can be explained based on the thermionic emission theory. The barrier height is defined by [10]

$$\phi = \frac{kT}{q} ln(AA^*T^2/I_s) \quad (2)$$

Where k is the Boltzmann constant, T is the operating temperature, A is the junction area, A* is the TiO_2 Richardson constant and I_S is the reverse saturation current. With $A^*=1200$ A.cm^2/K^2 for TiO_2 [15] and assuming V > 3kT/q, the effective barrier height is found to be 1.1 eV. The energy band diagrams of Ni-TiO_2 junction is shown in Figure 6. Based on this diagram the barrier height equals 1.2 eV that is about 10% wider than the theoretically predicted energy gap.

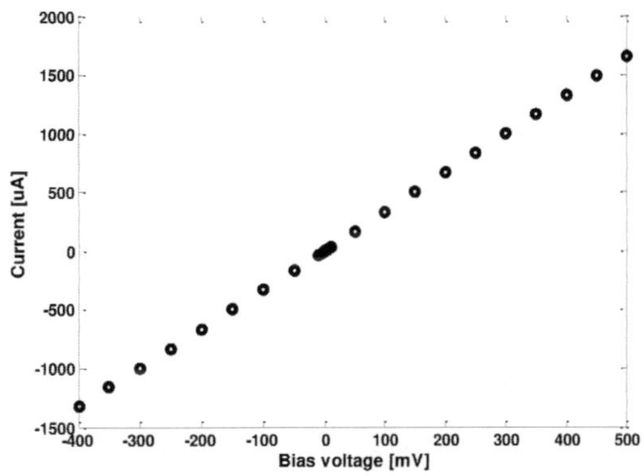

Figure 3. I-V characteristics of an as-fabricated Ni-TiO_2 diode at room temperature

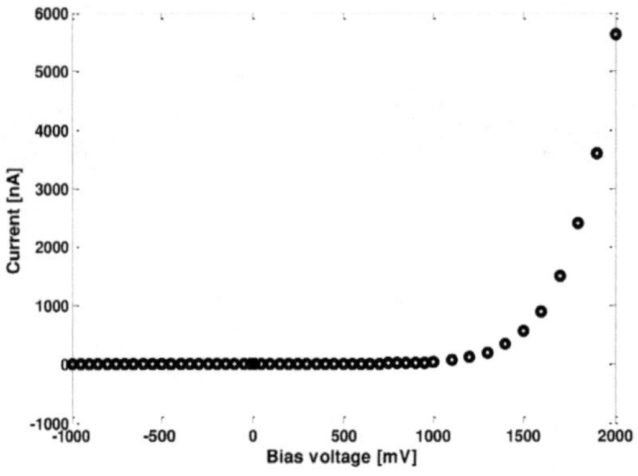

Figure 4. I-V characteristics of the Ni/TiO_2 diode after thermal annealing at 190°C in air.

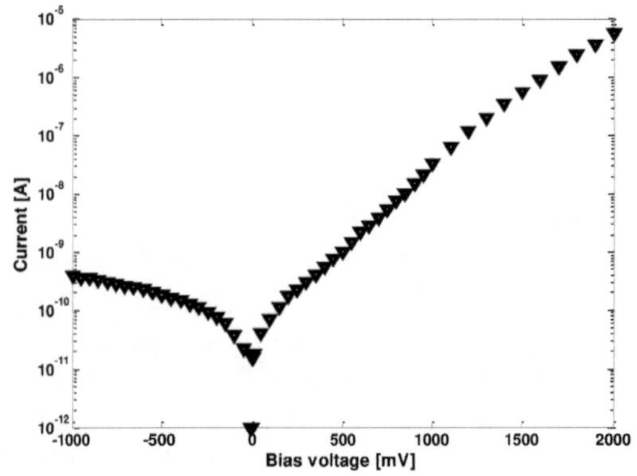

Figure 5. Semi logarithmic Current-voltage characteristics of the Ni-TiO_2 diode annealed at 190°C

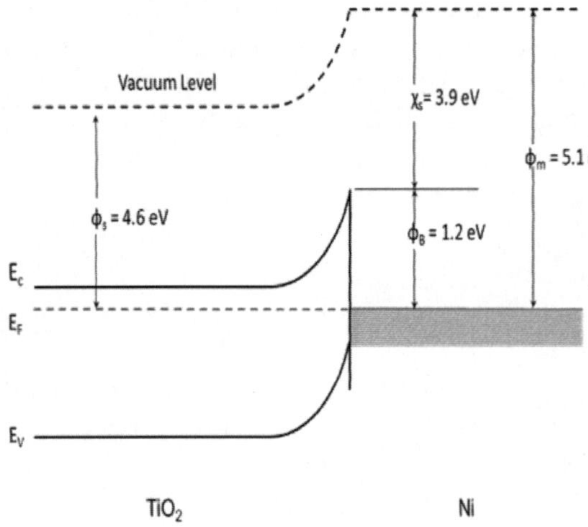

Figure 6. The proposed energy band diagram of the Ni/TiO_2 contact.

IOP Conf. Series: Materials Science and Engineering **108** (2016) 012031 doi:10.1088/1757-899X/108/1/012031

3.2. Photoelectric properties of Ni-TiO_2 contact

To study the photosensitivity of the diode, I-V characteristics are measured under varying light intensities. The results are presented in Figure 7. As expected, the reverse current of the diode, which is the sum of the thermal current and photo current, crosses the I-axis at negative values proportional to the light intensity. The relationship between voltage and current can be attributed to the presence of traps which release carriers at higher voltages [16]. Figure 8 is the spectral response measured at -300 mV biasing voltage for the TiO_2 photo detector with nickel Schottky contact. The photoresponse has a high magnitude over the band gap and a cut-off at wavelength around the absorption edge of the rutile phase TiO_2 ($\sim 405nm$) [17]. Because of high absorption coefficient at higher wavelengths, the photoresponse drops as the absorption result in generation of EHPs far away from the depletion region. These EHPs have a higher probability of recombination before participating in the photocurrent [18]. Therefore, this device shows the highest sensitivity at 365nm.

To study the stability of the optoelectronic features, the sample devices are exposed to UV light at wavelength of 365 nm continuously for 16 hours. It was found that the photoresponse of the device reaches to 98% of its primary state after 16 hours. The temporal response of the device is displayed in Figure 9. The device shows a fast time response with a response time of 10 ms and a recovery time of 120 ms. However, in Ni-TiO_2–Ni UV resistive detectors the rise and recovery times are reported to be 13 ms and 11000 ms, respectively [19].

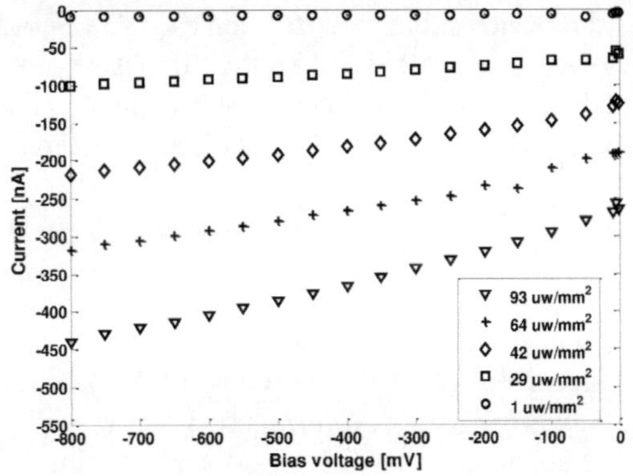

Figure 7. The reverse current of the Ni/TiO_2 diode under different UV light (at 390 nm) intensities and biasing voltages.

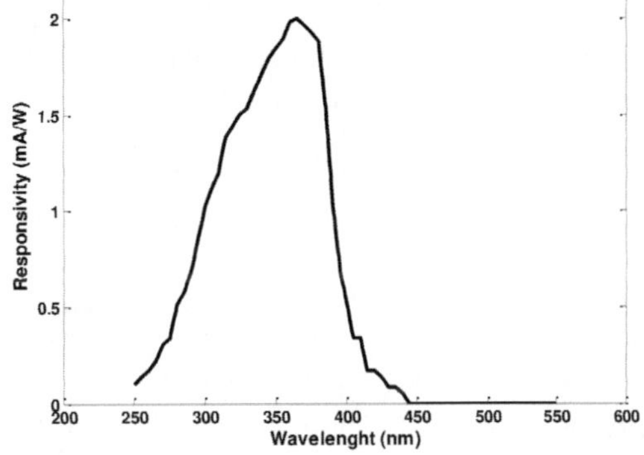

Figure 8. The spectral response of the Ni/TiO_2 photodiode.

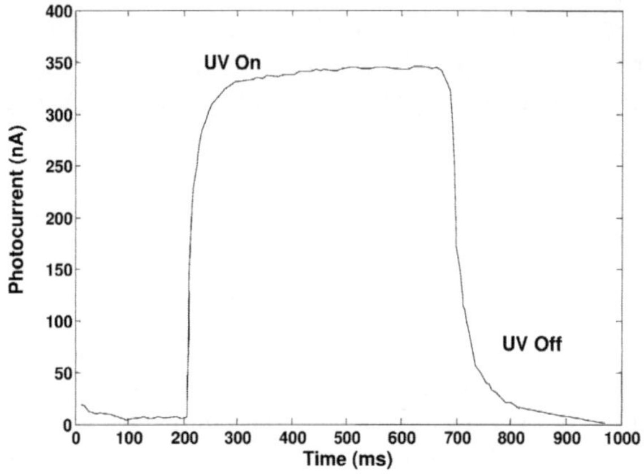

Figure 9. Transient response of the Ni/TiO_2 UV sensor.

4. Summary

The fabrication of a Ni-TiO_2 UV detector was reported. Our electrical measurements indicate that the as-fabricated Ni-TiO_2 contact has zero junction energy barrier. However, a thermal annealing process in air causes drastic changes in the device behavior. After annealing, a Shottky barrier height of 1.0 eV was observed, which is close to the 1.2 eV calculated by the difference of the metal work function and electron affinity of the oxide. The fabricated UV detector shows stable photoresponse. The maximum photoresponse is 2 mA/W at 365 nm. Fast rise and recovery times of 10ms and 120 ms were measured respectively.

References

[1] Hossein-Babaei F, Lajvardi M M and Boroumand F A 2012 *Sens Actuators A* **173** 116

[2] Lajvardi M M and Boroumand F A 2013 Environmental UV-A Level Monitoring Using an Ag-TiO_2 Schottky Diode *Key Engineering Materials* **543** 113-6

[3] Zeskind B J *et al.* 2007 Nucleic acid and protein mass mapping by live-cell deep-ultraviolet microscopy *Nature Methods* **4** 567-9

[4] Perera A G U *et al.* 2007 Performance improvements of ultraviolet/infrared dual-band detectors *Infrared physics & technology* **50** 142-8

[5] Ding H, Chen G, Majumdar A K, Sadler B M and Xu Z 2009 Modeling of non-line-of-sight ultraviolet scattering channels for communication *Selected Areas in Communications, IEEE Journal on* **27** 1535-44

[6] Nosaka A Y, Kojima E, Fujiwara T, Yagi H, Akutsu H and Nosaka Y 2003 Photoinduced changes of adsorbed water on a TiO_2 photocatalytic film as studied by 1H NMR spectroscopy *J. of Phys. Chem. B* **107** 12042-4

[7] Grätzel M 2001 Photoelectrochemical cells *Nature* **414** 338-44

[8] Goushcha A and Tabbert B 2007 Optical Detectors *Springer Handbook of Lasers and Optics* 503-62

[9] Hossein-Babaei F, Lajvardi M M and Alaei-Sheini N 2015 The energy barrier at noble metal/TiO_2 junctions *Appl. Phys. Lett.* **106** 083503

[10] Hossein-Babaei F and Rahbarpour S 2011 Titanium and silver contacts on thermally oxidized titanium chip: electrical and gas sensing properties *Solid-State Electronics* **56** 185-90

[11] Hossein-Babaei F and Rahbarpour S 2011 Separate assessment of chemoresistivity and Schottky-type gas sensitivity in M–metal oxide–M′ structures *Sens and Actuators B: Chemical* **160** 174-80

IOP Conf. Series: Materials Science and Engineering **108** (2016) 012031 doi:10.1088/1757-899X/108/1/012031

[12] Schierbaum K D, Kirner U K, Geiger J F and Göpel W 1991 Schottky-barrier and conductivity gas sensors based upon Pd/SnO_2 and Pt/TiO_2 *Sensors and Actuators B: Chemical* **4** 87-94

[13] Jakobi K, 3.1.2.4 Work Function Data, Ed. by G. Chiarotti, *Springer Materials, The Landolt–Börn stein Database* http://www.springermaterials.com DOI:10.1007/10086058_16

[14] Könenkamp R and Rieck I 2000 Electrical properties of schottky diodes on nano-porous TiO_2 films *Materials Science and Engineering: B* **69** 519-21

[15] Tang J, White M, Stucky G D and McFarland E W 2003 Electrochemical fabrication of large-area Au/TiO_2 junctions *Electrochemistry communications* **5** 497-501

[16] Frenkel J 1938 On pre-breakdown phenomena in insulators and electronic semi-conductors. *Physical Review* **54** 647

[17] Stamate M, Vascan I, Lazar I, Lazar G, Caraman I and Caraman M 2005 Optical and surface properties TiO_2 thin films deposited by DC magnetron sputtering method *J. Optoelectronics and Advanced Materials* **7** 771-4

[18] Tomás S A, Luna-Resendis A, Cortés-Cuautli L C and Jacinto D 2009 Optical and morphological characterization of photocatalytic TiO_2 thin films doped with silver *Thin Solid Films* **518** 1337-40

[19] Kong X *et al.* 2009 Metal-semiconductor-metal TiO_2 ultraviolet detectors with Ni electrodes *Appl. Phys. Lett.* **94** 123502

Fast electrochemical actuator

I V Uvarov[1], **A V Postnikov**[1] **and V B Svetovoy**[1,2]

[1] Institute of Physics and Technology, Yaroslavl Branch, Russian Academy of Sciencies, 150007, Yaroslavl, Russia

[2] MESA+ Institute for Nanotechnology, University of Twente, PO 217, 7500 AE Enschede, The Netherlands

E-mail: v.svetovoy@utwente.nl

Abstract. Lack of fast and strong microactuators is a well-recognized problem in MEMS community. Electrochemical actuators can develop high pressure but they are notoriously slow. Water electrolysis produced by short voltage pulses of alternating polarity can overcome the problem of slow gas termination. Here we demonstrate an actuation regime, for which the gas pressure is relaxed just for 10 μs or so. The actuator consists of a microchamber filled with the electrolyte and covered with a flexible membrane. The membrane bends outward when the pressure in the chamber increases. Fast termination of gas and high pressure developed in the chamber are related to a high density of nanobubbles in the chamber. The physical processes happening in the chamber are discussed so as problems that have to be resolved for practical applications of this actuation regime. The actuator can be used as a driving engine for microfluidics.

1. Introduction

The last decennium has witnessed an impressive trend to miniaturize systems of virtually any kind. This trend has many reasons: small systems are often cheaper to produce, they can have properties large systems have not, and they may facilitate use of large systems (cars, for example). An important and generic component in microsystems is the actuator. It plays role of a motor transforming electricity or other kind of energy into mechanical motion. In contrast with large scale systems, where effective engines are available (internal combustion or electromagnetic motors), microsystems suffer from lack of strong and fast actuators [1, 2]. Small electromagnetic motors cannot generate forces of useful magnitude but internal combustion engines [3] perform poorly due to increased heat losses via the volume boundary [4, 5] when the volume decreases.

Existing microactuators are using mostly two types of forces [1, 2, 6, 7]: electrostatic forces, which are weak, and those generated by thermal expansion, which are slow. Fast and strong piezoelectric elements are not compatible with microtechnology, need a high voltage for actuation, and have a small stroke. Electrochemical actuation was also discussed in many papers [8–14] but it is notoriously slow. One can produce a large amount of gas in a short time but it is impossible to get rid of this gas also fast.

Electrochemical decomposition of water is a well-known process but electrolysis performed in microsystems on a short-time scale brought unexpected surprises [15]. It was found that the local current density can be three orders of magnitude larger than that for the normal long-time (> 1 ms) electrolysis. The local relative supersaturations in the short-time regime ($1 - 100$ μs) was larger than 1000. In these conditions homogeneous nucleation of bubbles must happen and

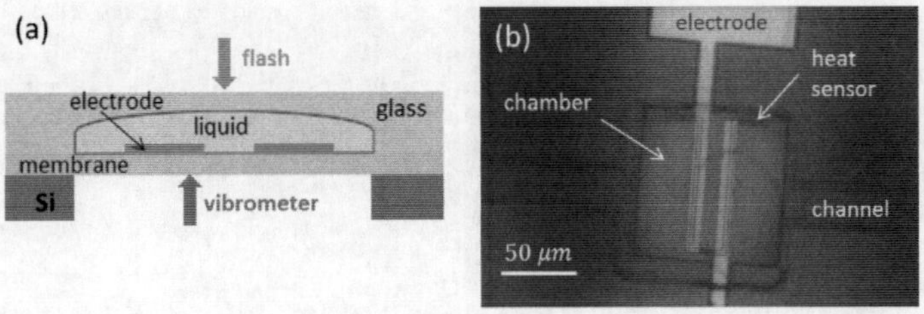

Figure 1. (a) Schematic representation of the chamber. (b) Top view of the actuator. The chamber, filling channels, and electrodes are visible.

was, indeed, observed [16]. Applying potential with the fast change of polarity (> 20kHz) visible production of gas disappeared but the current via the electrolyte practically did not change. A whole number of effects indicated that the gas disappearance is related to the reaction between hydrogen and oxygen that happens spontaneously in nanobubbles with the size smaller than 200 nm.

The phenomenon of gas combustion in nanobubbles puzzled many researches (see, for example, [17]). It is not clear why the reaction is ignited. High Laplace pressure in the bubbles at room temperature is not sufficient to ignite the reaction. Fast microseconds dynamics can be important but taken along it can hardly explain the ignition of the reaction. For the ignition one has to overcome a high energy barrier that seemed not possible at room temperature. Moreover, in contrast with the standard combustion theory [18] the reaction happens without significant temperature increase in the bubbles [15] although deposition of a large amount of energy is well visible. Combustion reactions inside of a small volume is an additional mystery of nanobubbles. The other well-known problem is unexpectedly long stability of surface [19] and bulk [20] nanobubbles.

In spite of poor understanding of the reaction mechanism the discovered phenomenon can be used to build a fast and strong actuator [21]. This actuator consisted of a microchamber covered with a flexible membrane and filled with an electrolyte. The electrolyses was performed by short voltage pulses of alternating polarity. The pressure in the chamber increased and pushed the membrane but no visible bubbles were formed. This is because most of the gas was packed in nanobubbles, which did not scatter light. When the pulses were switched off the pressure relaxed for 100 μs or so. It can be compared with the usual electrochemical actuators, for which the time scale is minutes [8, 9].

In this paper we describe a new actuation regime, for which a high pressure can be reached but it is relaxed just for 10 μs.

2. Experimental

We did experiments with the devices (see Fig. 1) fabricated on Si wafers covered with a layer of silicon nitride (530 nm thick). Platinum electrodes were deposited on top of this layer. Underneath of the electrodes there is a heat sensor made of polysilicon. Within the chamber area the nitride was made free by etching the Si wafer from the back side so that nitride layer played the role of a membrane. The chamber and filling channels were isotropically etched in borofloat glass. The silicon and glass wafers were anodically bonded. Nominal dimensions of the chamber are $100 \times 100 \times 5$ μm^3. The details of the design and fabrication were reported earlier [21]. The chamber was filled via the channel with 1 M solution of Na_2SO_4 in deionized water. The in/outlet openings of the channel were sealed after the filling.

Square voltage pulses of alternating sign were applied to the electrodes at frequencies

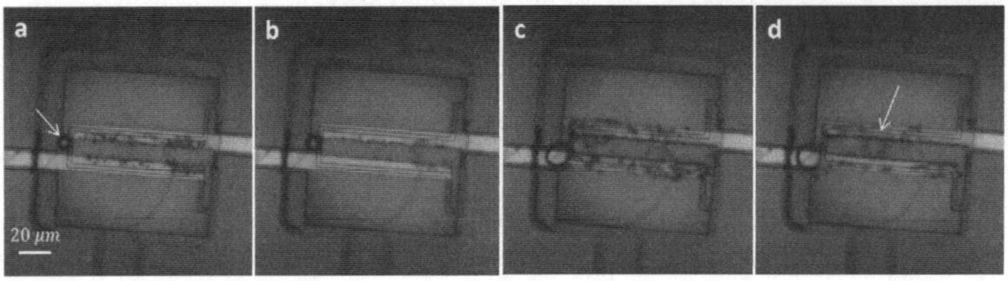

Figure 2. (a) and (b) Stroboscopic snapshots of the chamber at $t = 400\,\mu$s for a frequency of driving pulses $f = 150$ and 200 kHz, respectively. The arrow shows a pinned bubble, which existed before the process started. These images show the visible situation in the chamber during normal actuation. (c) and (d) The same parameters of driving pulses but the images were made at $t = 600\,\mu$s. The arrow in (d) indicates a bubble that appears in the chamber for a very short time. Appearance of the short-lived bubbles is the signal of the new actuation regime.

$f \sim 100$ kHz. To get high currents we use the electrochemical cell in the ohmic regime applying voltage amplitudes above 5 V [21]. Gas formation in the chamber was observed with a homemade stroboscope [22] (10 μs flashlight and a wavelength of $\lambda \approx 530$ nm).

Gas production in the chamber resulted in a pressure rise and a deflection of the membrane. Detailed information on the pressure was collected by observing deflection of the membrane nearby its center with a vibrometer (Polytec MSA-400). The laser beam ($\lambda = 633$ nm, size 1.5 μm) was focused on an opaque spot on the back side of the membrane to prevent possible scattering by occasional microbubbles formed in the chamber. The current and movement of the membrane were recorded in separate channels of the instrument. The membrane deflection d was calibrated by applying a static gas pressure, giving $\Delta P = 2.03d + 0.27d^3$ [21], where ΔP is the overpressure in bars and d is in μm.

3. Results

Normal actuation of the device was described in Ref. [21]. It was demonstrated that the pressure in the chamber can be as high $P = 4.6$ bar and the time for pressure relaxation can be as short as 100 μs. The actuator works well at frequencies $f > 20$ kHz. At high frequencies very few gas is visible in the chamber as one can see in Fig. 2(a) and (b), which correspond to $f = 150$ and 200 kHz, respectively. Both images were made at the time moment $t = 400\,\mu$s and the process run at the voltage amplitude $U = 8$ V. At lower frequencies the amount of visible gas increases and below 20kHz the chamber becomes completely filled with the gas and the actuation becomes impossible. A small amount of visible gas in the chamber exists in the form of small microbubbles located above the electrodes. New actuation regime manifests itself when the process run longer. In this case a faint contrast appears in between the electrodes. For example at $t = 600\,\mu$s it can be seen in Fig. 2(c) and (d), which also correspond to $f = 150$ and 200 kHz, respectively. This contrast resembles rather large microbubbles ($10 - 20\,\mu$m in diameter), which appear out of focus due to motion blur. This conclusion is supported by the observation of the chamber with a fast camera (180 000 frames per second) described in [23], where these bubbles are well visible. They appear in the chamber just for a few microseconds and accompanied by significant pressure jumps in the chamber (see below).

Typical response of the membrane on the electrical pulses with the amplitude $U = 9$ V at frequency $f = 100$ kHz is shown in Fig. 3(a). Well visible oscillations are superimposed on the monotonous increase of the membrane deflection. These oscillations are in phase with the driving pulses. As was explained in [21] they are related to the reaction happening in

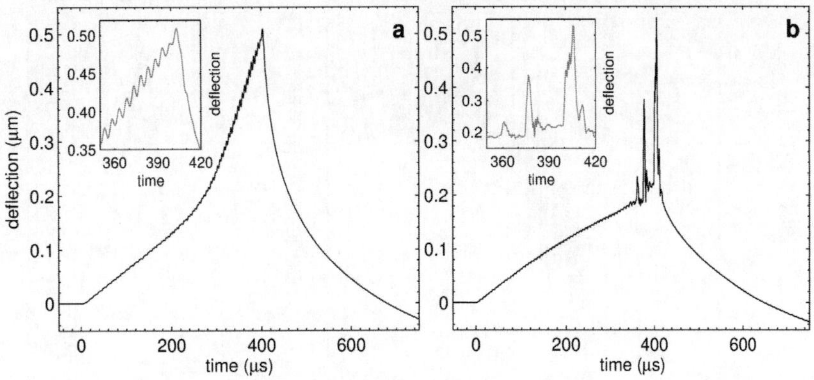

Figure 3. (a) Normal actuation regime. Membrane deflection as a function of time is presented as measured with the vibrometer. The frequency of the driving pulses is $f = 100$ kHz. The inset shows a detailed view around the point of maximal deflection. (b) Transition to the new actuation regime at higher frequency $f = 150$ kHz. High and short peaks are a characteristic feature of this regime.

nanobubbles containing a stoichiometric mixture of H_2 and O_2 gases. The monotonous deflection of the membrane is due unburned gas. For example, if a bubble contain only hydrogen or only oxygen the reaction does not happen and such a bubble will contribute to the pressure increase in the chamber. This gas is also collected in nanobubbles, which do not scatter light and cannot be observed in visible light. Very few bubbles visible in the chamber (see Fig. 2) cannot be responsible for a significant pressure increase. When the driving pulses are switched off the pressure drops down even faster than it was build up. It is explained by merging of non-stoichiometric bubbles with the following reaction between gases. The process is very sensitive to the external or internal temperature increase. Particularly, change of the slope in the membrane deflection in Fig. 3(a) is explained by the heat produced by the reaction of water formation.

If the same process run at higher frequency, for example, at $f = 150$ kHz as in Fig. 3(b), the pressure increases slower but then it jumps high for a very short time. These pressure jumps mark transition to the new regime. As was already explained these jumps are related to the short-lived microbubbles appearing in the chamber. The jumps never appear immediately after switching on the current. There must be an incubation period before the jumps can appear. The incubation time decreases with the increase of the voltage amplitude and decreases strongly even with a small temperature rise ($\sim 10\,°$C). We did not observe the pressure jumps at frequencies below 100 kHz. On the other hand, the amplitude of the jumps increases with the frequency above 100 kHz.

If the process run sufficiently long time, the pressure in the chamber reaches a saturation point on average but continues to fluctuate around this point. This situation is demonstrated in Fig. 4(a) for the amplitude $U = 8$ V and frequency $f = 300$ kHz. A typical pressure jump is shown in Fig. 4(b). One can see that the jumps are very short and highly energetic events. The width of a separate peak is estimated as $3\,\mu$s. The energy released in the pressure jump is $\delta E = \delta P V_{ch} \approx 3.5$nJ, where $\delta P \approx 0.7$bar (as follows from Fig.4(b)) is the amplitude of the jump and $V_{ch} = 5 \times 10^4\,\mu$m^3 is the volume of the chamber. Since the jump is related to the appearance and termination of one short-lived bubble with a size $D \sim 10\,\mu$m, the pressure change inside of this bubble can be estimated as $\Delta P_b = \delta E/V_b \sim 100$ bar, where $V_b = \pi D^2 h/4$ is bubble volume and $h = 5\,\mu$m is the chamber height. The only process which is able to provide this energy scale

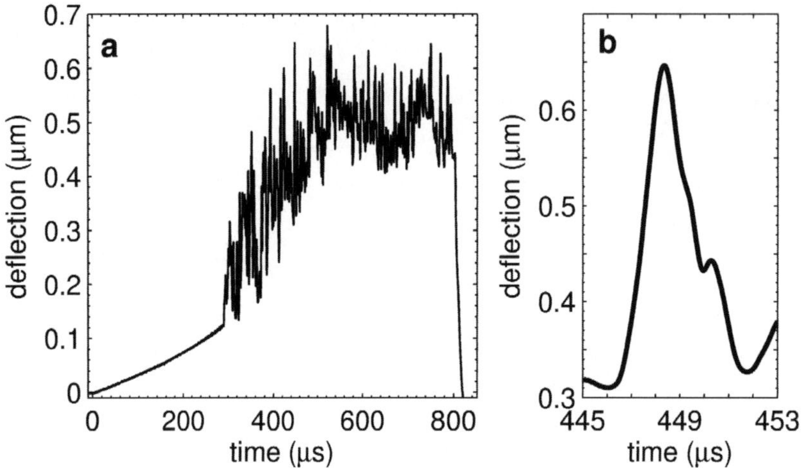

Figure 4. (a) Pressure fluctuations in the saturated regime. (b) Zoomed view of one separate peaks.

is the combustion between hydrogen and oxygen inside of the short-lived bubbles. We expect that these bubbles appear in the chamber due to merging of non-stoichiometric nanobubbles when the density of nanobubbles becomes critical [23]. The formed microbubble contains the stoichiometric mixture of gases, which are able to react. The reason for the ignition of the reaction is still not known. However, there are no doubts that the reaction happens because a huge amount of gas produced by the Faraday current does not show up in the chamber.

The most prominent feature of the new actuation regime is the time for the pressure relaxation when the driving pulses are switched off. As one can see in Fig. 4(a) it happens extremely fast just for $10\,\mu$s or so. Actually the pressure rises much slower than it is going down. This is a very important property for the fast actuation. Figure 5 demonstrates actuation of the membrane by three series of pulses separated by 200 μs pauses. The first series is 400 μs long. It is longer than the other two series, which are 200 μs long each, because of the incubation period. This first example of actuation in the new regime already demonstrates significant achievements. For quite a stiff SiN membrane the stroke is around 0.5μm and overpressure in the chamber is around 1 bar. Termination of the gas is not a problem for this actuation regime and the response time is limited by the production of gas.

4. Discussion

Long incubation period gives a restriction on the fast performance of the actuator. The physical reason for this delay time is that one has to produce a significant concentration of nanobubbles before they start to merge. This incubation time is very sensitive to the temperature and can be easily reduced a few times by increasing temperature on $10-20\,^{\circ}$C. It can be done, for example, with a heating element, which keeps the actuator at elevated temperature.

Fluctuations of pressure in the chamber are not favorable for smooth actuation. This problem can be solved by the proper design of the electrodes. The size of the short-lived microbubbles is related to the distance between the electrodes. Making this distance smaller and distributing electrodes more homogeneously in the chamber we can smooth out the fluctuations. The choice of working frequency is also an important parameter in this respect.

Combustion of hydrogen and oxygen is a highly exothermic process. For this reason the

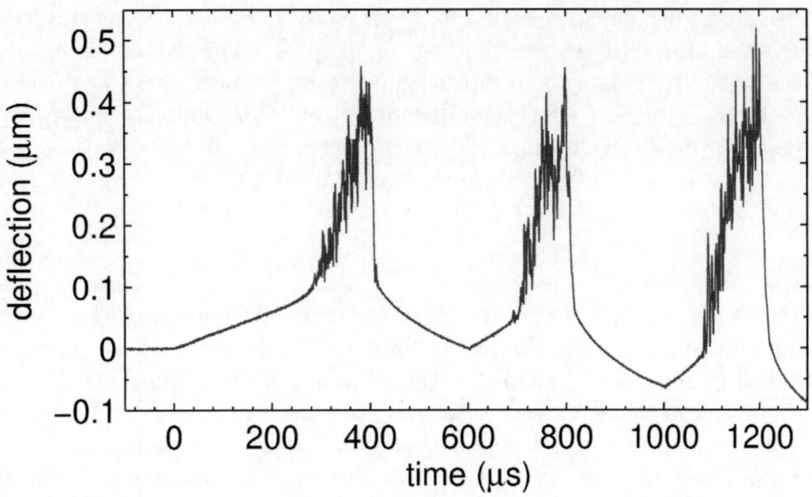

Figure 5. Actuation of the membrane by three series of pulses separated by 200 μs pauses. The first series is 400 μs long and the other two are 200 μs long. The driving pulses have the amplitude $U = 8$ V and frequency $f = 150$ kHz.

reaction between gases in nanobubbles, which are formed in close proximity to the electrodes, influences significantly on the electrodes resulting in their degradation with time. The effect is especially strong at relatively low frequencies $20-50$ kHz [15]. When the combustion happens in microbubbles one could expect faster degradation of the electrodes. However, we did not observe this effect in the new actuation regime. This is because most of the gas reacts in microbubbles located in between the electrodes. In this case highly energetic events of bubble termination do not harm the electrodes. Also, in this regime much higher frequencies are used to get a similar stroke of the membrane. The higher the frequency the smaller stoichiometric nanobubbles are formed resulting in a weaker influence on the electrodes. Nevertheless, a long-time stability of the electrodes was not analysed yet. Material for the electrodes is also an important issue. It was established that electrodes wear correlates with the material's yield strength [15]. The weakest effect was observed for tungsten and the strongest one for gold. Up to now we tested the actuators only with platinum electrodes.

One of the most important actuation parameter is the stroke. Current version of the actuator uses rather stiff silicon nitride membrane. For this membrane with the size 100×100 μm^2 the stroke that has been reached is ~ 1 μm. It can be increased further using softer material for the membrane. A convenient material in this respect is SU8. It has Young's modulus 100 times smaller than that for SiN and its thickness can be easily changed in a wide range to program a desired stroke. The stroke can be also increased due to higher pressure in the chamber. It is possible to do by increasing the Faraday current, which is related to the design of electrodes.

During operation of the actuator one or a few pinned bubbles often appear in the chamber as one can see in all four images in Fig. 2. In most cases pinning happens at the edge of the chamber at structural inhomogeneities. Sometime a pinned bubble can grow during the operation. In this case it becomes a real problem, which prevents normal functioning of the device. For practical application of the actuator one has to understand the reasons for pinning to be able to control it.

Summarising, the fast electrochemical microactuator has a number of very attractive features. It has a large stroke in comparison with its size, it needs rather low actuation voltage, it has a short response time and can develop a high pressure. Finally, the actuator is completely compatible with the standard microtechnological process. The main disadvantage is a low efficiency of the actuator. This is because the electrochemical process in general has the low efficiency. Then, only part of the gas produced electrochemically makes a useful work; a significant part of the gas is burned in stoichiometric nanobubbles without production of the mechanical work. In spite of this disadvantage the total power consumption is still below 100 mW.

5. Conclusions

We presented here a new actuation regime of the electrochemical actuator that is driven by short alternating polarity voltage pulses. Extremely short ($\sim 10\,\mu$s) termination time of the produced gas can be reached in this regime. This fast termination of gas is possible due to a very high density of nanobubbles in the chamber of the actuator. The high density of nanobubbles is also responsible for the high pressure developed by the actuator. The response time is limited by the rate of the gas production but not by the gas termination time as one could expect for an electrochemical actuator. We discussed problems and possible ways to improve performance of the actuator.

Acknowledgments

VBS thanks M. Elwenspoek and G. Krijnen for discussion of microactuators, S. Wildeman and D. Lohse for discussion of physical phenomena in the chamber. This work is financially supported by the Russian Science Foundation, Grant No. 15-19-20003. At earlier stage it was supported by the Dutch Technology Foundation, Grant No. 0785.

References

[1] Abhari F, Jaafar H and Yunus N A M 2012 *Int. J. Electrochem. Sci.* **7** 9765–9780
[2] Ashraf M W, Tayyaba S and Afzulpurkar N 2011 *Int. J. Mol. Sci.* **12** 3648–3704
[3] Maruta K 2011 *Proc. Combust. Inst.* **33** 125–150
[4] Veser G 2001 *Chem. Eng. Sci.* **56** 1265–1273
[5] Fernandez-Pello A C 2002 *Proc. Combust. Inst.* **29** 883–899
[6] Weiss L 2011 *Int. J. Therm. Sci.* **50** 639–647
[7] Volder M D and Reynaerts D 2010 *J. Micromech. Microeng.* **20** 043001
[8] Neagu C R, Gardeniers J G E, Elwenspoek M C and Kelly J J 1996 *J. Microelectromechan. Syst.* **5** 2–9
[9] Cameron C G and Freund M S 2002 *Proc. Natl. Acad. Sci. USA* **99** 7827–7831
[10] Hua S Z, Suchs F, Yang D X and Chopra H D 2002 *Anal. Chem.* **74** 6392–6396
[11] Ateya D A, Shah A A and Hua S Z 2004 *Rev. Sci. Instrum.* **75** 915–920
[12] Meng D D and Kim C J 2008 *Lab Chip* **8** 958–968
[13] Kjeang E, Djilali N and Sinton D 2009 *J. Power Sources* **186** 353–369
[14] Li P Y, Sheybani R, Gutierrez C A, Kuo J T W and Meng E 2010 *J. Microelectromechan. Syst.* **19** 215–228
[15] Svetovoy V B, Sanders R G P, Lammerink T S J and Elwenspoek M C 2011 *Phys. Rev.* E **84** 035302(R)
[16] Svetovoy V B, Sanders R G P and Elwenspoek M C 2013 *J. Phys.: Cond. Matter* **25** 184002
[17] Smart A G 2011 *Physics Today* **64** 17–18
[18] Lewis B and von Elbe G 1987 *Combustion, Flames and Explosions of Gases* (New York: Academic Press)
[19] Zhang X and Lohse D 2014 *Biomicrofluidics* **8** 041301
[20] Ohgaki K, Khanh N Q, Joden Y, Tsuji A and Nakagawa T 2010 *Chem. Eng. Sci.* **65** 1296–1300
[21] Svetovoy V B, Sanders R G P, Ma K and Elwenspoek M C 2014 *Sci. Rep.* **4** 4296
[22] van den Broek D M and Elwenspoek M 2008 *J. Micromech. Microeng.* **18** 064003
[23] Svetovoy V B, Wildeman S, Tas N and Lohse D (in preparation)

IOP Conf. Series: Materials Science and Engineering **108** (2016) 012033 doi:10.1088/1757-899X/108/1/012033

Ag-doped titanium dioxide gas sensor

Navid Alaei Sheini[1,a], Mahsa Rohani[1,b]
[1]Electronic Materials Laboratory, Electrical Engineering Department, K. N. Toosi University of Technology, Tehran 16317-14191, Iran

E-mail: [a]navid_alaei@ee.kntu.ac.ir, [b]mah.rohani@gmail.com

Abstract. Titanium dioxide has been utilized for the fabrication of oxygen sensitive ceramic bodies. In this work, disk-shaped TiO_2 pellets are fabricated by the sintering of the press-formed anatase powder at 1000°C. Two silver contacts are printed on one of the top base of each sample. Silver wire segments are connected to the printed electrodes. It is shown that the gradual diffusion of silver into titanium dioxide from the electrodes profoundly affects the resistive properties of the ceramic samples. SEM, XRD and EDAX analyses are carried out to determine the position of the silver diffused in the structure. At 350°C, before silver diffusion, the electrical resistance of the device decreases ten times in response to the presence of 3000 ppm ethanol contamination. Sensitivity (R_{air}/R_{gas}) to reducing gases is severely affected by the silver doping level in the titanium dioxide. The progress of silver diffusion continuously decreases the sensitivity till it become less than one. Further progress in silver diffusion brings the devices to the condition at which the resistance increases at the presents of reducing gases. In this condition, inverse sensitivities (R_{gas}/R_{air}) as large as 10^3 are demonstrated.
Keywords: Titanium dioxide, Gas sensor, Silver diffusion, Sensitivity, Reducing gas.

1. Introduction

Seiyama found gas sensitivity on metal oxides in 1962, and Taguchi brought metal oxide semiconductor-based gas sensors to the market [1,2]. ZnO, TiO_2 and SnO_2 are the most widely studied metal oxide semiconductors for gas sensing applications [3]. They are all wide-band-gap semiconductors whose charge carrier populations are determined either by doping with aliovalent cations [4] or oxygen nonstoichiometry [5]. In undoped metal oxide materials, oxygen deficient crystals present n-type semiconductivity since the formation of oxygen vacancies accompanies electron generation. In contrast, the p-type semiconductivity can be explained by the deficiency of metal ions in the material [6]. The effect of oxygen vacancy-related defects on gas-sensing properties of n-type metal oxide gas sensors has previously been investigated [7]. In n-type polycrystalline metal oxides the sensing mechanism mainly is related to the change in the barriers which formes between grains by imperfections and oxygen adsorption at working temperature [8]. Moreover, Schottky-type metal contacts can play important role in the sensing mechanism and in comparison to resistive gas sensors, Schottky types can potentially be more sensitive [9]. Although resistive gas sensors are popular owing to their reasonable prices and good durability [10], their selectivity and sensitivity are limited. These deficiencies are the main driving force for developing new materials and methods for superior gas detectors [10]. There are numerous papers focused on optimizing parameters of metal oxides to enhance these properties by controlling the size and the shape of the gas sensor's oxide grains [11] or introducing catalytically active additives in the sensing layer via doping or surface functionalization [12, 13]. In addition, the selectivity of these sensors have been improved by external methods like thermal modulation [14] and connection to microfluidic channels [15].

Titanium dioxide is known as n-type wide bandgap semiconductor, with 3.1 eV for rutile and 3.3 eV for anatase [16]. There are several methods to produce TiO_2 such as thermal oxidation [17], hydrothermal [18], and sol-gel [19]. Proper phase, morphology, grain size and porosity can be achieved by controlling the fabrication parameters in each method [20]. Both rutile and anatase phases of TiO_2 have been used for gas sensor [9, 17], UV detector [21] and solarcell fabrication [22]. In most of these cases, many noble metal-TiO_2 contacts are utilized for connecting them to the outside circuitry [23]. Besides, it is found that TiO_2 doped with noble metals, such as Ag, Au, Pt, and Pd can enhance device properties. Ag presents numerous interesting properties, such as high efficiency, low cost compared to other noble metals, and high oxygen adsorption reactivity. TiO_2 thin films doped with Ag have shown many attractive characteristics [24]. Moreover, Ag metallic ion is known as a high mobility cation, which has been used in resistive switching devices [25].

In the present study, we report on the electronic properties of the Ag/TiO_2/Ag structure and using the device as gas sensor. We show that silver diffusion into the TiO_2 layer, plays a significant role in determining sensor sensitivity. By gradual silver diffusion, the sensitivity of the device increases slightly, but with further progress of silver diffusion, the sensing mechanism changes and the sample resistance increases at the presence of the reducing gas. The change in the sensing mechanism is describe by TiO_2 acting like a p-type semiconductor when Ag particles present in the open structure of the polycrystalline TiO_2.

2. Experimental

TiO_2 pellets with 5 millimeters in diameter are prepared by addition of 2 milliliters of distilled water into 5 grams of anatase TiO_2 powder. The samples are dried and prepared for sintering at 1000°C. The grains grow during sintering, and this phenomena lead to larger grains which, evidently, reduce the population of the smaller pores. In this process, samples are placed in an electric muffle furnace at definite temperature for 60 minutes. The atmosphere of the furnace is air.

Samples are fabricated by silver electrode deposition on the thermally oxidized titanium foil chips. As shown in figure 1, a microheater is attached beneath the substrate and provides elevated temperatures up to 500°C. Large area ohmic connections to Ag/TiO_2 structure were prepared by silver paste printing. After settling and hardening of the silver paste, the samples are placed at 300°C for 15 minutes.

As shown in figure 2, the measurement system comprises of a signal generator to provide the input ac voltage, Sanwa PC-5000 multimeter for continuous monitoring of the sensor output, and a fine type-S thermocouple placed on the sample surface to controls the operating temperature of the sensor. For measuring the sample resistance, a sinusoidal voltage with frequency of 80 Hertz is applied to the sample and a series resistor is used for monitoring the output voltage.

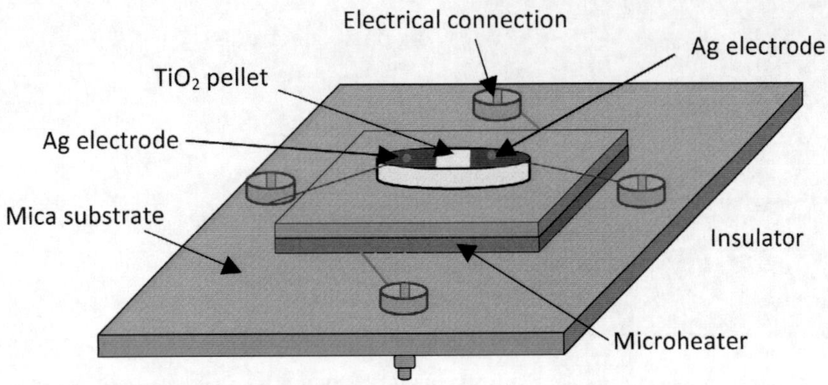

Figure 1. The schematic diagrams fabricated device comprising sensing oxide, silver electrodes on the top surface, and the microheater.

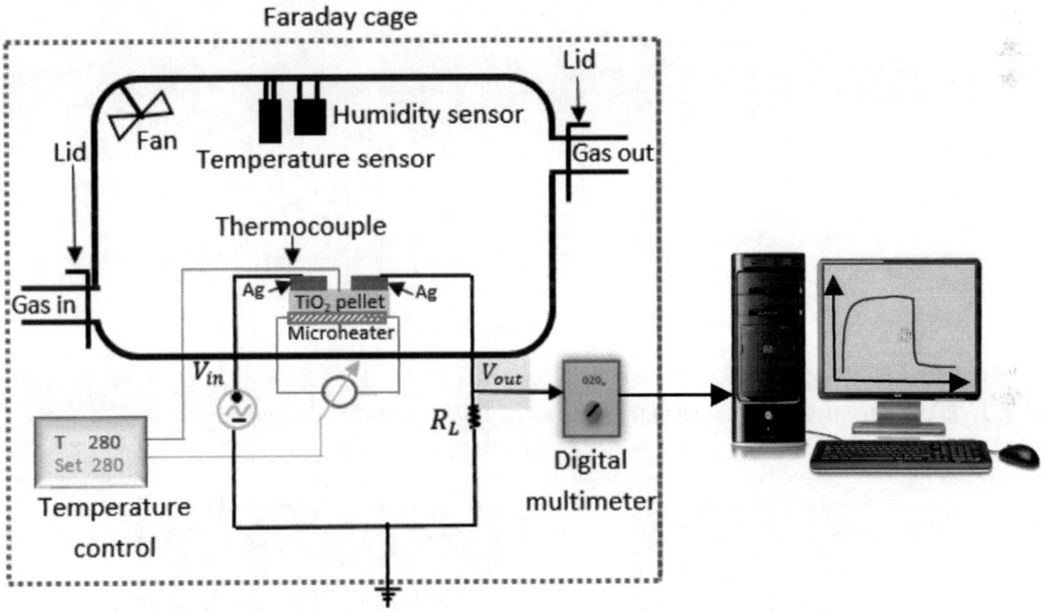

Figure 2. The schematic diagram of the experimental setup used for recording sample resistance at presence of gas contaminant.

Silver paste electrodes are connected to external circuit with silver wire segments (figure 3). XRD pattern shows that the TiO_2 samples sintered at 1000°C for 60 minutes are totally of anatase phase.
The transient response of the sample to the presence of 1000 ppm ethanol contamination at 320°C is a 6-fold change in conductance, as shown in figure 4. The transient response of pure TiO_2 shown in figure 4 is consistent with the TiO_2 disk produced being an n-type semiconductor. The increase in conductivity of pure TiO_2 at the presence of reducing gases is due to the increased oxygen vacancy and decreased potential barriers established at the grainboundaries.

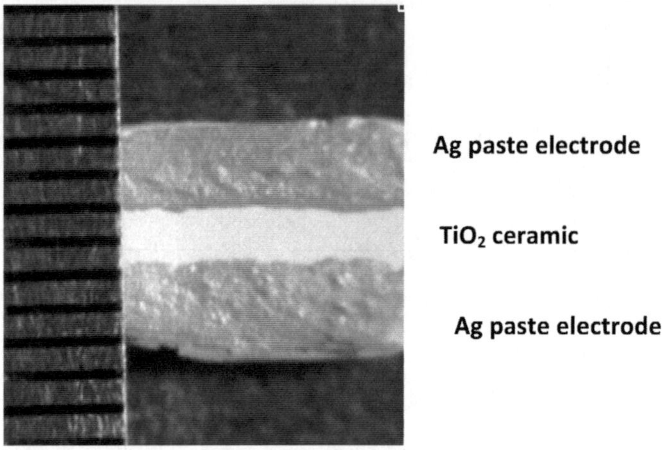

Figure 3. Optical microscope image of the pure titanium dioxide with two parallel silver electrodes deposited on it.

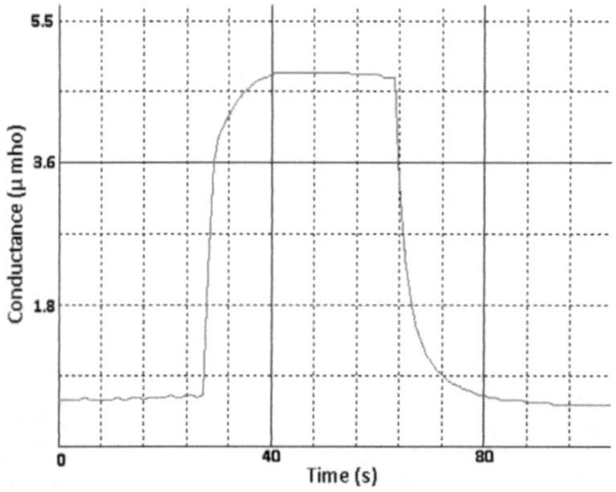

Figure 4. Transient response of the sample to the presence of 1000 ppm ethanol contamination at 320°C.

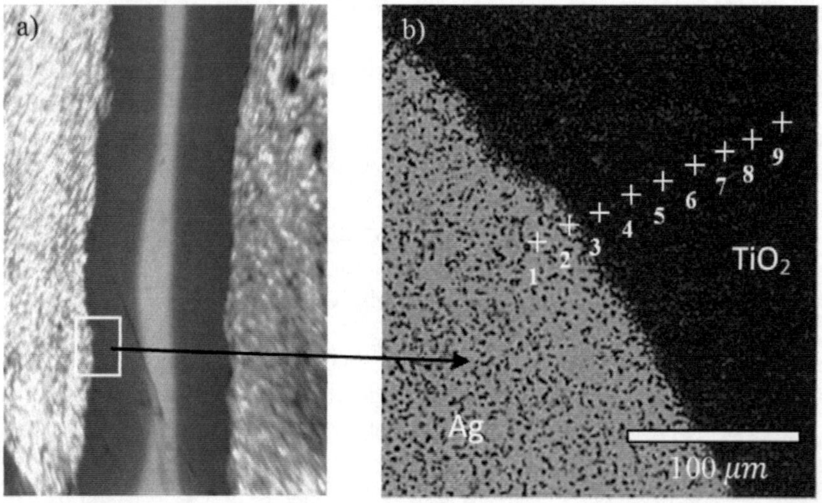

Figure 5. Micrographs illustrating silver diffusion from the silver electrode into the titanium dioxide: a) the silver halo created around the silver electrodes by silver diffusion, b) SEM image of the area selected in (a).

Silver diffusion into TiO_2 body is investigated at elevated temperatures up to 400°C. The established electrical field helps silver diffusion. Microscopic images of an Ag-TiO_2 contact after silver diffusion are given in figure 5.

3. Results and discussion

The concentration of silver in points 2 to 9 shown in figure 5 is analyzed by EDAX surface analyzer. The results are plotted vs. distance in figure 6.

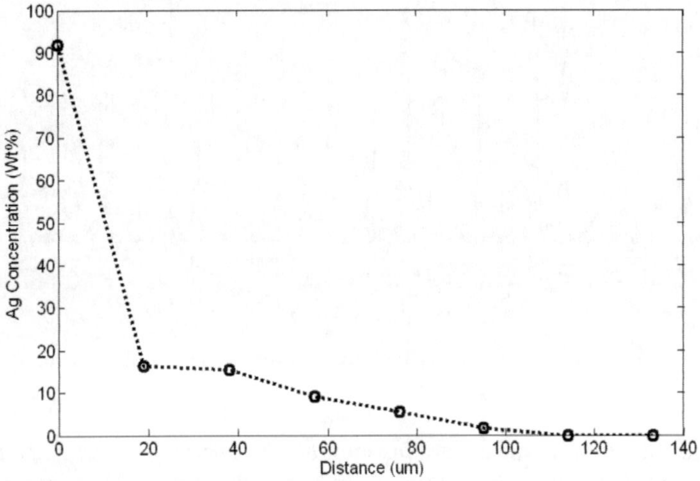

Figure 6. Silver concentration in points marked in figure 5-b.

The XRD analysis is carried out on the TiO_2 body after sintering. The pure TiO_2 body is provided and sintered in 1000°C for 60 minutes. The surface of a sintered body is completely covered with the silver paste. It is, then, placed in an electric muffle furnace and annealed for 30 minutes at 500°C. After extraction from the furnace, silver layer is removed from the surface of the sample body. XRD was, subsequently, carried out on this sample. As shown in figure 7, in addition to picks of anatase phase, there are 2 little picks related to rutile phase. With respect to this reality, pure TiO_2 is in anatase phase, it is evident from patterns that presence of Ag in pure TiO_2 even in 500°C converts anatase phase to rutile.

The XRD pattern in figure 7 shows that our TiO_2 sample is a mixture of anatase, and rutile phases. Ag and Ag_2O are present in the sample and there is no sign of possible alloy phases between AgO and TiO_2. Although the XRD analysis system used, claims 5% accuracy, it appears that Ag has not diffused into the TiO_2 crystal structure significantly. The silver particles appear to be mostly metallic residing within the pores and grainboundaries.

Silver diffusion into TiO_2 body severely affects the conductance and response of the sample to the presence of reducing gases. Experiment of silver diffusion from contacts is carried out on several samples and all of them have the same pattern in changing conductance and respond to reducing gas.

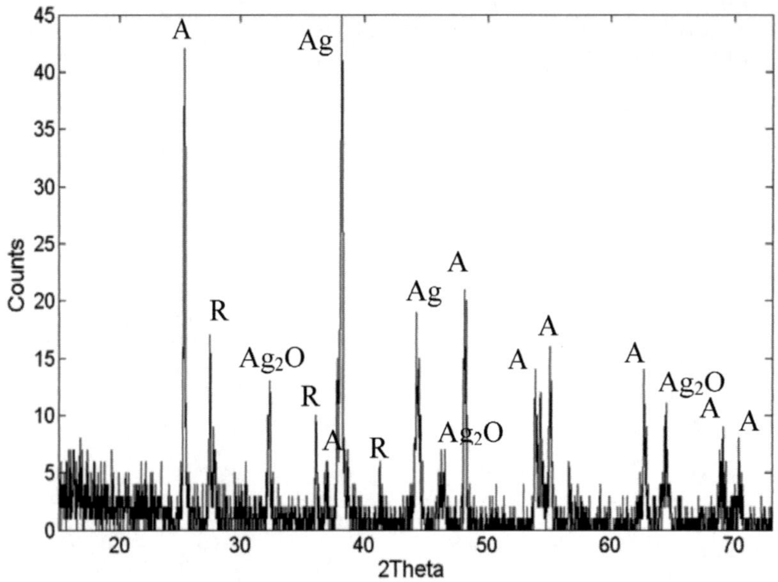

Figure 7. XRD pattern for titanium dioxide after silver diffusion shows peaks of anatase, rutile, Ag and Ag_2O.

4. Conclusion

The Effect of silver diffusion from the electrodes into the TiO_2 pellet of a gas sensor on the responses of the gas sensing properties of the device was experimentally investigated. It was concluded that the sensitivity can increase significantly by diffusing silver to the titanium dioxide body. Before silver diffusion, sensor's response to 3000 ppm ethanol contamination was conductance increase of 10-fold, but after silver diffusion the sensor showed conductance decrease as response to ethanol presence. Silver diffusion into the TiO_2 ceramic body and the sensitivity changes before and after silver diffusion at different conditions requires further studies. The attractions of Ag-TiO_2-Ag structure as a potentially high sensitivity and low cost gas sensing device are clear and make it a good candidate for more investigations.

References

[1] Seiyama, T., and A. Kato. "A new detector for gaseous components using semiconductor thin film." *Anal. Chem* 34.11 (1962): 1502-1503.More references

[2] Simon, Isolde, et al. "Micromachined metal oxide gas sensors: opportunities to improve sensor performance." *Sensors and Actuators B: Chemical* 73.1 (2001): 1-26.

[3] Hossein-Babaei, F. "Porous silver-TiO_2 Schottky-type chemical sensor fabricated on thermally oxidised titanium." *Electronics letters* 44.2 (2008): 161-162.

[4] Szczuko, D., et al. "XPS investigations of surface segregation of doping elements in SnO_2." *Applied Surface Science* 179.1 (2001): 301-306.

[5] Kılıç, Çetin, and Alex Zunger. "Origins of coexistence of conductivity and transparency in SnO_2." *Physical Review Letters* 88.9 (2002): 095501.

[6] Kim, Hyo-Joong, and Jong-Heun Lee. "Highly sensitive and selective gas sensors using p-type oxide semiconductors: Overview." *Sensors and Actuators B: Chemical* 192 (2014): 607-627.

[7] Ahn, M-W., et al. "Gas sensing properties of defect-controlled ZnO-nanowire gas sensor." *Applied physics letters* 93.26 (2008): 263103.

[8] Barsan, Nicolae, and Udo Weimar. "Conduction model of metal oxide gas sensors." *Journal of Electroceramics* 7.3 (2001): 143-167.

[9] Hossein-Babaei, F., and S. Rahbarpour. "Separate assessment of chemoresistivity and Schottky-

type gas sensitivity in M–metal oxide–M′ structures." *Sensors and Actuators B: Chemical* 160.1 (2011): 174-180.

[10] Smulko, Janusz, et al. "New approaches for improving selectivity and sensitivity of resistive gas sensors: A review." *8th International Conference on Sensing Technology (ICST 2014), Liverpool, UK, 2-4 September*. 2014

[11] Franke, Marion E., Tobias J. Koplin, and Ulrich Simon. "Metal and metal oxide nanoparticles in chemiresistors: does the nanoscale matter?." *Small* 2.1 (2006): 36-50.

[12] Hyodo, Takeo, et al. "Gas-sensing properties of ordered mesoporous SnO_2 and effects of coatings thereof." *Sensors and Actuators B: Chemical* 93.1 (2003): 590-600.

[13] Velmathi, Guruviah, Sanganeni Mohan, and Rabinder Henry. "Analysis of Factors for Improving Functionality of Tin Oxide Gas Sensor." *IETE Technical Review* ahead-of-print (2015): 1-8.

[14] Hossein-Babaei, Faramarz, and Amir Amini. "A breakthrough in gas diagnosis with a temperature-modulated generic metal oxide gas sensor." *Sensors and Actuators B: Chemical* 166 (2012): 419-425.

[15] Hossein-Babaei, Faramarz, and Vahid Ghafarinia. "Gas analysis by monitoring molecular diffusion in a microfluidic channel." *Analytical chemistry* 82.19 (2010): 8349-8355.

[16] Linsebigler, Amy L., Guangquan Lu, and John T. Yates Jr. "Photocatalysis on TiO_2 surfaces: principles, mechanisms, and selected results." *Chemical reviews* 95.3 (1995): 735-758.

[17] Hossein-Babaei, F., and S. Rahbarpour. "Titanium and silver contacts on thermally oxidized titanium chip: electrical and gas sensing properties." *Solid-State Electronics* 56.1 (2011): 185-190.

[18] Pasang, T., et al. "Tuning of band gap in TiO_2 and ZnO nanoparticles by selective doping for photocatalytic applications." *Materials Research Innovations* 19.1 (2015): 73-80.

[19] Miao, Zheng, et al. "Electrochemically induced sol-gel preparation of single-crystalline TiO_2 nanowires." *Nano Letters* 2.7 (2002): 717-720.

[20] Ogale, Satishchandra Balkrishna, T. Venky Venkatesan, and Mark Blamire, eds. *Functional Metal Oxides: New Science and Novel Applications*. John Wiley & Sons, 2013.

[21] Hossein-Babaei, Faramarz, Mehdi Mohamadzade Lajvardi, and Farhad Akbari Boroumand. "Large area Ag–TiO_2 UV radiation sensor fabricated on a thermally oxidized titanium chip." *Sensors and Actuators A: Physical* 173.1 (2012): 116-121.

[22] McFarland, Eric W., and Jing Tang. "A photovoltaic device structure based on internal electron emission." *Nature* 421.6923 (2003): 616-618.

[23] Hossein-Babaei, F., Mehdi M. Lajvardi, and Navid Alaei-Sheini. "The energy barrier at noble metal/TiO_2 junctions." *Applied Physics Letters* 106.8 (2015): 083503.

[24] Bensouici, F., et al. "Synthesis, characterization and photocatalytic behavior of Ag doped TiO_2 thin film." *Superlattices and Microstructures* (2015).

[25] Waser, Rainer, et al. "Redox-based resistive switching memories–nanoionic mechanisms, prospects, and challenges." *Advanced Materials* 21 (2009): 2632-2663.

Physico-chemical mechanism for the vapors sensitivity of photoluminescent InP quantum dots

P. Prosposito[a1], R. De Angelis[a], F. De Matteis[a], F. Hatami[b], W.T. Masselink[b], H. Zhang[c], M. Casalboni[a]

[a] Industrial Engineering Department and INSTM, University of Rome "Tor Vergata", via del Politecnico 1, 00133 Rome, Italy.

[b] Department of Physics, Humboldt University, Berlin, Newtonstr. 15, D-12489, Germany.

[c] Van't Hoff Institute for Molecular Sciences, University of Amsterdam, Science Park 904, 1098 XH Amsterdam, The Netherlands.

E-mail: prospositooma2.infn.it

Abstract. InP/InGaP surface quantum dots are interesting materials for optical chemical sensors since they present an intense emission at room temperature, whose intensity changes rapidly and reversibly depending on the composition of the environmental atmosphere. We present here their emission properties by time resolved photoluminescence spectroscopy investigation and we discuss the physico-chemical mechanism behind their sensitivity to the surrounding atmosphere. Photoluminescence transients in inert atmosphere (N_2) and in solvent vapours of methanol, clorophorm, acetone and water were measured. The presence of vapors of clorophorm, acetone and water showed a very weak effect on the transient times, while an increase of up to 15% of the decay time was observed for methanol vapour exposure. On the basis of the vapor molecule nature (polarity, proticity, steric hindrance, etc.) and of the interaction of the vapor molecules with the quantum dots surface a sensing mechanism involving quantum dots non-radiative surface states is proposed.

1. Introduction

The optical properties of many semiconductor quantum dots (QDs) such as broad absorption band and size-dependent spectral emission, together with a high effective surface area available for interaction with target chemicals are attractive for applications as chem-/bio-sensors and lab-on-chips[1-3]. So far, most of the reports of optical chemical sensing with QDs are based on colloidal quantum dots, however recently a new approach has been introduced which is based on uncapped Surface QDs (SQDs) grown by epitaxy on solid substrates[4,5]. In these cases it is possible to achieve sufficient quantum confinement and radiative emission in QDs even without a capping layer[6,7]. In this case a direct contact with the external environment takes place and the short-range interaction between dots and environment allows their exploitation as sensor (e.g. for organic molecules).

A strong sensitivity to the surface environment has been reported for SQDs made of InAs[8], InGaAs[4] and InP[5]. These QDs are very attractive for chem-/bio-sensor and lab-on-chip design since they can be grown on a conventional wafer. On this regard, InP SQDs have some key advantages. Their emission

[1] To whom any correspondence should be addressed.

intensity is only 2-5 times lower than buried ones[6], which is quite outstanding compared to analogous systems where the intensity for uncapped QDs is usually orders of magnitude lower than for capped QDs[9]. Moreover, their near infrared emission, matching fibre optics and biological tissue transparency windows, make them suitable for a wide range of applications.

Recently, we characterized the methanol sensitivity of InP SQDs by continuous wave photoluminescence (CW-PL) spectroscopy[5,10]. We observed a rapid and reversible increase in the PL intensity, proportional to the methanol concentration. However, a model of the sensing mechanism is still lacking and would be beneficial in order to fine tuning the property of the dots and the whole semiconductor heterostructure to achieve methanol sensing with resolution suitable for practical application.

In this work, we report on the sensing mechanism investigated by time resolved photoluminescence (TR-PL) spectroscopy. We compare the response of the InP SQDs in nitrogen atmosphere to that in vapor of methanol, water, acetone and chloroform solvents. We found that the decay transients of the PL in nitrogen environment can be well fitted by a three exponential decay function having characteristic times of 0.5, 2.0 and 7.0 ns. When the gas environment is changed introducing fixed amount of solvent vapour an increase of the fast and intermediate times was observed. Such an effect is particularly evident for methanol where a 15% enhancement of the fast decay time has been measured. We discuss here a physico-chemical model based on the saturation effect of the QD non-radiative surface states by the vapor organic molecules which is fully compatible with the observed experimental behaviour.

2. Experimental

Self-assembled epitaxial $InP/In_{0.48}Ga_{0.52}P$ SQDs were grown by gas source molecular beam epitaxy on (100) GaAs substrate. The sample synthesis is described in detail elsewhere[5,10]. Structural characterization was accomplished with atomic force microscopy. TR-PL spectra were recorded using a Ti:Sapphire laser Chameleon ULTRA-II (Coherent) equipped with an optical parametric oscillator tuned at 488 nm. The emitted light was analyzed with a Time-Correlated Single Photon Counting technique for the lifetimes measurements and with a spectrograph equipped with a CCD for the fluorescence surface images. Measurements were performed at 760 Torr and 25 °C in a sealed flow cell with an optical window where the atmosphere can be changed from pure N_2 to a mixture of solvent vapor and N_2. Such mixtures were produced by bubbling the solvent in a Drechsel bottle to obtain the saturated vapor of the solvent with N_2 in a controlled amount. The solvent concentration was estimated with an accuracy of 5%. The gas mixture was injected into the cell and the TR-PL measurements were taken in static atmosphere. The initial condition was restored flowing pure N_2 for 5 min. The liquid solvents used to produce the vapors were Sigma-Aldrich analytical reagent grade methanol (MeOH) 99.8%, chloroform (CF) 99.8% and acetone (Ac) 99.5% used without further purification.

3. Results and discussion

Figure 1 shows an AFM micrograph of the sample. The investigated SQDs have lens shaped with a lateral size of about 20 - 30 nm and a height in the range of 2 - 6 nm. The dot density is 9×10^{10} dots/cm^2 and by a statistical analysis of the size and the shape of about 30 individual dots measured from a 600×600 nm^2 AFM micrograph we observed a bimodal Gaussian distribution for the QD heights, with peak positions of 2.8 and 4.3 nm.

In previous studies[5,10] we found that the composition of the surrounding gaseous environment strongly affects the PL intensity of InP SQDs. In particular, we observed that the exposure of samples to methanol vapor results in an enhancement of the PL intensity while spectral shape and position remain unchanged, as shown in the inset of figure 2. Such findings pointed out the need of a model of the sensing mechanism that at present is still lacking. On this regard we performed TR-PL measurement with and without vapors. Figure 2 shows the PL transients measured at 1.47 eV (the peak of the emission spectrum) in N_2 and methanol, and its initial response function (IRF). Taking into account the IRF, the transients can be well fitted with a 3-exponential decay law with an initial fast (τ_F), an intermediate (τ_I) and a slower (τ_L) decay time of about 0.5 ns, 2 ns and 7 ns, respectively.

Figure 1. AFM image of InP SQDs.

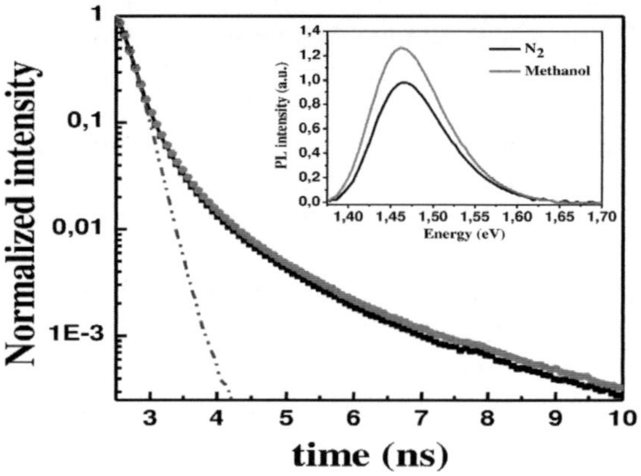

Figure 2. PL transients, red curves are related to methanol vapor while black ones are in inert atmosphere of nitrogen. Blue dashed line is the IRF. Inset shows the InP SQDs CW emission.

The fast and intermediate times are due to the emission of confined states in QDs. For InP SQDs with large lateral size the main parameter responsible for the emission energy and radiative lifetime is the dot height. Likely, the presence of two different radiative lifetimes is a direct consequence of the bimodal height distribution in this particular sample. The longest component can be attributed to QDs coupling. A recent theoretical investigation showed the relevance of InP SQDs lateral coupling in determining emission energy and optical spectra[11]. According to this study, the appearance of molecular terms in the optical spectra are likely to occur in our samples. When the sample is exposed to methanol vapor the initial and middle components of the PL transient lengthen while for the long component we did not observe any significant change. To test the sensitivity and selectivity of the InP SQDs we performed a series of experiments with other common solvents, namely chloroform, acetone and water with fixed concentration of 2×10^4 ppm. The results are summarized in table I.

Table I. InP SQDs lifetimes with N_2, methanol (MeOH), chloroform (CF), acetone (Ac) and water (H_2O) solvents.

Environment	Pure N_2	MeOH	CF	Ac	H_2O
τ_F (ns)	0.53	0.61	0.56	0.56	0.56
τ_I (ns)	1.83	2.03	1.93	1.94	1.91
τ_L (ns)	7.41	7.40	7.44	7.65	7.54

We evaluated the lifetime relative change induced by the presence of the vapor solvents as $\Delta\tau_i/\tau_i\% = (\tau_{solvent}-\tau_{N2})/\tau_{N2}\ \%$, where i=F,I,L indicates the fast, intermediate and long decay times, respectively. $\tau_{solvent}$ represents the decay time when the sample is exposed to solvent vapor and τ_{N2} is the decay time in pure N_2. Methanol presents a consistent stretching of the decay times with values of 15% and 11% for τ_F and τ_I, respectively. The other solvents produce only a slight increase of both τ_F and τ_I of approximately few percents irrespective of the solvent type.

The time resolved photoluminescence results strongly support the sensing mechanism model we previously hypothesized[5]. The target molecules present in the gas environment, passivating the non-radiative decay channels related to the QDs surface states, produce at the same time an enhancement of the PL signal in the steady state and an increase in the luminescence decay time. In our case the surface states are mostly represented by the dangling bonds that could be both positive In^+ and negative P^- ions. In particular, we believe that the involved defect type are in prevalence negatively charged (donor). Indeed, the molecules employed for the experiments present differences regarding the polarity, charge distribution and steric hindrance. Chloroform is a solvent with a weak polarity, its interaction with defects may occur by weak London dispersion force and therefore its effectiveness in the passivation of the radiative pathway is limited. Acetone is a polar aprotic species having a positive charge distribution rather spread on the carbon moiety with respect to the negative one (around oxygen) so it preferentially interacts with positive defects. Since it has a very limited effect on our system we suggest that the involved defects are not positively charged. Moreover, its low response can be ascribed both to its low polarity and its large steric hindrance. Methanol produces the most noticeable effect and it is a polar and protic solvent. In this case the positive charge is rather localized on the hydrogen of the OH group which tends to interact with negatively charged species on the surface. On the other side, using water which is a very polar and protic molecule the effect we measured is very low. To explain such apparent inconsistency, a more complex representation has to be taken into account where the solvent-solvent intermolecular interactions play an important role. Water molecules tend to interact strongly with each other forming a complex network of hydrogen bonds that interacts with the QD surface in a less effective way with respect to methanol that instead tends to be adsorbed on the surface as a monolayer. This could explain its higher effectiveness in luminescence intensity enhancing.

4.Conclusion

We studied the chemical sensitivity of InP SQDs by TR-PL spectroscopy. PL transients with and without vapors were fitted with a 3-exponential decay law. A very weak effect was found for acetone, chloroform and water while a consistent increment of the decay times was measured for methanol. This result suggests a sensing mechanism based on the ability of the organic molecules to saturate non-radiative QDs surface states. The model we propose can satisfactorily account for both the observed experimental effects since by hindering the non-radiative channels a more intense CW-luminescence signal and a longer lifetime have been found.

Acknowledgments

We acknowledge the financial support of LASERLAB-EUROPE (no. 284464, EC's VIIth F. P.) and the “Uncovering Excellence 2014” program of the University of Rome Tor Vergata.

References

[1] Medintz I L, Uyeda H T, Goldman E R, Mattoussi H 2005 *Nature Mater.* **4** 435
[2] Costa-Fernández J M, Pereiro R, Sanz-Medel A 2006 *Trends Anal. Chem.* **25** 207
[3] Frasco M F, Chaniotakis N 2009 *Sensors* **9** 7266
[4] Milla M J, Ulloa J M, Guzmán A 2012 *Appl. Phys. Lett.* **100** 131601-1
[5] De Angelis R, Casalboni M, Hatami F, Ugur A, Masselink W T, Prosposito P 2012 *Sens. Act. B* **162** 149
[6] Hestroffer K, Braun R, Ugur A, Tomm J W, Röder B, Hatami F 2013 *J. Appl. Phys.* **114** 163510-5
[7] Ellstrëm C, Trägårdh J, Samuelson L, Seifert W, Pistol M-E, Lemeshko S, Pryor C, *Appl. Phys. Lett.* 2006 **89** 033111-1
[8] Lin A, Liang B L, Dorogan V G, Mazur Y I, Tarasov G G, Salamo G J, Huffaker D L 2013 *Nanotechnolgy* **24** 075701-1
[9] Wu H-B, Xu S J, Wang J 2006 *Phys. Rev. B* **74** 205329-1
[10] De Angelis R, D'Amico L, Casalboni M, Hatami F, Masselink W T, Prosposito P 2013 *Sens. Act. B* **189** 113
[11] Barettin D, Auf der Maur M, De Angelis R, Prosposito P, Casalboni M, Pecchia A 2015 *J. Appl. Phys.* **117** 094306-1

IOP Conf. Series: Materials Science and Engineering **108** (2016) 012035 doi:10.1088/1757-899X/108/1/012035

Topology optimization of 3D compliant actuators by a sequential element rejection and admission method

R Ansola[1,3], E Veguería[1], C Alonso[1] and O M Querin[2]
[1] Department of Mechanical Engineering, University of the Basque Country, Alda. Urquijo s/n, 48013 Bilbao, Spain.
[2] School of Mechanical Engineering, University of Leeds, Leeds LS2 9JT, UK.

E-mail: ruben.ansola@ehu.es

Abstract. This work presents a sequential element rejection and admission (SERA) method for optimum topology design of three dimensional compliant actuators. The proposed procedure has been successfully applied to several topology optimization problems, but most investigations for compliant devices design have been focused on planar systems. This investigation aims to progress on this line, where a generalization of the method for three dimensional topology optimization is explored. The methodology described in this work is useful for the synthesis of high performance flexure based micro and nano manipulation applications demanding for both sensing and control of motion and force trajectories. In this case the goal of the topology optimization problem is to design an actuator that transfers work from the input point to the output port in a structurally efficient way. Here we will use the classical formulation where the displacement performed on a work piece modelled by a spring is maximized. The technique implemented works with two separate criteria for the rejection and admission of elements to efficiently achieve the optimum design and overcomes problems encountered by other evolutionary methods when dealing with compliant mechanisms design. The use of the algorithm is demonstrated through several numerical examples.

1. Introduction

Compliant actuators obtain their mobility from flexibility of their parts as opposed to classic rigid body mechanisms. Therefore they can be built using fewer parts, require fewer assembly processes and need no lubrication. An important application of compliant mechanisms lies in Micro Electro Mechanical Systems (MEMS) design, where due to the small size, hinges and bearings cannot be used. As a result these types of devices must be built and designed as compliant actuators etched out of a single piece of material. The design of compliant mechanisms was initially accomplished by trial and error. However, the idea of introducing more systematic design procedures captured the mind of researchers [1].

Two different design approaches were considered: lumped and distributed compliant mechanisms. In the first approach [2], rigid body mechanisms were converted into partially compliant mechanisms composed of small flexible pivots and rigid links. In the second approach, distributed compliant mechanisms were obtained with the use of topology optimization techniques, where optimum designs were automatically obtained for prescribed design domains, boundary conditions and functional

[3] To whom any correspondence should be addressed.

specifications. The main advantage of this approach was that there was no need to predetermine the number of links or the location of the flexural joints [3]. Since then, topology optimization has been successfully applied to optimize compliant sensors and actuators in many practical engineering designs with the use of finite element analysis, since this technique enables systematic design directly from the behavioural specifications. The structural topology design problem is formulated as a material distribution problem within a given design domain, where material should be placed and connected to some portions of the boundary with some number of holes inside to optimize an objective function.

Although the design goals for structures and compliant actuators are different, the same topology optimization methods can be adapted to design both types of elements, introducing some modifications to specifically suit the functional requirement of compliant devices. We should bear in mind that adequate flexibility is deemed essential to afford the required displacement of the actuator at the point of interest. Additionally, a compliant mechanism also needs to be stiff enough to be able to sustain external loads. Therefore, an optimum balance between the two requirements of flexibility and stiffness is essential in the synthesis of compliant actuators.

The pioneering topology optimization method used to design compliant mechanisms was the homogenization method [4]. This technique considered a material model with infinite number of microscopic cells containing varying degrees of solid and void. The optimum topology was obtained by seeking the optimal porosity of the domain. Although this method pioneered the design of compliant mechanisms with topology optimization, the most widely used method for compliant mechanisms is the SIMP parameterization [5]. In this approach, material properties were considered constant within each element and densities were the design variables. The effective property of each element consisted of its density raised to a power and multiplied to the material properties of the solid material. This method was applied to a wide variety of compliant actuators design problems [6]. A number of heuristic or intuition based methods have been also applied to the design of compliant mechanisms (genetic Algorithms [7], Level Set Methods [8] and the Evolutionary Structural Optimization method [9]). More recently the bi-directional evolutionary structural optimization (BESO) method was developed for topological design of compliant mechanisms [10], where solid and void elements are all grouped in one single list and a penalty function is mandatory to reach convergence. The sequential element rejection and addition (SERA) method [11] that the authors propose in this work does not need any artificial material interpolation scheme, and a penalization scheme is not required to find a pure discrete solution.

The Sequential Element Rejection and Admission (SERA) method adds and removes material from the design domain using separate criteria for the update of solid and void elements. This strategy takes into account the current material status and controls how elements change their density, avoiding the algorithm to enter a loop where elements may change randomly from real to virtual and back to the previous material model in each iteration [12]. The optimization problem will be defined as the maximization of the Mutual Potential Energy [13], equivalent to the displacement at the output port. The ratio between input and output stiffness is controlled using a spring model [14], which model the input actuator behaviour and the stiffness of the work piece located at the output port. Examples of these type of actuation principles are electrothermal heating, piezoelectric actuation, shape memory alloys, etc. A well known filtering technique [15] is used to avoid the formation of checkerboard patterns, giving the method the necessarily mesh-independency. The validity of the proposed method for topology optimization of three dimensional actuators is demonstrated with the use of two benchmark problems, where motion is induced by an input force applied directly: an inverter device and a crunching actuator. Future work should consider the application of the SERA method for optimum topology design of more complex thermal and electrothermal devices.

IOP Conf. Series: Materials Science and Engineering **108** (2016) 012035 doi:10.1088/1757-899X/108/1/012035

2. Topology optimization formulation for three dimensional compliant mechanisms

In order to produce the specific displacement when the load in the input port is applied the compliant mechanism is required to be strong and flexible. Figure 1 shows such a compliant mechanism, occupying a three dimensional domain Ω. It is subjected to a force F_{in} at the input port P_{in} and is supposed to produce an output displacement Δ_{out} at the output port P_{out}.

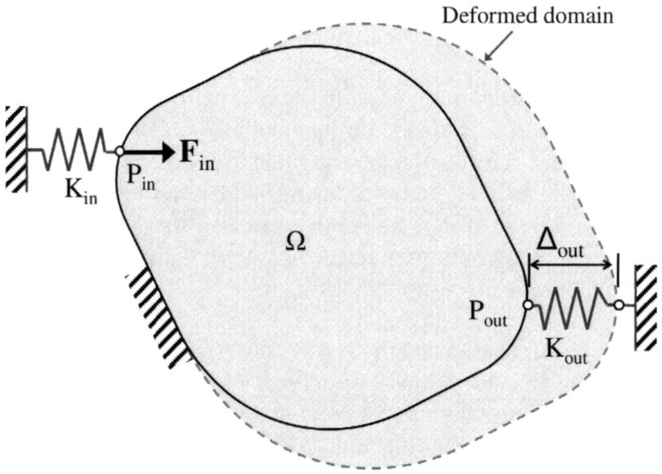

Figure 1. Problem definition

Obtaining the optimum design that converts the input work into an output displacement in a predefined direction is the aim of the topology optimization procedure implemented here for 3D compliant actuators design. The mathematical formulation of the required output displacement is expressed as the Mutual Potential Energy (MPE). Then the problem can be written as the maximization of the Mutual Potential Energy (MPE) subjected to a target volume fraction V^*.

$$\max \text{MPE} \tag{1}$$

$$\text{s.t. } \sum_{e=1}^{N} \rho_e \leq V^*, \ \rho_e = \{\rho_{min}, 1\}, \ \ e=1,\ldots,N \tag{2}$$

where ρ_e is the density of the eth finite element, N is the number of finite elements, V_e is the volume of the eth element, V_{Tot} is the total volume for the domain and ρ_{min} is the minimum density considered, a typical value of which is 10^{-4}.

The Mutual Potential Energy in equation (3) was defined as the deformation at a prescribed output port in a specified direction. To obtain the MPE, two load cases are calculated:
1) The Input Force Case, where the input force F_{in} is applied to the input port P_{in}, named with the subscript 1 in equations (3) and (4) and figure 2a;
2) The Unit Force Case, where a unit force is applied at the output port P_{out} in the direction of the desired displacement, named with the subscript 2 in equations (3) and (5) and Figure 2b.

$$\text{MPE} = \boldsymbol{U}_2^T \cdot \boldsymbol{K} \cdot \boldsymbol{U}_1 \tag{3}$$

$$\boldsymbol{K} \cdot \boldsymbol{U}_1 = \boldsymbol{F}_1 \tag{4}$$

$$\boldsymbol{K} \cdot \boldsymbol{U}_2 = \boldsymbol{F}_2 \tag{5}$$

where **K** is the global stiffness matrix of the structure; $\mathbf{F}_1$ is the nodal force vector which contains the input force F_{in}; $\mathbf{F}_2$ is the nodal force vector which contains the unit output force F_{out}; and $\mathbf{U}_1$ and $\mathbf{U}_2$ are the displacement fields due to each load case.

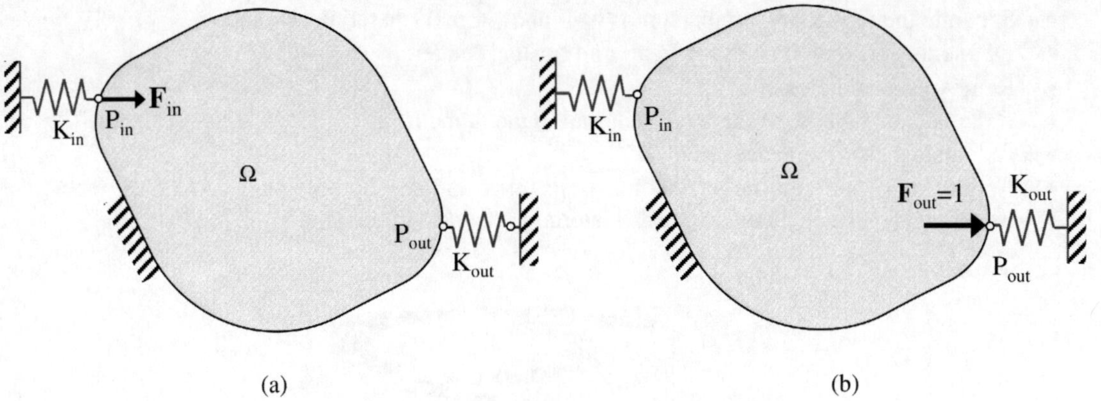

Figure 2. Representation of two load cases

In order to control the displacement amplification it is possible to specify different values of the output spring, K_{out}, that simulates the resistance in the output displacement. The spring model of figure 1 is used to define the stiffness ratio between the input and output ports. The artificial input spring K_{in} together with a spring force F_{in} simulates the input work of the actuator.

3. Sensitivity number computation

A sensitivity analysis is carried out as part of the optimization process to provide information on how sensitive the objective function is to small changes in the design variables. This sensitivity number in each element determines which elements are removed or added so that the objective function is maximized.

$$\alpha_e = -\boldsymbol{U}_{1e}^T \cdot \Delta\boldsymbol{K}_e \cdot \boldsymbol{U}_{2e} \tag{6}$$

where $\mathbf{U}_{1e}$ is the displacement vector of element e due to the applied load $\mathbf{F}_1$; $\mathbf{U}_{2e}$ is the displacement vector of element e due to the output load vector $\mathbf{F}_2$; and $\Delta\mathbf{K}_e$ is the variation of the elemental stiffness matrix.

4. Sequential Element Rejection and Admission method for three dimensional compliant mechanisms

The method considers two separate material models, 'Real' and 'Virtual' material and two separate criterions of rejection and admission of elements allow material to be introduced and removed from the design domain by changing its status from 'virtual' to 'real' and vice versa (figure 3). The 'real' material present at the end of the optimization will be the optimized topology. The twelve steps that drive the SERA method for compliant mechanisms are given below:

- Define the design problem. The maximum design domain must be defined and meshed with finite elements. All boundary constraints, loads and the target volume fraction V^* must also be specified.
- Assign 'real' and 'virtual' material properties to the initial design domain.
- Calculate the variation of the volume fraction in the ith iteration which consists of the volume fraction to be added $\Delta V_{Add}(i)$ and removed $\Delta V_{Remove}(i)$.
- Carry out a Finite Element Analysis of the two load cases to produce the displacement vectors $\boldsymbol{U}_1$ and $\boldsymbol{U}_2$. The elemental and global stiffness matrixes, $\boldsymbol{K}_e$ and $\boldsymbol{K}$, are also calculated as part of the FEA.
- Calculate the elemental sensitivity numbers.

- Apply a mesh independent filtering to the sensitivity numbers.
- Separate the sensitivity numbers into 'real' and 'virtual' materials, α_R and α_V.
- Define the threshold values for 'real' and 'virtual' material, α_R^{th} and α_V^{th}.
- Remove and add elements.
- Calculate the volume of the 'real' material in the domain.
- Calculate the convergence criterion ε_i.
- Repeat steps until the target volume is reached and the optimization converges. The final topology is represented by the 'real' material in the design domain.

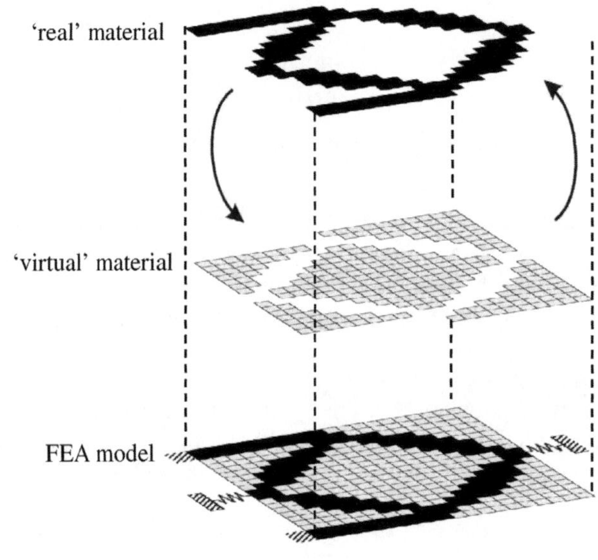

Figure 3. The SERA material models

4.1. Definition of the starting design domain

The SERA method can start from a full design domain (all elements are 'real' material), from a void design domain (all elements are 'virtual' material), and also with any amount of material present in the domain. In this last case, a combination of 'real' and 'virtual' material is the starting point. For any of these cases, the material present in the domain is assigned the 'real' material properties and material not present in the domain is assigned the 'virtual' material properties. The method converges toward the optimum topology regardless of the initial design domain. However, the initial design domain determines the number of iterations needed to achieve that optimum. The examples shown at the end of the paper were all solved starting from a full design domain. Nevertheless, expressions for both cases are included in the following subsections for the sake of completeness.

4.2. Calculation of volumes to add $\Delta V_{Add}(i)$ and remove $\Delta V_{Remove}(i)$

In the first stage of the process different amounts of material are added and removed in each iteration until the target volume fraction V^* is reached. In the second stage, once the target volume fraction is reached, material re-distribution takes place by both adding and removing the same amount of material until the problem converges.

4.2.1. First stage: real material volume fraction and target volume fraction are different. Two starting domain cases exist:

a) When the design starts with a volume fraction higher than the target fraction or a full domain $V_F(0)$ is the volume in the design domain at the beginning of the process The volumes in the following iterations will be calculated using the equation (7). In each iteration the variation of the volume will be calculated with equation (8) and then the amount of material to add and remove will be calculated in equations (9) and (10), where: PR is the Progression Rate, with typical values ranging between 0.01 and 0.05; SR is the Smoothing Ratio, with typical values in the range between 1.2 and 1.4.

$$V_F(i)=\max\left(V_F(i-1)\cdot(1-\mathrm{PR}),V^*\right) \tag{7}$$

$$V(i)=|V_F(i)-V_F(i-1)| \tag{8}$$

$$\Delta V_{Add_F}(i)=\Delta V(i)\cdot(\mathrm{SR}-1) \tag{9}$$

$$\Delta V_{Remove_F}(i)=\Delta V(i)\cdot\mathrm{SR} \tag{10}$$

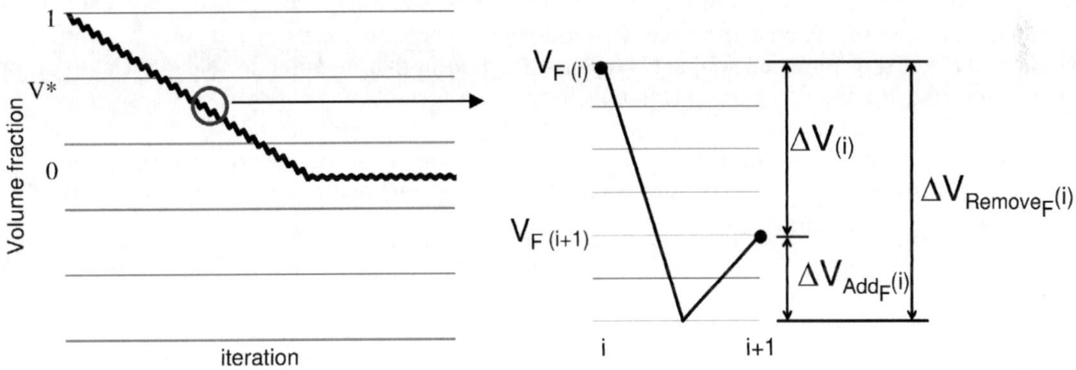

Figure 4. Removal and addition of material from a full domain

b) When the design starts with a volume fraction lower than the target fraction or a void domain $V_v(0)$ is the volume in the design domain at the beginning of the process. The volumes in the following iterations will be calculated using the expression (11). In each iteration the variation of the volume will be calculated with equation (12) and then the amount of material to add and remove will be calculated using equations (13) and (14).

$$V_V(i)=\min\left(V_V(i-1)+\mathrm{PR},V^*\right) \tag{11}$$

$$V(i)=|V_V(i)-V_V(i-1)| \tag{12}$$

$$\Delta V_{Add_V}(i)=\Delta V(i)\cdot\mathrm{SR} \tag{13}$$

$$\Delta V_{Remove_V}(i)=\Delta V(i)\cdot(\mathrm{SR}-1) \tag{14}$$

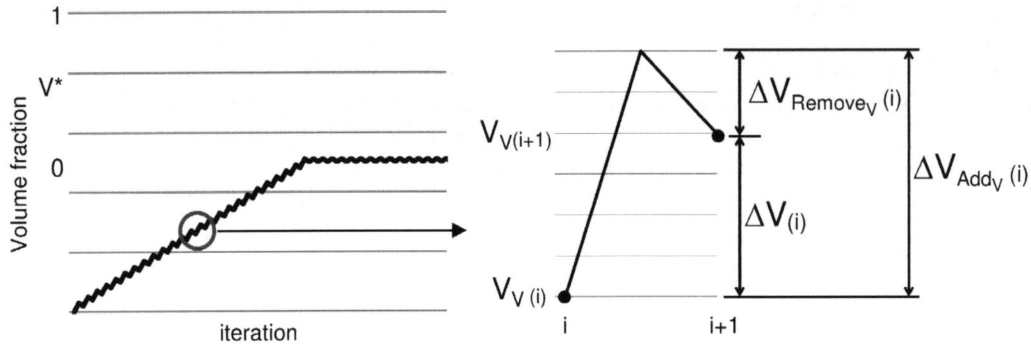

Figure 5. Removal and addition of material from a void domain

A graphical representation of the removal and addition of elements in each iteration for case (a) is given in figure 5 and for case (b) in figure 6. In both cases, each iteration consists of two sub-steps which add and remove material from the design domain. The difference depends on the amount of material to be added or removed so that the volume fraction in that iteration decreases for case (a) or increases for case (b).

4.2.2. Second stage: real material volume fraction and target volume fraction are similar. The process of material re-distribution takes place: this consists of both adding and removing the same amount of material from the design domain:

$$\Delta V_{Add_{F\ or\ V}}(i)=\Delta V_{Remove_{F\ or\ V}}(i)=\beta\cdot V^{*} \tag{15}$$

where β is the material re-distribution fraction, with typical values ranging between 0.001 and 0.005.

4.3. Removal and addition of elements

The sensitivity number for the eth finite element α_e (6) is a function of the variation between two iterations in the stiffness matrix of that element $\Delta \mathbf{K}_e$ (16).

$$\Delta \boldsymbol{K}_e=\boldsymbol{K}_e(i)-\boldsymbol{K}_e(i-1) \tag{16}$$

where $\mathbf{K}_e(i)$ is the stiffness matrix in the ith iteration for the eth finite element; and $\mathbf{K}_e(i-1)$ is the stiffness matrix in the $(i-1)^{th}$ iteration for the same finite element. If an element is added, $\mathbf{K}_e(i) = \mathbf{K}_e$ and $\mathbf{K}_e(i-1) \approx 0$, so the variation of the elemental stiffness matrix is $\Delta \mathbf{K}_e = \mathbf{K}_e$. But if an element is removed, $\mathbf{K}_e(i) \approx 0$ and $\mathbf{K}_e(i-1) = \mathbf{K}_e$, and $\Delta \mathbf{K}_e = -\mathbf{K}_e$.

Then, the elemental sensitivity numbers for the 'real' and 'virtual' material are given by equations (17) and (18), respectively.

$$\alpha_{eR}=\boldsymbol{U}_{1e}^{T}\cdot\boldsymbol{K}_e\cdot\boldsymbol{U}_{2e} \tag{17}$$

$$\alpha_{eV}=-\boldsymbol{U}_{1e}^{T}\cdot\boldsymbol{K}_e\cdot\boldsymbol{U}_{2e} \tag{18}$$

As the objective is to maximize the MPE, the elements with the higher values of sensitivity number are the ones to be added and removed (figure 6). Before the addition or removal of elements according to this sensitivity values, this values will be smoothed with a mesh independency filter. The threshold values α_{th}^{R} and α_{th}^{V} are the sensitivity values that remove or add the amount of volume $\Delta V_{Remove}(i)$ and $\Delta V_{Add}(i)$ defined for each iteration are given in figure 6.

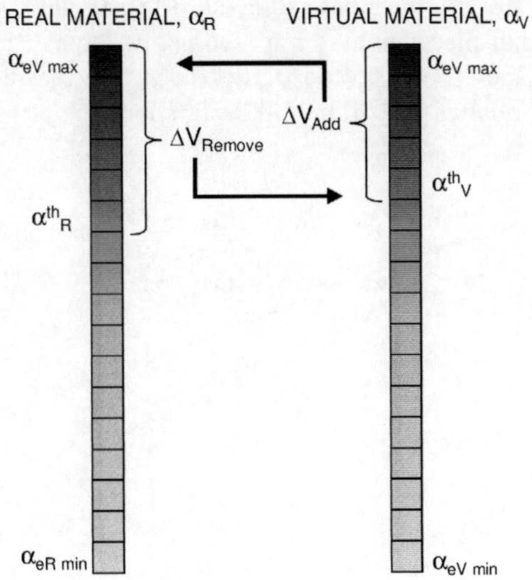

Figure 6. Lists of real and virtual material and the volumes to be removed and added

4.4. Convergence criterion

The convergence criterion is defined as the change in the objective function in the last 10 iterations (19), which is considered an adequate number of iterations for the convergence study. It implies that the process will have a minimum of 10 iterations as the convergence criterion is not applied until the iteration number has reached this minimum number of iterations.

$$\varepsilon_i = \frac{\left|\sum_{i=9}^{i=5} MPE_i - \sum_{i=4}^{i} MPE_i\right|}{\sum_{i=4}^{i} MPE_i} \tag{19}$$

where ε_i is the convergence criterion, with typical values ranging between 0.001 and 0.01.

5. Numerical examples

Three dimensional inverter and crunching benchmark actuators are presented to demonstrate the efficiency of the proposed method. Many examples of compliant mechanisms, such as inverters and grippers, can be found in the literature. Here, some of them have been re-designed using the proposed method and illustrating the ability of the design method for 3D applications. The material properties used in all examples are the following: Young's modulus is taken to be E = 200 GPa and the Poisson's ratio is v = 0.3. The density of the virtual material is taken to be $\rho_{min} = 10^{-4}$, which is equivalent to 0.01 % of the stiffness of a real material. The algorithm has been applied for simple compliant actuators optimization under directly applied input loads. The topologies presented in this section were modelled using geometrical nonlinear elastic analysis, because otherwise the results may be useless in some cases as large displacement mechanisms, even if obtained results do not show any locking or related problems.

5.1. Inverter mechanism

The design domain for the inverter mechanism is shown in figure 7. The aim of this topology optimization problem is to design an actuator that converts an input displacement on the left edge to a displacement in the opposite direction on the right edge. The domain consists of a parallelepiped of size 40×40×5 mm discretized using cubic eight node finite elements. The boundary is fixed at the upper and bottom left corners and a standard input force F_{in} = 1 kN is applied. A stiffness ratio of K_{out}/K_{in} = 1 is defined for the springs that model the input actuator and the output workpiece. The target volume fraction V* is 0.4 and the filter radius used to prevent chequerboard is r_{min} = 1.5 mm. Both the top face and the side displacement are imposed taking advantage of the symmetry; i.e., nodes can only move within the planes shown in the figure. Therefore the results shown in the following subsections correspond to the quarter part of the full design domain.

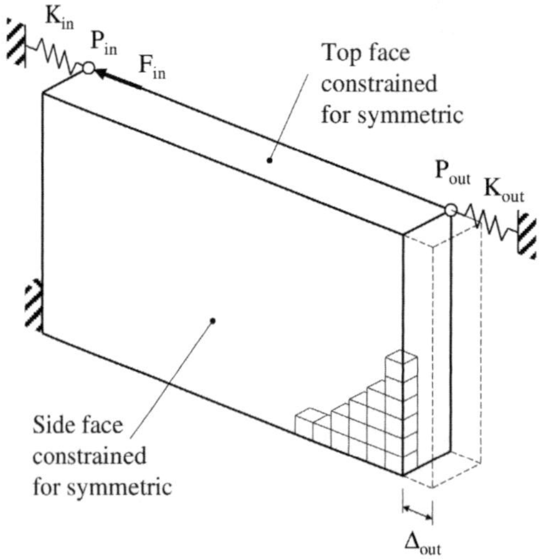

Figure 7. Design domain and boundary conditions for the inverter mechanism

The optimized topology was obtained after approximately 150 iterations. Figure 8 shows the optimum topology of the analyzed quarter part of the domain. It can be observed that the obtained material distribution and the connections of the supports with the input and output ports are basically similar to the classical planar inverter mechanism with slight differences. This example agrees well also with the solution offered by the SIMP parametrization strategy. One can notice that the resulting mechanism is not truly compliant because there is a moment free one-node connected hinge. This hinge should be avoided performing a post processing of the resulting topology, where it can be substituted with a long slender compliant hinge. An alternative solution is to applied diverse methods proposed in literature to alleviate the problem from the very beginning of the optimum design problem formulation.

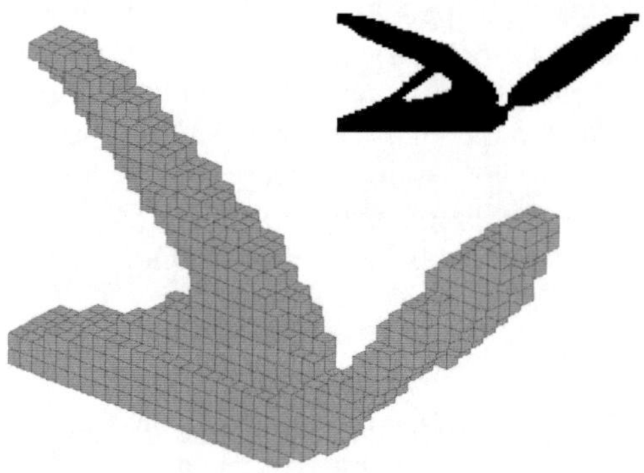

Figure 8. Optimum topology for the inverter mechanism

In figure 9 the output displacement evolution is plotted as a function of the iteration number and snapshots of the actuator topologies at the different stages are shown. As it can be observed in the evolution history, the objective function does not have positive values until a well defined mechanism is constructed with material joining the input and output ports. This is because the objective function represents the displacement in the output port and therefore positive values can only be obtained when the material distribution promotes the displacement transmission through the hinges created during the optimization process. Initial displacement at the output port was 0.9603 mm inwards, while final displacement is 3.742 mm outwards, this time along the desired direction. Figure 9 describes the gradual formation of the optimum design and illustrates how the material rejection and addition takes place, using an image sequence captured during the process. The optimal topology is the same regardless of the path of the optimization, even if it can take different number of iterations until the topology efficiently grows towards the optimum, depending on the initial material distribution and the final volume fraction.

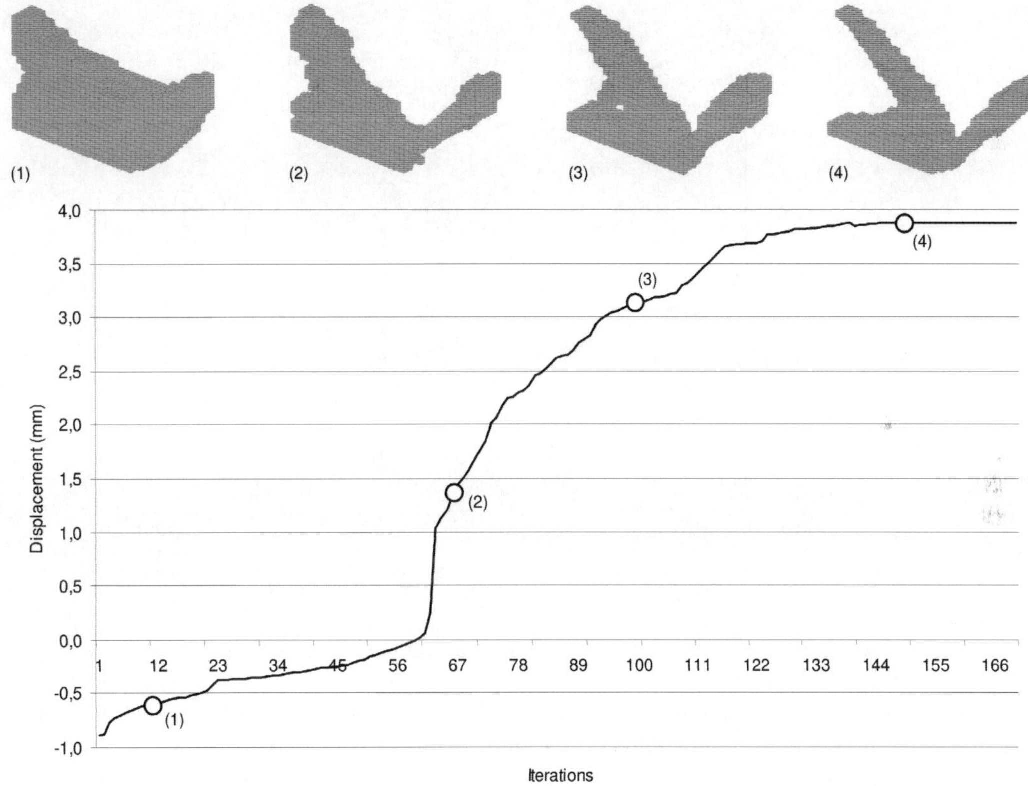

Figure 9. Evolution history for the inverter mechanism.

5.2. Crunching mechanism

The design domain for a crunching mechanism is shown in figure 10. Again it is a square of size 40 × 40 mm with a thickness of 5 mm and subdivided using 1 mm cubic eight node finite elements. The mechanism is supported at the left side and is subject to a vertical input load F_{in} = 1 kN at the upper and lower right corners, where due to the symmetry of the problem only a quarter of the full design domain is modelled. The objective is to maximize the horizontal displacement on the output port in the outward diretions by distributing material in the design domain area. The allowable amount of material for this example is taken to be 40% of the full design domain. The spring based formulation used in this work allows the displacement amplification to be controlled by specifying different values of the input and output springs as part of the definition of the problem. As in the previous example a ratio of $K_{out}/K_{in} = 1$ has been adopted.

Figure 11 shows the obtained optimum topology for the crunching mechanism, which resembles the solutions obtained by other authors using different material models and optimization algorithms for two dimensional problems such as SIMP. The same figure includes the optimized topology for the planar actuator optimum design problem. The resulting three dimensional topology is comparable with the 2D solution, obtained with the SERA procedure described in this work. Although topologies are not exactly the same, the most important features of the design are present in both cases. It must be noted as well that the angle between the bars and the location of holes depends heavily on the smoothing of sensitivities and stiffness of the structure and springs, which reasonably are different in both cases.

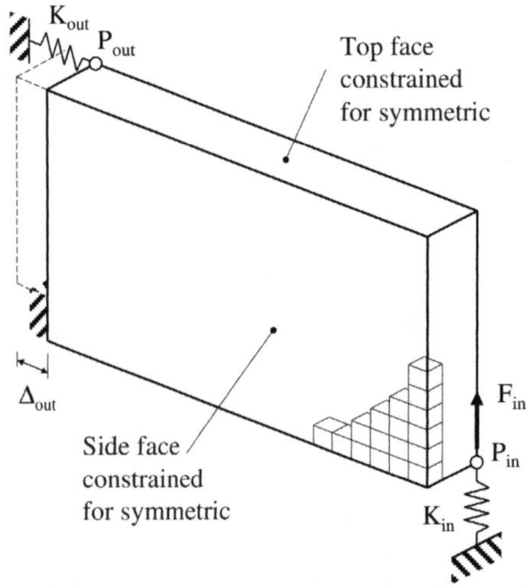

Figure 10. Design domain and boundary conditions for the crunching mechanism

6. Conclusions

An extension of the sequential element rejection and admission (SERA) method for topology optimization of three dimensional compliant actuators has been presented in this work. This method overcomes the issues noticed evolutionary methods are used for the design of compliant mechanisms. The main difference of this method with respect to other bi-directional methods which add and remove elements from the design domain is that solid and void materials are treated separately so that the addition and removal of elements have separate criteria. The problem of designing compliant mechanisms is defined here as the maximization of the mutual potential energy and a spring model is used to control the input and output stiffness. This formulation meets the flexibility and stiffness requirements necessary to design compliant devices that satisfy the kinematic requirements and at the same time withstand the applied loads. The solved numerical examples agree well with solutions offered by the SIMP interpolation scheme and demonstrate that the SERA method is both an efficient method of designing 3D compliant actuators. Optimum topologies resemble the solutions obtained with the same method for two dimensional problems. It should be mentioned that both SIMP and SERA methods show the same issues concerning one-node connected hinges, so techniques to avoid them should be explored and implemented. The generalization of the method for more complex electrothermal actuators in three dimensions is left for future work.

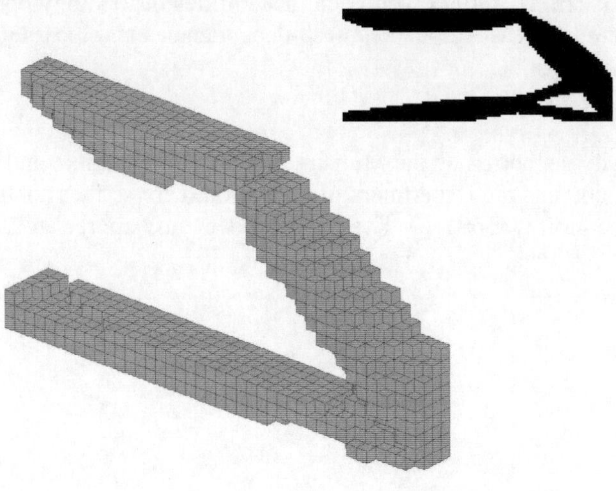

Figure 11. Optimum topology for the crunching mechanism.

7. References

[1] Howell L 2001 *Compliant mechanisms* (New York: Wiley)

[2] Howell, L L and Midha A 1994 A Method for the Design of Compliant Mechanisms With Small-Length Flexural Pivots *ASME J. Mech. Des.* **116** 280–90

[3] Bendsoe M P and Sigmund O 2003 *Topology optimization: theory, method and application* (Berlin: Springer)

[4] Ananthasuresh G K, Kota S and Gianchandani Y 1993 Systematic synthesis of micro compliant mechanisms: preliminary results *Proc. 3rd national conf of appl mech and robot (Cincinnati)* vol 2 p 82

[5] Bendsoe M P 1994 Optimal shape design as a material distribution problem *Struct. Optim.* **1** 193–202

[6] Sigmund O 2001 Design of multiphysics actuators using topology optimization. Part I: One-material structures *Comput. Meth. Appl. Eng.* **190** 6577–604

[7] Parsons R and Canfield S 2002 Developing genetic programming techniques for the design of compliant mechanisms *Struct. Multidisc. Optim.* **24** 78–86

[8] Yulin M and Xiaoming W 2004 A level set method for structural topology optimization and its applications *Adv. Eng. Softw.* **35** 415–441

[9] Ansola R, Vegueria E, Canales J and Tárrago J 2007 A simple evolutionary topology optimization procedure for compliant mechanism design *Finite Elem. Anal. Des.* **44** 53–62

[10] Huang X, Li Y, Zhou S W, Xie Y M 2014 Topology optimization of compliant mechanisms with desired structural stiffness *Engineering Structures.* **79** 13–21

[11] Rozvany G I N and Querin O M 2002 Theoretical foundations of Sequential Element Rejections and Admissions (SERA) methods and their computational implementations in topology optimisation *Proc. 9th AIAA/ISSMO symposium on multidisciplinary analysis and optimization (Georgia)* p 5521

[12] Alonso C, Querin O M and Ansola R 2013 A Sequential Element Rejection and Admission (SERA) method for compliant mechanisms design *Struct. Multidisc. Optim.* **47** 795–807

[13] Shield R and Prager W 1970 Optimal structural design for given deflection *J. Appl. Math. Phys.* **21** 513–23

[14] Ananthasuresh G K, Kota S and Gianchandani Y 1994 A methodical approach to the design of compliant micromechanisms *Proc. Solid state sensor and actuator workshop* p 189-192

[15] Sigmund O and Petersson J 1998 Numerical instabilities in topology optimization: a survey on procedures dealing with checkerboards, mesh-dependencies and local minima *Struct. Multidisc. Optim.* **16** 68–75.

Acknowledgments
This work was financially supported by the Ministry of Education, Culture and Sport in Spain through the project DPI2012-36600 and the Department of Education, Universities and Research of the Basque Government through the project IT691-13. Their support is greatly appreciated.

IOP Conf. Series: Materials Science and Engineering **108** (2016) 012036 doi:10.1088/1757-899X/108/1/012036

LED-based NDIR natural gas analyzer

Sergey Fanchenko[1,2], Alexander Baranov[1], Alexey Savkin[1], Vladimir Sleptsov[1]

[1] "MATI" -Russian State Technological University, Moscow, 121552, Russia
[2] NRC "Kurchatov Institute", Moscow, 123182, Russia
E-mail: fanchenko.sergey@kiae.ru

Abstract. A new generation of the light-emitting diodes (LEDs) and photodiodes (PDs) was used recently to develop an open path non-dispersive infrared (NDIR) methane analyzer. The first open path detector prototype was constructed using LEDs for measurement and reference channels, accordingly, and first measurements for methane gas have been performed using optical paths of the order of several meters [3]. The natural gas consists of several first alkanes, mainly methane, and it is important to have a possibility of measuring all of them. In the present work we report the results of NDIR measurements for propane-butane mixture and new measurements of methane using LEDs for measurement and reference channels at 2300 and 1700 nm wavelengths, accordingly. The necessity of the double beam scheme is demonstrated and obtained results for methane and propane-butane mixture are compared.

1. Introduction

Low cost and low power consuming methane detection is needed for control of large areas both for pipeline inspection and for pollution control. Mainly point sensors have been developed for natural gas detection – catalytic, semiconductor and optical ones [1-3]. However this kind of sensors provides the measurements only in the place were they are mounted. To enable the large area monitoring these sensors should be joined in a sensor network (or wireless sensor network [4]). The number of sensors in a wireless network for combustive gas monitoring should be tens or even hundreds, but still a chance exists that the natural gas leakage won't be detected. Therefore optical open path detection systems have been developed using different light sources – mainly xenon lamps, LEDs and lasers in order to provide long distance measurements. Gas sensors based on optical absorption offer fast responses (time constants below 1 s), minimal drift and high gas specificity, with zero cross response to other gases [3]. Xenon lamps and LEDs are used for non-dispersive measurements with rather broad light spectra with the width of the order of hundreds of nm, while laser diodes and lasers - for tunable spectroscopy with very narrow spectra with the width of tenth of a nm. But laser systems are rather expensive, while xenon lamps systems are rather high power consuming ones and besides need protecting shields, enlarging essentially the size of the constructions. So the best candidates for low cost and low power consuming systems are LED-based systems. Recently qualitative technological break-through in growing of narrow band-gap heterostructures was achieved [5]. High effective LEDs for the spectral range 1600-5000 nm and spectral matched photodiodes were developed. For developing of LEDs and PDs lattice-matched heterostructures with GaInAsSb quaternary solid solutions in the active layer were used. It became possible to create gas sensors that meet all demands of the mass market: compactness; low electrical power consumption; high selectivity; high sensitivity; long life-time without additional calibration and low cost in mass production. Average optical power up to 3.5 mW and peak power up to 170 mW was achieved for 1600-2400 nm LEDs. Typical LED spectral widths are comparable with the width of methane absorption bands. It was shown recently that for methane concentrations of the order of Low Explosive Limit (LEL) the highest sensitivity could be achieved by using absorption band at 2300 nm wavelength (for high values of the product of concentration by optical path the light transmission for the strongest absorption bans goes to zero) [6].

NDIR transmission is described by integral formulation of Beer-Lambert low:

[1] fanchenko.sergey@kiae.ru

$$A = \frac{\int d\upsilon I_0(\upsilon)F(\upsilon)e^{-LK(\upsilon)}}{\int d\upsilon I_0(\upsilon)F(\upsilon)}$$

where υ is the frequency, $K(\upsilon)$ is the frequency dependent linear absorption coefficient, $I_0(\upsilon)$ is the light source spectra and $F(\upsilon)$ is the sensitivity of the photodetector, L – the optical path in the absorbing media. Coefficients $K(\upsilon)$ are usually obtained from HITRAN database [7]. For large values of the argument in the exponent the transmission for the strongest absorption band becomes too low and it is more efficient to work with a weaker absorption band. It's important only for open path detectors measuring concentrations near LEL level, for trace gas detectors, measuring concentrations on ppm or lover level, the most efficient for measurements is the strongest absorption band

One of the peculiarities of NDIR detection scheme is the use of two channels – a measurement and a reference one. Emission wavelength of the reference beam should be shifted from the measurement one, so that the reference signal won't be absorbed by the specified gas. The total measured intensity is normalized to the intensity of the reference channel in order to exclude the scattering and absorption effects that may occur along the whole optical path due to atmospheric precipitations or other obstacles.

The NDIR gas analyzer was invented and first used in Germany in the 1930's. Recent breakthrough in LEDs technology permitted one to use nonconventional IR light sources for gas detectors. Firstly LEDs/PDs pairs were used in portable trace gas analyzers [8] and a bit later - in open path ones, the construction of an open path IR hydrocarbon gas detector, based on heterostructure PDs and LEDs, being reported in [9]. The 3400 nm measurement wavelength was used for methane main absorption band detection, the reference wavelength being available in two options – 3070 and 3850 nm and the optical path being of the order of 40 m. Now LED-based NDIR detectors are mainly used as trace gas analyzers [3,10]. Considering distances much more than 40 m, we are developing natural gas NDIR detector using absorption band at 2300 nm. The only problem in this case is in choosing the appropriate LED for the reference channel with a spectrum close to the spectrum of the measurement LED and outside the methane and water absorption bands at 2300 and 1900 nm, accordingly. Also one should take into account that for existing LEDS there is a gap in wavelength from about 2.4 to 2.8 μm due to existence of miscibility region for GaInAsSb based solid solutions which depends on the epitaxy temperature and the compound composition. Considering the currently available LEDs the only possibility to solve the problem with reference beam is to use the LED with 1700 nm peak, nevertheless it crosses the methane absorption band at 1650 nm (Figure 1). This cross-reference can be excluded by renormalization of calibration curve due to the fact that this band is much weaker than that at 2300 nm. The first preliminary results of methane concentration measurements by LED-based NDIR prototype analyzer in measurement channel were reported recently [6]. Now the measurements in the reference channel are also possible.

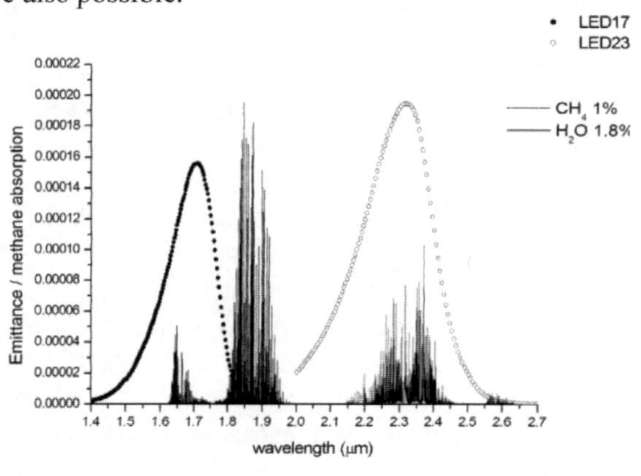

Figure 1. LED spectra and methane and water absorption bands

The natural gas consists of several first alkanes, mainly – methane, absorption bands being rather similar for all of them. LELs of the main components of the natural gas are given in Table 1. For most combustible gas sensors the default alarm is set as 10 % of LEL with a full range 0 – 100 % LEL.

Table 1. LEL values for natural gas components

Component	LEL (vol. %)
Methane	4.4
Ethane	2.5
Propane	1.7
i-Butane	1.4
n-Butane	1.3
Pentane	1.4

Structurally, the alkanes are quite simple, having no functional groups. They form a homologous series differing only in the number of methylene (–CH_2–) segments. The first and lowest-weight alkane in the series is methane, CH_4, which is quite atypical of the series because the carbon atom is tetrahedrally coordinated to the four hydrogen atoms, and it contains no C-C bonds [11]. Alkanes IR spectra display only C-C and C-H bond vibrations, being somewhat self-similar: the same absorption bands are present in essentially the same positions, differing only in such less obvious yet still diagnostic characteristics as absorption depth and band width. As a result the cross sensitivity is present for practically all low number alkanes. Nevertheless the sensitivity of NDIR method depends on the particular alkane, and we have performed NDIR measurements of propane-butane mixture in order to compare the results with data for methane. One of the ways of improving NDIR selectivity is the use of additional selective reflector [12], it could be rather easily incorporated in an open path gas analyzer if needed, but in the current paper we are considering only the simplest NDIR scheme, in which the absorption coefficient integrated over the whole emitting band width is measured.

2. Experimental

Prototipe NDIR analyzer consists of two independent units – emitting and detecting blocks mounted on tripods (Figure 2 a, b). It's construction permits one to use at a time only one channel (measurement channel at 2.3 μm or reference one at 1.7 μm), so that full NDIR measurements should be performed successively. In optical scheme one collimating lens is used in the emmiting block and one focusing one – in the detecting block.

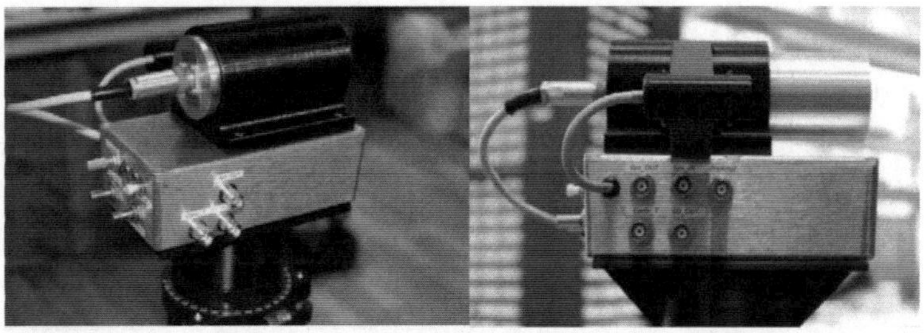

a b

Figure 2. Photo of the prototype NDIR analyzer blocks (a – emitter, b – receiver)

In emitting and receiving blocks "LED Microsensor NT" LEDs and PDs are used [13]. These LEDs are characterized by compact size of 0.35·0.35 mm, low power consumption (<1 mW), short response time – 10-50 ns, operation temperatures up to +150°C, lifetime 80 000 hours. LEDs are mounted into a package with a thermoelectric module (Peltier element) enabling LED chip temperature stabilization. Besides there is a possibility to arrange multi-element arrays enabling to obtain multi-wavelength emitters in single compact packages. The PD characteristics are following: cut off at wavelength 2.45 μm (Figure 3), sensitive area diameter 0.3 mm, detectivity 2-5·10^{10} cm·$Hz^{0.5}W^{-1}$ [10].

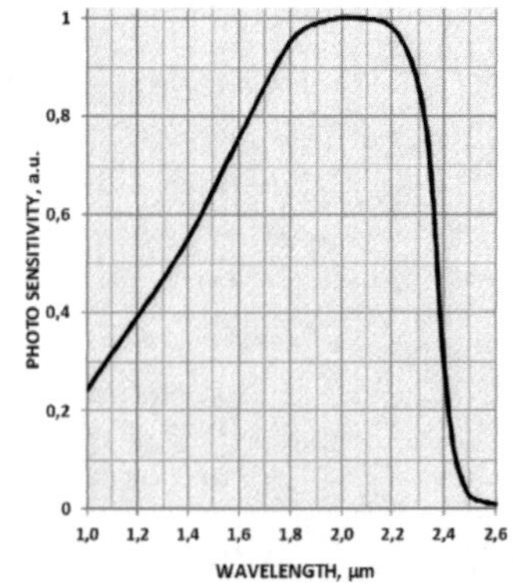

Figure 3. PD spectral response [13]

Propane-butane mixture (30% propane + 70% butane) was chosen for NDIR analyzer testing. Firstly the measuring channel at 2.3 μm was checked, the gas cell of the length of 10 cm was filled with propane-butane mixture) and then - with air. In Figure 4 it is shown that the NDIR intensity is lowered when the gas cell is filled with testing gas and is restored very well after it is filled once more with air, the noise being on the level of several tenths of a per cent.

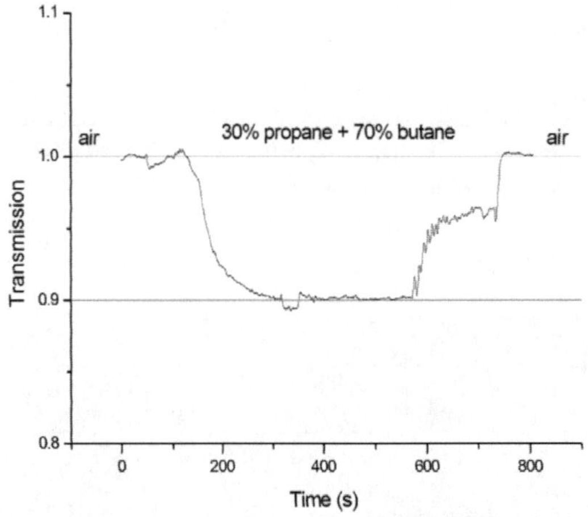

Figure 4. Transmission in propane-butane mixture (optical path 10 cm)

Secondly, the reference channel at 1.7 μm was checked, and it was shown, that the variation of the reference signal due to absorption is much less than of the measurement one, so that the signals ratio could be used in NDIR double beam scheme as calibration curve (Figure 5, curve 23/17).

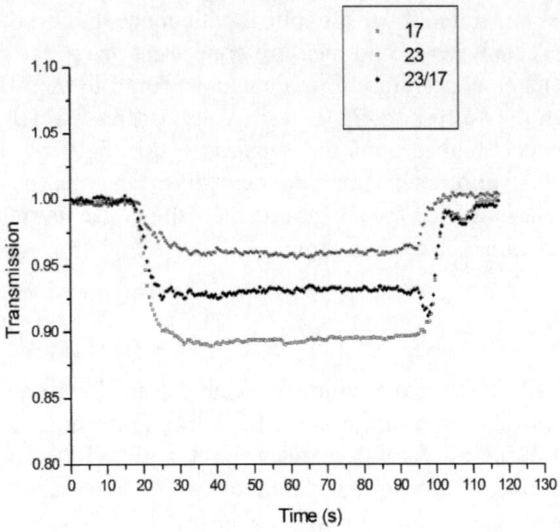

Figure 5. Transmission in propane-butanr mixture (23 – measurement channel, 17 – reference one, 23/17 - calibration curve)

Thirdly, the necessity of a double beam scheme was demonstrated for methane measurements. The gas cell was filled with 100% methane and successive measurements were performed both for measurement and reference channels. The results are depicted in Figure 6 together with the theoretical lines calculated using HITRAN database [7]. It is seen that both experimental curves are lower than theoretical ones the difference being practically the same. This difference is about 15 % for reference channel and is associated with some additional scattering or absorption in the gas cell. It's much stronger than the noise and its influence can be excluded only by normalizing the measurement channel intensity to reference channel one. The comparison of renormalized NDIR transmittances shows, that in methane the absorption effect is by 70% stronger than in propane-butane mixture.

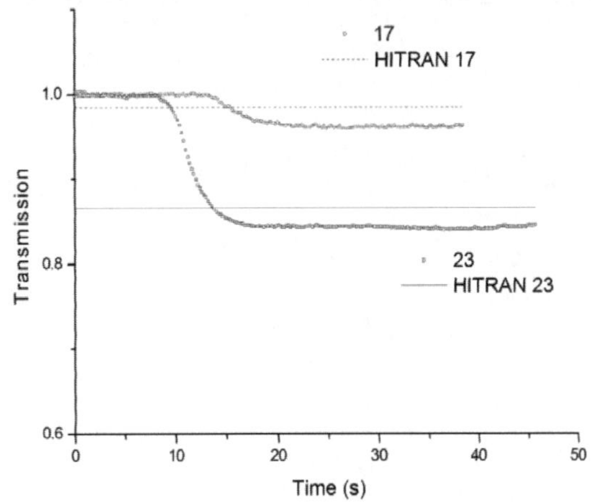

Figure 6. Transmission in 100% methane (23 – measurement channel, 17 – reference one, hitran 23 and HITRAN 17 – corresponding HITRAN-calculated curves)

One should note that methane LEL is equal to 4.4% (table 1). The usual unit in open path measurements is LEL/m (the integrated over all optical path concentration divided by LEL value and one meter).The open path gas analyzer should measure concentrations in the range 0 – 100 % LEL/m. The dangerous level for methane concentration in air is considered to be 50% of LEL/m or 2.2%·m. For half a meter optical path the corresponding LEL/m value is 25% which results in absorption on the level of 3.3%, being much higher than the sensitivity threshold of the constructed analyzer prototype (the noise is on the level of a tenth of a per cent). It means that the constructed gas analyzer could be used as a reliable alarm device for dangerous methane concentrations detection in the range from several to hundred % LEL/m.

3. Conclusions

It is shown that LED-based NDIR detectors with a 2-channel optical scheme are very perspective for alkane open path analysers of gas concentrations near LEL. For the first time LED with a peak emission at 2.3 μm is used in the measuring channel and 1.7 μm LED – in the reference one. It is shown that the influence of alkane absorption band near to the reference emitter spectrum can be excluded by renormalization of the calibration curve. The first prototype of an open path LED-based detector was constructed and the first measurements have shown that the methane concentration could be measured with a bit better sensitivity than for other alkanes.

From figure 6 it is seen that in the double beam scheme for optical path 10 cm the pure absorption in 100% methane is rather low (of the order or 15%), while the absorption band at 2.3 μm is rather strong by itself. It results from the fact, that LED linewidth is wider than the absorption band width and that absorption spectrum is discrete while the emitting spectrum is continuous one (figure 1). As a result essential part of light is not absorbed at all propagating through methane. To enhance the absorption effect the filters, narrowing LEDs spectra, could be used or a selective reflector before the photodiode could be added [11]. Such modifications should improve the sensitivity of an open path analyzer enhancing the signal to the background ratio.

Acknowledgments

This work was supported by the grant No. RFMEFI57714X0022 from the Ministry of Education and Science of Russian Federation

References

[1] Karpov E E Karpov E F. Suchkov A Mironov S Baranov A Sleptsov V Calliari 2013 Energy efficient planar catalytic sensor for methane measurement *Sensors and Actuators A: Physical* **194** 176–180
[2] Samotaev N N Vasiliev A A Podlepetsky B I Sokolov A V Pisliakov A V 2007 The mechanism of the formation of selective response of semiconductor gas sensor in mixture of CH4/H2/CO with air *Sensors and Actuators B: Chemical* **127** 242–247
[3] Hodgkinson J and Tatam R 2013 Optical gas sensing: a review *Meas. Sci. Technol.* **24** 012004
[4] Somov A Baranov A Spirjakin D Spirjakin A Sleptsov V Passerone R 2013 Deployment and evaluation of a wireless sensor network for methane leak detection *Sensors and Actuators A: Physical* **202** 217–225
[5] Stoyanov N D Salikhov K M Kalinina K V Zhurtanov B E Kizhaev S S 2012 Middle infrared LEDs: key element for new generation chemical sensors *Proc. SPIE 8257 Optical Components and Materials* **IX** 82571E
[6] Fanchenko S Baranov A Savkin A Petukhov A Kalinina K Zhurtanov B Velikotny M 2015 Non-dispersive LED-based methane open path detector capabilities *Proc. EESMS 2015* in press

IOP Conf. Series: Materials Science and Engineering **108** (2016) 012036 doi:10.1088/1757-899X/108/1/012036

[7] Rothman L S Gamache R R Goldman A Brown L R Toth R A Pickett H M Poynter R L Flaud J M Camy-Peyret C Barbe A Husson N Rinsland C P and Smith M A H 1987 The HITRAN database: 1986 Edition *Appl.Opt.* **26** 4058-4097

[8] Aleksandrov S Gavrilov G Kapralov A Karandashov S Matveev B Sothikova G and Stus N 2002 Portable optoelectronic gas sensors operating in the mid-IR spectral range (λ = 3–5 μm) *Proc. SPIE* **4680**188–94

[9] Lugovskaya A I Login S A. Balashov O B Kuznetsov A A Vasiliev E V Chernyak E Y 2003 Remote Infrared Detector hydrocarbon gases *Products and Systems. Control, monitoring, diagnostics* **6** 55-59

[10] Gibson D and MacGregor C. 2013 A Novel Solid State Non-Dispersive Infrared CO2 Gas Sensor Compatible with Wireless and Portable Deployment *Sensors* **13** 7079-7103

[11] Clark R N Curchin J M Hoefen T M Swayze G A 2009 Reflectance spectroscopy of organic compounds: 1. Alkanes *J. Geophys. Res.* **Vol. 114** E03001

[12] Makeenkov A, Lapitskiy A, Somov A, Baranov A 2015 Flammable gases and vapors of flammable liquids. Monitoring with infrared sensor node *Sensors and Actuators* **B 209** 1102–1107

[13] http://lmsnt.com/datasheets/catalogue/LED_Microsensor_NT _catalogue_rev310315.pdf (June, 2015).

Precise sensing and selection of molecules by the interface between the metal nanocluster and the oxide support

S A Khubezhov[1], I V Silaev[1], I V Tvauri[1], G S Grigorkina[1], Z S Demeev[1], A.G. Ramonova[1], D V Kuznetsov[2], T I Radchenko[3], T T Magkoev, Jr[1], S Ogura[4], D Sekiba[4], K Fukutani[4] and T T Magkoev[1 5]

[1]North Ossetian State University, Nanophysics and Nanotechnology Research Center (CKP), 44-46, Vatutina street, Vladikavkaz 362025, Russia

[2]National University of Science and Technology “MISIS”, Leninsky av. 4, Moscow 119049 , Russia

[3]Municipal Budget Educational Institution “MBOU SOSH 26”, pr. Kosta 221, Vladikavkaz 362020, Russia

[4]Institute of Industrial Science, The University of Tokyo, Komaba, Meguro-ku, Tokyo 153-8505, Japan

E-mail: t_magkoev@mail.ru

Abstract. Molecules being adsorbed on the surface dramatically change the physics and the chemistry of the substrate, thus offering an opportunity for precise sensing of the molecules via monitoring the transformation of the state of the adsorbent. Combination of different types of substrates, forming the surface interface border, generally exhibits synergistic effect dramatically enhancing sensitivity factor. In this regard, the aim of the present work is to demonstrate how the metal/oxide interface, formed in-vacuo (10-10 Torr) by deposition of metal nanoclusters on oxide support, affects behavior of NO and CO molecules being adsorbed on the surface. Coadsorption of NO and CO molecules on the Ni clusters deposited on MgO(111) film formed on Mo(110) crystal has been studied by reflection–absorption infrared spectroscopy (RAIRS) and temperature programmed desorption (TPD). Observed high sensitivity of metal/oxide interface to the molecular behaviour suggests an opportunity for design of new molecular sensors based on the metal/oxide nanostructures.

1. Introduction

Nanoparticles, especially those supported by solid surfaces, thus forming interface borders, exhibit extremely high sensitivity to a variety of physical and chemical phenomena. Among these, molecular adsorption and surface reaction are one of the key issues attracting considerable interest of researches to such nanostructures [1]. For instance, combining metal nanoparticle with the solid oxide support provides synergistic effect dramatically enhancing molecular sensitivity and selectivity, thus offering an opportunity of design of new high-sensitive molecular sensors [2,3]. In this case, however, it is necessary detail atomic-level understanding of the physics and chemistry of molecular interaction at the metal/oxide interface. In relation to this the aim of the present study was to design in-vacuo (10^{-10} mbar) the model metal/oxide nanostructure consisting of Ni nanoparticles supported on MgO(111)

[5] To whom any correspondence should be addressed.

thin film and investigate its sensitivity and selectivity to carbon monoxide (CO) and nitric oxide (NO) molecules. One of the key features of the designed metal/oxide nanostructure is that the 0.2 nm thick MgO(111) polar film formed on Mo(110) bulk crystal exhibits high electrostatic surface field dramatically affecting molecular structure of adsorbing molecules, thus producing new reaction pathways and sensitivity features.

2. Experimental

Measurements have been carried out in ultra-high vacuum chamber (base pressure: 10^{-10} mbar) equipped by conventional Fourier transform infrared spectrometer adapted to measure normal vibrations at surfaces at grazing incidence and detection angles (Fig. 1), and quadrupole mass-spectrometer.

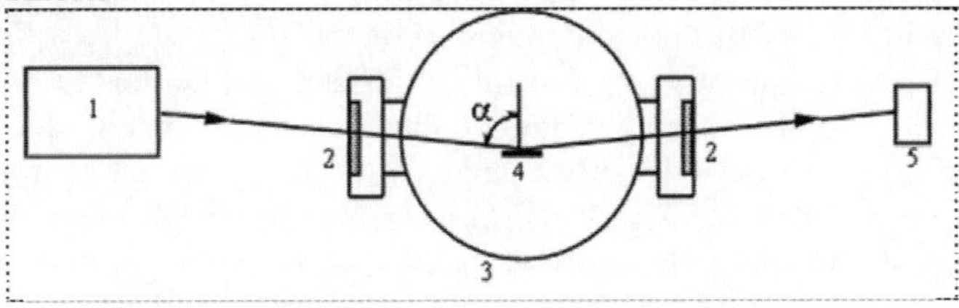

Figure 1. RAIRS scheme. 1-interferometer, 2-CaF_2 windows, 3–UHV chamber, 4–sample, 4–CdHgTe detector

Additional techniques included Auger electron spectroscopy (double pass cylindrical mirror analyser with coaxial electron gun), work function measurements (Anderson method) and low energy electron diffraction (four-grid rear view optics). Preparation and control of MgO(111) film on Mo(110) was done in situ by a well-known procedure of oxidation of Mg vapour in an oxygen ambient at a substrate temperature of 900 K. This polar surface, usually unstable in case of bulk crystal, forms in the thin film due to the ordering influence of the hexagonal atomic symmetry of the underlying Mo(110) support. In the measurements, the thickness of MgO film was 5 monolayers, i.e. 0.2 nm. Nickel overlayers on MgO film were deposited by thermal evaporation at a substrate held at 600 K. According to Auger uptake curves the three-dimensional islands are formed in this case. Calibration of Ni flux was done by monitoring the characteristics of a well-studied Ni/Mo(110) system: Mo Auger signal attenuation with Ni coverage, superstructure formation and work function change. The average size of the clusters was estimated by measuring the adsorbate (Ni(LVV)/substrate (O(KVV) Auger ratio versus Ni coverage plot. Prior to CO and NO adsorption the formed Ni/ MgO(111) system was cooled down to 85 K. Adsorption of gas molecules was done by backfilling the UHV chamber to a partial pressure not exceeding 5 x 10^{-8} mbar (1L= 10^{-6} mbar x s).

3. Results and discussion

Series of infrared spectra obtained upon initial CO adsorption (curve 1) and subsequent NO dosing (curves 2-7) on Ni/MgO(111) are shown in Fig. 2(a). The spectrum of pure CO consists of two absorption bands at 1837 and 2026 cm^{-1}. Assuming structural ordered faceting of Ni clusters formed at elevated substrate temperature one can compare this lines to those which are formed on Ni(111) bulk crystal and were initially assigned to species bound in two-fold bridge and atop sites, respectively. Upon NO dosing to CO/Ni/MgO(111) a new band at 1479 cm^{-1} appears, increases in intensity and shifts to higher wavenumbers with increasing exposure (curves 2-7). Such behaviour of the band was assigned to the change of NO adsorption geometry in two-fold bridge sites from tilted at low coverage to upright at saturation. Up to the NO exposure of 1.5L the CO vibration bands do not change substantially: One can notice a slight increase in the high-frequency CO band and corresponding decrease for the low-frequency band with almost no change of the wavenumber (curves 2-4). Dramatic change in the CO absorption bands occurs at NO exposure of 2L (curve 5). The intensity of these bands considerably decreases accompanied with their blue shift by about 20 cm^{-1}, and appearance of a new band at a 2131 cm^{-1} (curve 5). Further increase of NO exposure results in the blue shift up to 2150 cm^{-1} and an intensity increase of the new line labelled as 'A' and decrease of the intensity of the line initially centred at 2026 cm^{-1} (labelled as 'B') (Fig. 2(b), curves 5-7). At the same time, coverage behaviour of the NO absorption band shows little sensitivity to the presence of pre-adsorbed CO: Qualitatively this behaviour is about the same as for NO on pure nickel. This allows to assume that mutual influence of adsorbed molecules on each other is quite different: adsorbed CO undergoes dramatic changes whereas NO remains almost intact, regarding the vibration properties. To find out the origin of the new fine 'A' (Fig. 2(b)) the isotopically labelled ^{13}CO has been used to measure IR spectra. It was found that all the fines, shown in Fig. 2(b) were systematically shifted to lower wavenumbers by 35 cm^{-1}. This can be viewed as an evidence that the line 'A' is rather due to CO molecules, than to other species which could form on the surface. In this case, however, it seems surprising that the wavenumber of the fine 'A' at NO exposure of 4L (Fig. 2(b), (curve 7) is higher than that for CO molecules in the gas phase: 2140 cm^{-1}. Additional information on the properties of the systems under consideration has been obtained by temperature programmed desorption (TPD) measurements. The ^{13}CO TPD spectra of the systems corresponding to IR spectra (curve 1 and 7, (Fig. 2(a)) are shown in Fig. 2(c) (curves 1 and 2, respectively). It can be seen that despite the fact that the CO IR intensity dramatically decreases after NO dosing (compare curves 1 and 7 in Fig. 2(a)), the amount of CO molecules on the surface does not noticeably change as follows from the comparable TPD peak areas in these two cases (curves 1 and 2 in Fig. 2(c)). At the same time the shape of the CO desorption spectra changes with NO post-adsorption: Intensity of the high temperature peak 'B' at 480 K decreases and a new peak 'A' at about 380 K appears. It is reasonable to assign the latter one to those CO species which give rise to the vibrational band at 2130-2150 cm^{-1}, since both of the fines labelled as 'A' in RAIRS and TPD appear after NO adsorption.

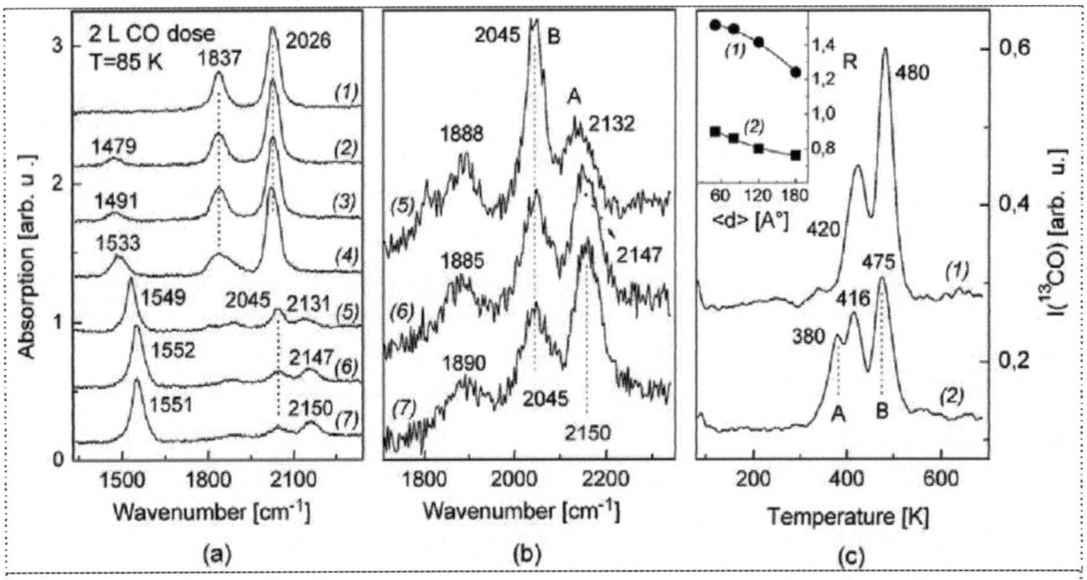

Figure 2. RAIRS (a, b) and TPD (c) spectra for NO+CO/Ni/MgO(111) system. MgO film thickness: c.a. 20 A. Average Ni cluster size (d): 120 A. CO exposure: 2 L. NO exposure (L), (a), (b): 1-0; 2-0.5; 3-1.0; 4-1.5; 5-2.0; 6-2.5; 7-3; (c): 1-0; 2-3. CO and NO adsorption temperature: 85 K. Inlay in (c) shows dependence of the ratio (R) of intensities of RAIRS (curve 1) and TPD (curve (2) peaks 'A' and 'S' (R=A/B) on the value of the average Ni cluster size (d). High-frequency part of curves 5,6 and 7 in (a) is shown in a bigger scale in (b).

The TPD data suggest that the observed decrease in CO IR intensity caused by NO adsorption is not due to the decrease of concentration of CO species on the surface. Therefore, the IR intensity decrease can be attributed to at least two reasons: (1) Decrease of the CO dynamical dipole moments perpendicular to the surface (RAIRS method in the configuration used is sensitive to these modes), and (2) Change in the adsorption site and/or geometry of CO molecules. Vibrational intensity decrease suggests that the CO species which initially were aligned along the surface normal become tilted as NO is added. This change occurs only after achieving of the NO coverage corresponding to the exposure of 1.5L (Fig. 2(a), curve 5). At the same time, a new high-frequency species ‘A’ appear (curves 5-7). Until the exposure of 1.5L the CO vibrational spectrum does not considerably change (curves 1-4). This relative stability of CO IR bands with account of the behaviour of NO IR band, which resembles that on pure nickel, suggests that up to 1.5L of NO exposure, the CO and NO adsorbed species do not considerably influence each other. Adsorption of NO does not cause the change of the CO adsorption site, which would otherwise lead to the considerable change of the wavenumber. The dipole-dipole interaction between these species results only in some change of the wavenumber of CO species (by 4 cm^{-1}) and slight intensity variation by 1% (Fig. 2(a), curves 1-4). Dramatic change of the CO IR. band at NO exposure of 2L and higher is an evidence of substantial transformation of the properties of CO molecules. Intensity decrease of the IR line “B” by the order of magnitude (Fig. 2(a), curve 5) can be attributed to the fact that at higher exposures, when the density of NO layer grows, the CO molecular axis becomes tilted with respect to the surface normal due to the increased dipole repulsion. Appearance of a high-frequency band ‘A’ (Fig. 2(a) and (b)) can be reconciled if one assumes the change of the adsorption site by a fraction of CO molecules which is caused by subsequent NO adsorption. Taking into account that the observed value of the wavenumber (2150 cm^{-1}) is higher than that for CO adsorbed on Ni crystal planes and which is even higher than that for CO in the gas phase, it is possible to expect that this might be due to CO bound to the cluster/oxide perimeter interface. To check this possibility the RAIRS and TPD A/B peak ratio R has been measured as a function of cluster average size (d) (Fig. 2(c), inlay). Observed increase of the R value for both RAIRS and TPD with decreasing cluster size provides an evidence of the validity of the above assumption, because the total perimeter of the cluster/oxide interface increases with decreasing cluster size, thus increasing the relative population of CO species bound to this interface. Such an effect might be stimulated by the quite high local electric field existing at the border of the cluster and the polar oxide surface [4].

4. Conclusion
Carbon monoxide molecules adsorbed on the surface of Ni clusters deposited onto MgO(111) thin film change their properties after subsequent adsorption of nitric oxide molecules evidenced by the dramatic change of the CO vibrational properties and thermal desorption spectra. These changes occur only after achieving a certain coverage of NO corresponding to the exposure of about 2L. At the same time vibrational behaviour of NO molecules shows little sensitivity to the presence of initially adsorbed CO. The observed features in RAIRS and TPD can be reconciled if one assumes a change of the adsorption geometry and site by a fraction of CO molecules displaced by NO adsorption from the cluster surface inner area to the cluster/oxide perimeter interface. This high molecular site-specificity and sensitivity can be viewed as a fundamental background of technology of design of new precise molecular and gas sensors and catalysts.

Acknowledgements
The work was supported by The Ministry of Education and Science of Russian Federation (Goszadanie SOGU) and Government of RSO-Alania. The reported study was funded by RFBR and Russian-Japanese (RFBR-JSPS) grants.

References
[1] Cuenya B R and Behafarid F 2015 *Surf. Sci. Reports* **70** 135
[2] Kulkarni S C and Patil D S 2015 *Sensor Lett.* **13** 294
[3] Raghav R and Srivastava S 2015 *Sensor Lett.* **13** 254
[4] Brown W A and King D A 2000 *J. Phys. Chem.* **B104** 2578

IOP Conf. Series: Materials Science and Engineering **108** (2016) 012038 doi:10.1088/1757-899X/108/1/012038

Highly sensitive multi-layer pressure sensor with an active nanostructured layer of an organic molecular metal

V Laukhin[1,2,4] V Lebedev[2], E Laukhina[2,3], C Rovira[2,3] and J Veciana[2,3]
[1] Institució Catalana de Recerca i Estudis Avançats (ICREA), Barcelona, Spain
[2] CIBER de Bioingeniería, Biomateriales y Nanomedicina (CIBER-BBN), Spain
[3] Institut de Ciencia de Materials de Barcelona (ICMAB-CSIC), Campus UAB, Bellaterra, Spain

E-mail: vladimir@icmab.es

Abstract: This work addresses to the modern technologies that need to be instrumented with lightweight highly sensitive pressure sensors. The paper presents the development of a new plain flexible thin pressure sensor using a nanostructured layer of the highly sensitive organic piezoresistive metal β-$(BEDT\text{-}TTF)_2I_3$ as an active component; BEDT-TTF=bis (ethylenedithio)tetrathiafulvalene. The original construction approach permits one to operate the developed sensor on the principle of electrical resistance variations when its piezoresistive layer is elongated under a pressure increase. The pressure sensing element and a set of gold electrodes were integrated into one compact multi-layer design. The construction was optimized to enable one generic design for pressure ranges from 1 to 400 bar. The pressure tests showed that the sensor is able to control a small pressure change as a well definite electrical signal. So the developed type of the sensors is very attractive as a new generation of compact, lightweight, low-cost sensors that might monitor pressure with a good level of measurement accuracy.

1. Introduction

Engineering flexible all-organic sensing materials with electrical detection principle brings great opportunities in the field of physical sensors for their applications in hi-techs [1-3]. A need for flexible pressure sensors that can be easily located at the body tissue, synthetic and natural textiles, artificial skins, catheters, etc. is constantly on the rise across many hi-techs and industries [2, 4]. In this context, our interest in the use of piezo-resistive organic molecular metals as active polycrystalline layers in sensor engineering has grown steadily over the last decade due to their important material properties, such as electrical conductivity, high strain (pressure) resistance sensitivity, long-term stability, and weightless [5a, 6]. Additionally, it is worthwhile mentioning that the covering process of polymeric films with the oriented highly piezo-resistive polycrystalline layer of the molecular conductor β-

[4] To whom any correspondence should be addressed

$(ET)_2I_3$ (ET=BEDT-TTF=bis(ethylenedithio)tetrathiafulvalene, figure 1) is a simple two step procedure carried out at moderated temperature [7, 8].

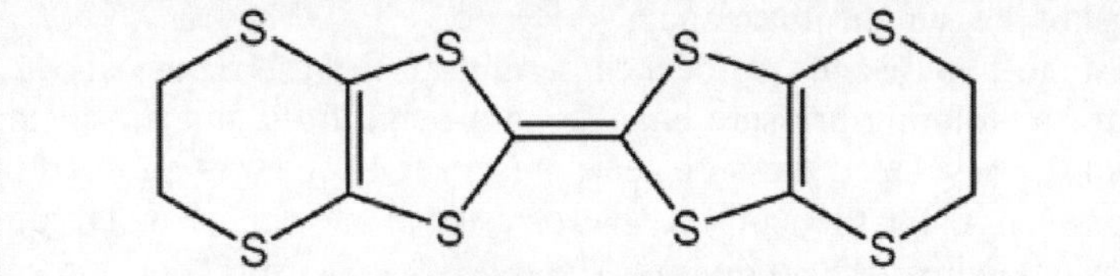

Figure 1. Skeletal formula of BEDT-TTF

The polycrystalline conductive covering layer reveals attractive mechanical properties such as excellent adhesion to a polymeric substrate and high fracture toughness. We showed that the electro-mechanical behaviour of these films is conditioned by the softness of the lattice of the BEDT-TTF-based molecular conductors [5]. The maximum strain (ε_{max}) before the fracture of the conductive layer is about 1 % [9] that is much greater than the same characteristic for conventional metals and semiconductors, in which very little deformation (ε_{max}≅0.2 %) occurs before the point, at which the stress is sufficient to induce failure [10]. The tests demonstrated that BL (bi-layered) films are able to control very small strain (relative strain 10-3 %) in a wide range of deformation. Haematoxylin-Eosin staining of the tissues adjacent to the sensing BL film and to the silicon band showed the less inflammatory reaction to the above mentioned BL film as compared with the reaction to a silicone band [11]. So, the developed sensing BL films have a higher biocompatibility than standard silicone-based materials used in surgery. The BL film-based membranes with sensitivity to pressure changes being 13.3 Ω/mmHg have been developed and the first simple prototypes of devices for monitoring the body movements (breathing rhythms [5a], blood pulse [12], and intraocular pressure [11]) have been designed and fabricated. It should be note that all developed prototypes operated on the principle of electrical resistance variations when the plans of sensitive BL films were elongated/deformed under load. The BL film dilatation results in stretching the conductive crystallographic *ab* plane of the "*c*" oriented sensing crystallites of the β-$(BEDT\text{-}TTF)_2I_3$ molecular metal and in its turn increases the distances between the BEDT-TTF molecules layered along the *ab*-plane (figure 2). This leads to a decrease of the intra- and inter- BEDT-TTF molecular stack transfer integrals with an enormous increase of the BL film resistance [6].

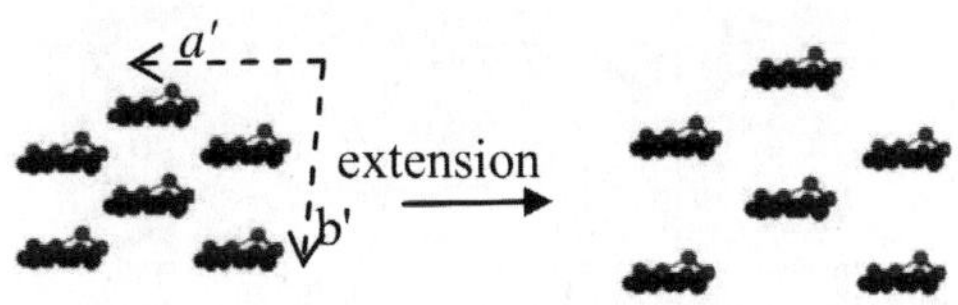

Figure 2. Schematic representation of the extension of the crystallographic *ab* plan of the sensing crystallites of the piezo-resistive metal β-$(BEDT\text{-}TTF)_2I_3$ under 2D extension

The tests of devices [5a, 11, 12] demonstrated that the BL film-based sensors are able to replace conventional sensors in biomedical technologies that need to control any very delicate bending as well as quite large deformation. However, it remains a challenge to develop film sensors being capable of measuring quite large pressure values. In this connection we focus our study on the development of design approaches enable the use of delicate highly strain sensing BL films for monitoring hundreds of bars with a high level of measurement accuracy.

Here we demonstrate two design approaches permitting the use of the above mentioned piezo-resistive BL films for monitoring pressure changes in a range from ambient conditions up to a few hundred bars. One of the developed pressure sensors (sensor 1) operates on the principle of electrical resistance decrease when the lattice of the piezo-resistive molecular metal β-$(BEDT\text{-}TTF)_2I_3$ is compressed under quasi-hydrostatic conditions. The other one (sensor 2) operates on the principal of electrical resistance increase when the conductive crystallographic ab plane of the β-$(BEDT\text{-}TTF)_2I_3$ crystallites is stretched in all directions due to the pressure induced deformation

2. Fabricating compact flexible multi-layer pressure sensors with the β-$(BEDT\text{-}TTF)_2I_3$ as a piezo-resistive component

2.1. Preparing flexible piezo-resistive BL-based gauge

In line with the early reported method [7, 8] we first prepared a 25 μm thick polycarbonate (PC) films that contain a 2 wt. % of BEDT-TTF which is a precursor for the molecular metal β-$(BEDT\text{-}TTF)_2I_3$. The films were cast on glass supports at 130 °C from a 1,2-dichlorobenzene solution of PC and BEDT-TTF.

In order to cover the film with the $(BEDT\text{-}TTF)_2I_3$-based layer, we exposed the film surface to the vapors of a solution of iodine in dichloromethane. The surface of a polycarbonate film easily swells under its exposure to dichloromethane vapors; this swelling facilitates a migration of BEDT-TTF molecules from the film bulk to the swollen film surface where the part of donor molecules are oxidized to radical cations by I_2, which penetrates in the film surface together with dichloromethane vapors. This redox process induces the rapid nucleation of the highly insoluble $(BEDT\text{-}TTF)_2I_3$ molecular metal and the metallic facing layer is formed. The treatment of the film surface with iodine/dichloromethane vapors resulted in the formation of the covering layer of the α-phase of $(BEDT\text{-}TTF)_2I_3$. The prepared polycarbonate/α-$(BEDT\text{-}TTF)_2I_3$ BL film was annealed at 150°C during 30 min; this annealing permitted us to prepare the highly piezo-resistive polycarbonate/β-$(BEDT\text{-}TTF)_2I_3$ BL film with gauge factor 20 [5, 8].

The surface analysis on a micro scale, performed using "Quanta FEI 200 FEG-ESEM" scanning electron microscope (SEM) showed that the crystallites are of submicron sizes (Figure 3).

Figure 3. SEM image of the conductive covering layer based on β-$(BEDT\text{-}TTF)_2I_3$

The formation of the β-$(BEDT\text{-}TTF)_2I_3$-based sensing layer was confirmed by the X-ray analysis. As figure 4 shows, the X-ray diffraction pattern presents the family of *(00l)* reflections of β-$(BEDT\text{-}TTF)_2I_3$ [5a,b]; therefore, the β-$(BEDT\text{-}TTF)_2I_3$-based layer of the polycarbonate/β-$(BEDT\text{-}TTF)_2I_3$ BL film is formed by c^* oriented crystals that is in a good agreement with early reported data [7, 8]. At this orientation the c^* axis of the linked crystals is mainly perpendicular to the film surface and, consequently, their conductive crystallographic *ab* plans are parallel to it.

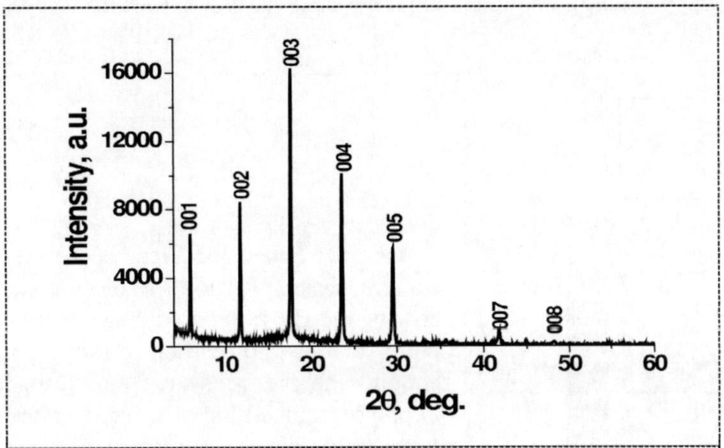

Figure 4. X–ray powder diffraction pattern of the piezoresistive covering layer of β-$(BEDT)_2I_3$

The BL film-based gauges with an area of ca. 12 × 5 mm^2 were cut and then mounted as a pressure sensing component in two different compact flexible multi-layer pressure sensors (hereafter 1 and 2).

2.2. Setting up pressure sensor 1

Pressure sensor 1 was packaged using a two-step process (figure 5) in which the BL film-based gauge was integrated with 4 gold electrodes into a compact flexible two layer structure (figure 6). This design involves the use of the electrical response of the lattice of β-$(BEDT\text{-}TTF)_2I_3$ to pressure when it is compressed under quasi hydrostatic conditions; high surface friction of polycarbonate layers permits their use as a pressure-transmitting medium.

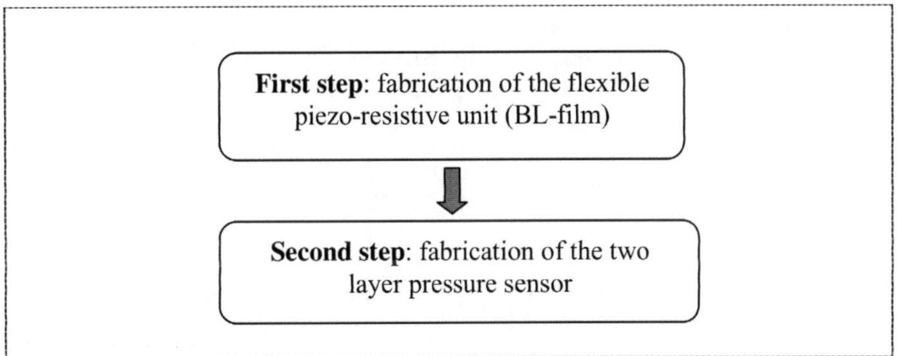

Figure 5. Scheme of the fabrication of pressure sensor 1

To fabricate pressure sensor 1, the BL film-based gauge of 25 μm thick with an area of ca. 12 × 5

mm^2 was glued with the conductive side up to a polycarbonate-based supporting film of 400 μm thick and the area 30 × 30 mm^2 as it is schematically shown in figure 6, top. Then, another polycarbonate film of 200 μm thick with the plan area of 30 × 50 mm^2 on which surface the four gold electrodes of 1 μm thick were preliminarily deposited was placed over the BL film-based gauge to make electrical contacts with the sensing part of the piezo-resistive unit; the film with gold electrodes was fixed using glue as it is shown in figure 6, bottom. To complete pressure sensor 1, four Cu wires were attached to the gold electrodes using graphite paste; to remove any stress from the gold electrodes, Cu wires were additionally attached (using glue) to the polycarbonate film with the plan area of 30 × 50 mm^2 as it is shown in figure 6, bottom.

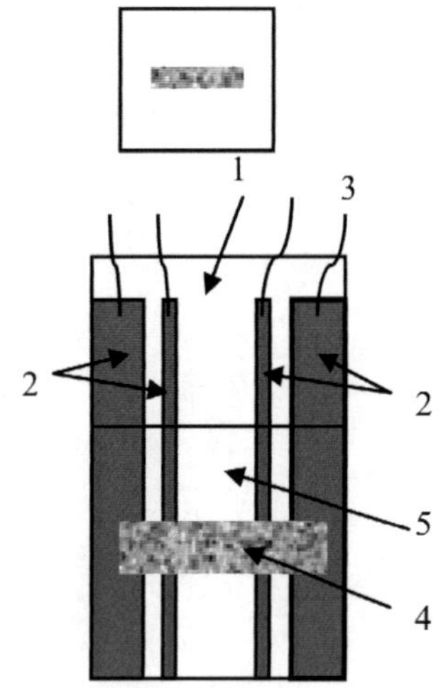

Figure 6. Schematic representation of the pressure sensor 1: (top) a piezo-resistive unit showing the BL film-based gauge location on the centre of the polycarbonate supporting film; (bottom) the piezo-resistive measuring element: (1) polycarbonate film, on which surface the gold electrical electrodes (2) were deposited; (3) four Cu wires attached to the gold electrodes using graphite paste; (4) BL film-based gauge between the polycarbonate films of 200 and 400 μm thick with a plane areas 30×50 and 30×30 mm^2, respectively (5)..

2.3. Setting up pressure sensor 1.

Pressure sensor 2 was packaged using a three-step process (figure 7), in which the BL film-based gauge was integrated with 4 gold electrodes into a compact flexible six layer structure.

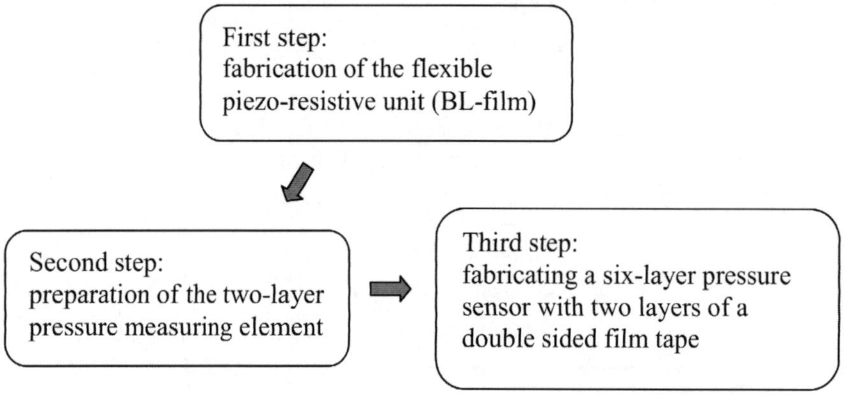

Figure 7. Scheme of the fabrication of multi layer pressure sensor 2

A key design issue for pressure sensor 2 is the use of two layers of a double sided film tape that characterized with small surface friction. This design proposes the dilatation of the plan of the flexible pressure measuring element shown in figure 6, top due to small surface friction when pressure is load perpendicularly to its plane. The piezo-resistive unit and flexible two layer pressure measuring element were manufactured in series as it was described in the previous section (figure 6). At the final step, the flexible pressure measuring element was sandwiched between two Kapton layers of 130 μm thick using a double sided film tape (figure 8).

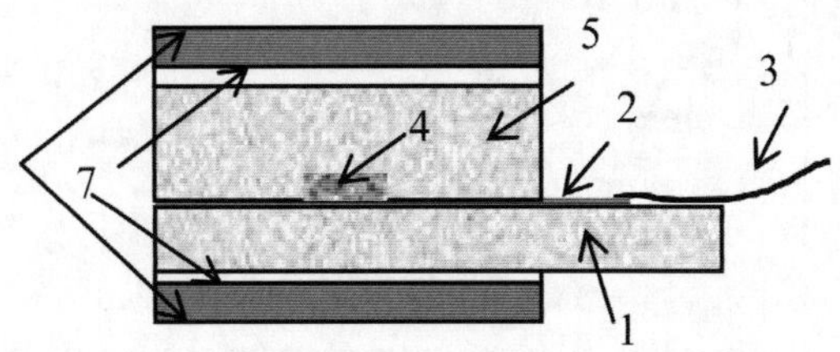

Figure 7. Schematic view of the multi-layer pressure sensor 2: (1) polycarbonate film on which the gold electrical contacts (2) were deposited; (3) four Cu wires attached to the gold contacts using graphite paste; (4) piezo-resistive BL film-based gauge; (5) polycarbonate film of 400 μm thick with a plane area 30x30 mm; (6) two Kapton films of 130 μm thick with a plane area 30x30 mm glued to the polycarbonate films with a double sided film tape (7)

3. Pressure tests

To study the effect of pressure on the electrical resistance of the developed multi-layer pressure sensors, the sensors 1 and 2 were placed on the piston of the hydraulic press ("Unipress", Poland) in turn. Sensors were subjected to different pressure load cycles by hydraulic press that pushes the piston with the pressure sensor against the stationary disc; the electrical response of the sensor was simultaneously measured by Multimeter Agilent 34970A in a four contacts configuration. Here it should be noted that the developed design permits 1-D pressure load on the area 30x30 mm^2 of the multi-layer sensors excluding the area were Cu wires are attached to the gold electrodes (figure 9).

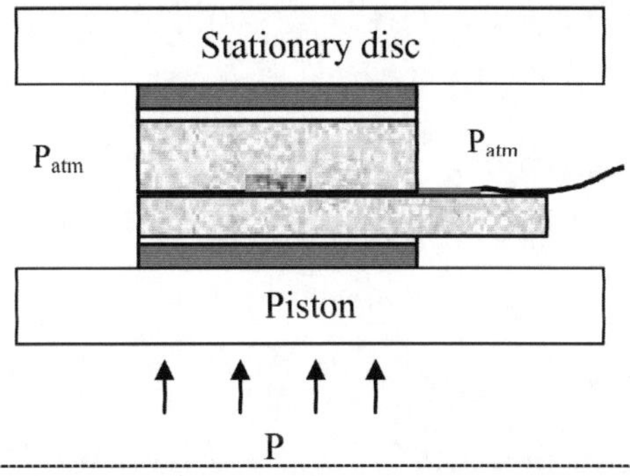

Figure 9. Schematic representation of pressure loading at cyclic pressure load tests of developed multi layer sensors

3.1. Electrical response of sensor 1 to pressure

As figure 10 shows that the electrical response of sensor 1 to the multi step cyclic load is a reversible well-detected signal; the maximum pressure loading was 545 bars

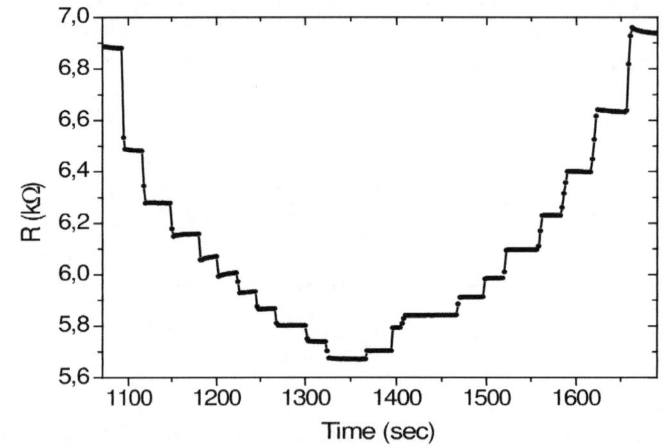

Figure 10. Electrical resistance response of sensor 1 to the pressure variation in the range 0-545-0 bar; each step up and down is 55 bars.

For better visualization of the electro-mechanical behaviour of sensor 1, its electrical resistance response to step-like cyclic pressure loading is plotted as relative resistance changes versus pressure in figure 11. With reference to this figure it can be seen that the resistance of sensor 1 significantly decreases when pressure increases from 0 up to 545 bars.

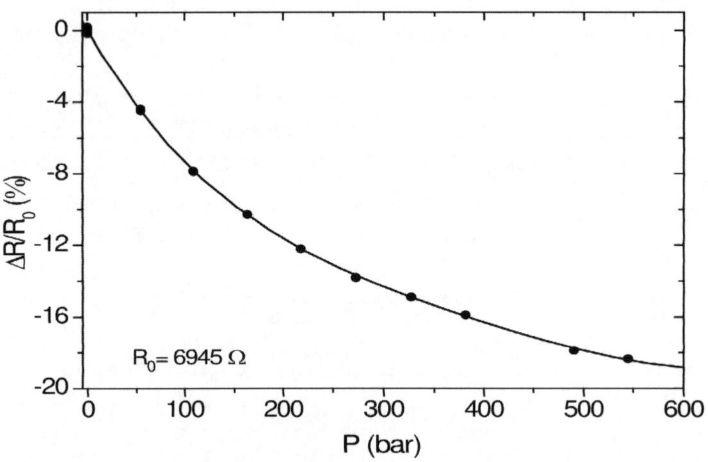

Figure 11. Electrical resistance response of sensor 1 to pressure loading

Therefore, as it was suggested in section 2.2., this design approach allowed us to operate the sensor on the principle of electrical resistance variations when the lattice of piezo-resistive crystallites of β-(BEDT-TTF)$_2$I$_3$ is compressed under load. Under compression of the β-(BEDT-TTF)$_2$I$_3$ lattice the distances between the BEDT-TTF molecules decreases (figure 12) that in turn increases the intra- and inter- BEDT-TTF molecular stack transfer integrals that results in decreasing of the BL film resistance. As the pressure-dependent compression of the lattice parameters monotonically slows down when pressure increased from 0 up to 545 bars, the sensor electrical response to pressure significantly deviates from the linearity. This nonlinar behavior is one of the disadvantages of sensor 1 that is able to impose significant restrictions for sensor applications.

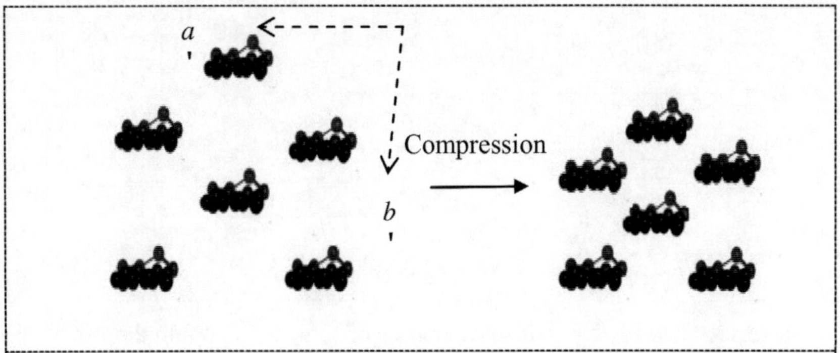

Figure 12. Schematic representation of the crystallographic *ab* plan of the sensing crystallites of the piezo-resistive metal β-(BEDT-TTF)$_2$I$_3$ under quazi hydrostatic compression.

3.2. Electrical response of sensor 2 to pressure
In this section we will show that the optimized multi layer construction of the sensor 2, whose operated principle is based on the expansion of the piezo-resistive β-(BEDT-TTF)$_2$I$_3$ layer under pressure load is able to overcome the above mentioned problem.

Multi-layer pressure sensor 2 was subjected to different pressure load cycles using the hydraulic press as it is schematically shown in figure 9. The pressure tests reveal that electrical response of the multi-layer sensor 2 to pressure is very strong and well reproducible (figure 13).

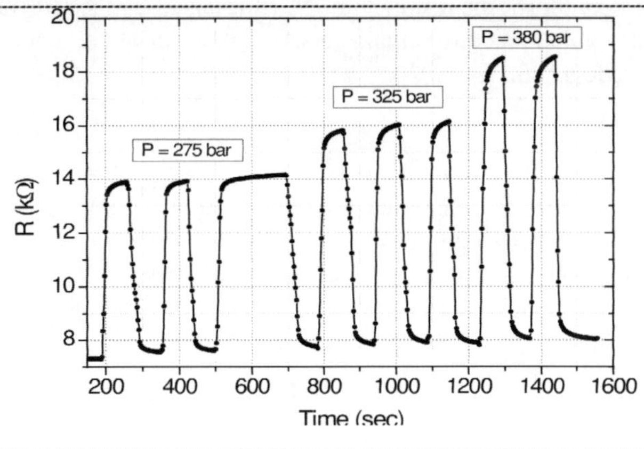

Figure 13. Illustration of repeatability of the respond of the sensor 2 at different pressure cycles: 0–275-0 bar, 0-325-0 bar and 0–380-0 bar.

For better visualization, the electrical resistance data, which were collected for the second loop of multi cyclic pressure load in the range 0-275 bar, are plotted as relative resistance changes versus pressure in figure 14. The figure shows that in contrast to sensor 1 the electrical response of sensor 2 to pressure does not deviate essentially from the linearity.

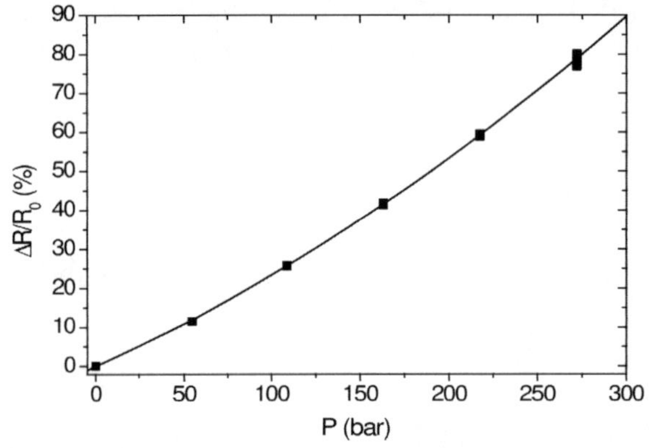

Figure 13. Normalized resistance change of sensor 2 *vs* applied pressure for the pressure range 0-275 bar.

The presence of two layers of a double sided film tape is one of the key solutions in this sensor construction that permits the sensor to solve the problem of the non linear pressure-dependent compression of the lattice parameters. Due to low surface friction of the tape layers, the plane of the piezo-resistive BL film-based gauge is stretched when pressure is applied. Under extension, the distances between the BEDT-TTF molecules in the lattice of the β-(BEDT-TTF)$_2$I$_3$ crystallites increases (figure 2) that in turn decreases the intra- and inter- BEDT-TTF molecular stack transfer integrals and as a result the resistance of the β-(BEDT-TTF)$_2$I$_3$ layer increases (figure 13, 14). The sensitivity of the senor, calculated as the ratio between the relative resistance change and pressure change was found to be around 0.3%/bar. Taking into account the initial resistance value of the sensor (7.6 kOhms), it is easy to estimate that the sensor sensitivity corresponds to 23 Ohms/bar. Thus, the developed multi-layer design allows the use of thin BL film-based gauges for fabricating compact flat pressure sensors which are able to control small pressure changes in the wide pressure range (0-400 bar) as a well defined electrical signal.

4. Summary

This study showed that the BL films: polycarbonate/(001) oriented β-(BEDT-TTF)$_2$I$_3$ layer may be easily mounted in pressure sensors as a flexible pressure measuring element.

The construction of a compact flexible multi-layer pressure sensor was designed and optimized to enable one generic design for pressure ranges at least up to 400 bars.

The pressure tests demonstrated that the developed type of the sensors is very attractive as a new generation of durable, low-cost, all-organic pressure sensors that could monitor quite large pressure values with a good level of measurement accuracy.

5. Acknowledgements

The authors acknowledge the financial support from Instituto de Salud Carlos III, through “Acciones CIBER.” The Networking Research Center on Bioengineering, Biomaterials and Nanomedicine (CIBER-BBN), an initiative funded by theVINational R&D&I Plan 2008–2011, Iniciativa Ingenio 2010, Consolider Program, CIBER Actions and financed by the Instituto de Salud Carlos IIIwith assistance from the European Regional Development Fund. The authors also appreciate the financial support through the projects: BE-WELL (CTQ2013–40480-R) granted by DGI (Spain), and GenCat

(2014-SGR-17) financed by DGR (Catalunya), the European Commission's Seventh Framework Programme for Research under contracts FP7-OCEAN-2013-614155.

6. References

[1] Yirish S, Kirinaki N V, Myshkin I L 2005 World Sensors and MEMS Markets: Analysis and Trends *Sensors & Traducers Magazine (S&T e-Digest)* **62** Issue 12 456-461

[2] Jacobs I S, Axisa F, Schmitt P M, Gehin C, Delhomme G, McAdams E, Dittmar A 2005 Flexible technologies and smart clothing for citizen medicine, home healthcare and disease prevention *IEEE Transactions on Information Technology in Biomedicine* **9** (3) 325-336

[3] Someya T, Sekitani T, Iba S, Kato Y, Kawaguchi H, Sakurai T 2004 Conformable, flexible, large-area networks of pressure and thermal sensors with organic transistor active matrixes, *Proc. Natl. Acad. Sci. USA* **101** 9966–9970

[4] Leonardi M,. Pitchon E M, Bertsch A, Renaud Ph, Mermoud A 2009 Wireless contact lens sensor for intraocular pressure monitoring: assessment on enucleated pig eyes *Acta Ophthalmol.* **87** 433–437

[5] (a) Laukhina E, Pfattner R, Ferreras L R, Galli S, Mas-Torrent M, Masciocchi N, Laukhin V, Rovira C, Veciana J 2010 Ultrasensitive Piezoresistive All-Organic Flexible Thin films, *Adv. Mater.* **22** 977-991; (b) Laukhina E 1997 Systematic Study of the $(ET)_2I_3$ Reticulate Doped Polycarbonate Film: Structure, ESR, Transport Properties and Superconductivity *J. Phys. I France* **7** 1665-1675

[6] Kondo Y R, Higa M, Kagoshima S, Hoshino H, Mori T, Mori H 2006 Electrical and Structural Properties of θ-type BEDT-TTF Organic Conductors under Uniaxial Strain *J. Phys. Soc. Jpn.***75** 044716-7

[7] . Jeszka J K, Tracz A 1992 A new method of growing crystalline networks in polymer matrices by simultaneous CT complex formation and in situ crystallization *Polym. Adv. Technol.* **3** 139–142

[8] Laukhina E, Rovira C, Ulanski J 2001 Organic metals as active components in surface conducting semi-transparent films *Synth. Met.* **21** 1407-8

[9] Laukhina E, Pfattner R, Mas-Torrent M, Rovira C, Veciana J, Laukhin V 2011 Film-based Sensors with Piezoresistive Molecular Conductors as Active Components: Strain Damage and Thermal Regeneration *Sensors & Transducers Journal* **10** 1-12

[10] a) Bridgman P W 1952 *Physics of High Pressure* G. Bell & Sons Ltd., London, 257, b) Hoffmann K 1987 *Einfu¨hrung in die Technik des Messens mit Dehnungsmessstreifen* Hottinger Baldwin Messtechnik GmbH, Darmstadt 62

[11] Laukhin V et al 2011 Prototype for Noninvasive IOP Monitoring with a Nanostructured Polymeric Sensor Embeded in a Contact Lens *Sensors and Actuaters A* **170** 36-43

[12] Laukhina E, Lebedev V, Laukhin V, Oncins G, Pfattner R, Rovira C, Veciana J 2011 Lightweight biocompatible physical sensors: polymeric films self-metallized with organic molecular conductors *Transducers'11* Beijing 1978-1981

TiO_2-TiO_2 composite resistive humidity sensor: ethanol cross-sensitivity

Milad Ghalamboran[1,a], **Yasin Saedi**[1,b]

[1]Electronic Materials Laboratory, Electrical Engineering Department, K. N. Toosi University of Technology, Tehran 16317-14191, Iran

[a]miladghalamboran@gmail.com, [b]saedi.yasin@gmail.com

Abstract. The fabrication method and characterization results of a TiO_2-TiO_2 composite bead used for humidity sensing along with its negative cross-sensitivity to ethanol vapor are reported. The bead shaped resistive sample sensors are fabricated by the drop-casting of a TiO_2 slurry on two Pt wire segments. The dried bead is pre-fired at 750°C and subsequently impregnated with a Ti-based sol. The sample is ready for characterization after a thermal annealing at 600°C in air. Structurally, the bead is a composite of the micron-sized TiO_2crystallites embedded in a matrix of nanometric TiO_2particle aggregates. The performance of the beads as resistive humidity sensors is recorded at room temperature in standard humidity level chambers. Results evince the wide dynamic range of the sensors fabricated in the low relative humidity range. While the sensor conductance is not sensitive to ethanol vapor in dry air, in humid air, sensor's responses are negatively affected by the contaminant.

Keywords: TiO_2, Sol-gel impregnation, Composite, Humidity sensor, Dynamic range, Cross-sensitivity,

1. Introduction:

Next to temperature, humidity is the most frequently measured environmental parameter. In addition to monitoring relative humidity (RH) levels in our living spaces, humidity sensors help keep RH levels optimal for industrial production. Particularly, humidity measurement and control in low RH range is of technical significance in many advanced technologies [1-5]. Humidity-sensitive materials, i.e. materials whose electrical properties, mainly conductivity [6, 7] and/or dielectric constant [8, 9], are affected by humidity, are classified into three main categories of electrolytic [10], ceramic [11] , and polymeric [12]. Ceramic humidity sensors are typically considered superior because of their thermal, mechanical, and chemical stability, as well as their longevity and recyclability [11, 13]. Among ceramic type humidity sensors, the thin film metal oxide capacitive sensors operates based on the change in the dielectric constant of the oxide material with respect to the change in the atmospheric conditions [14]. Porous ceramics are usually used in production of humidity sensors [2]. Some of the common ceramics are $ZnCr_2O_4$, $MgAl_2O_4$, SnO_2 [15] and TiO_2 [2], Al_2O_3 [16]. Making humidity sensors from other materials, such as TiO_2, $BaTiO_3$ and α-hematite, has also been reported in the literature [17, 18].
Conductance measurement on the sensing elements of these sensors requires forming at least two metal-metal oxide contacts [19-21] stable during the normal lifetime of the device. Considering the harsh environmental conditions and the electrochemical processes taking place, such contacts are made utilizing noble metals [19]. Forming of metal contacts to polycrystalline semiconductors is an important and intricate part of the gas and humidity sensor fabrication [22]. Making metallic connections to polycrystalline TiO_2 has been the subject of many investigations [23, 24]. Environmentally hazardous gaseous components, particularly volatile organic compounds, have

negative effects on the performance and accuracy of the humidity sensors [25]. Similarly, humidity changes in the atmosphere affect the responses of gas sensors [26]. Ethanol vapor, for instance, affects the responses of both capacitive and resistive humidity sensors, and the ideal device is the one whose responses to humidity change is not influenced by such contaminants in the surrounding atmosphere. Here, we investigate the significance of the pore size distribution in the TiO_2-TiO_2 composite resistive humidity sensors, and investigate their dynamic range and cross-sensitivity to ethanol vapor.

2. Experimental:

T Rutile TiO_2 powder is mixed with distilled water for slurry preparation; 12 mg of the produced viscous slip is dropped onto the junction of two d= 60 μm pure platinum wire segments. After drying, the sample is sintered at 750°C for 30 minutes. Ten similar samples are prepared. Five samples, after cooling in air down to room temperature are impregnated with titanium alkoxide solution in ethanol. The impregnated samples are dried at room temperature and subsequently annealed at 600°C which converts the injected sol to titanium dioxide. The fabricated sample is shown in Figure 1. In this sensor structure need for metal deposition on the oxide ceramic is removed and the suspension wires, i.e. the Pt wire segments, operate as the electrical contacts. Further information regarding the fabrication process is given in references [19].

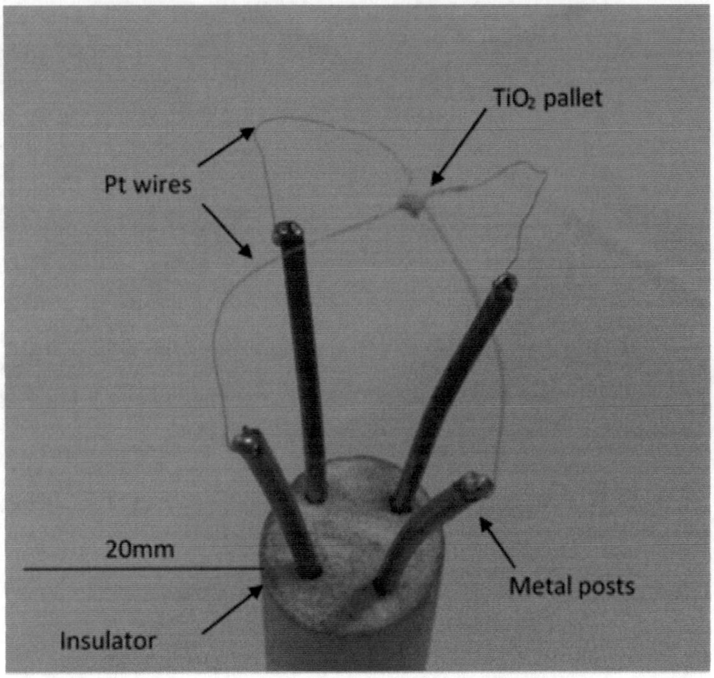

Figure 1. The TiO2-TiO2 composite bead formed on two crossed platinum wires.

Standard salts (see Table 1) are used for providing test chambers of calibrated relative humidity. Philips XL30 scanning electron microscope (SEM) is used for imaging; a micrograph of the sample sintered at 750°C is shown in Figure 2.

Table 1. Humidity level of standard salts at 25°C

Standard salt	Humidity level at 25°C
Silica gel	4%
Magnesium chloride	25%
Sodium chloride	75%
Potassium chloride	85%

Figure 2. SEM image of the TiO2 ceramic body produced.

The pore size distribution is determined by line averaging method on the obtained SEM micrographs. The population of the different size range pores is graphically presented in Figure 3.

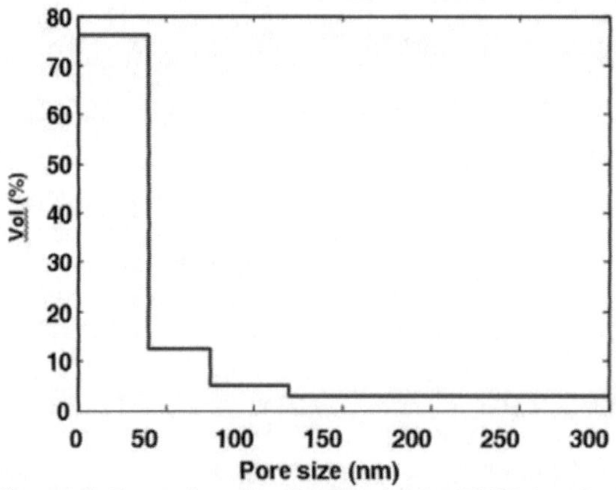

Figure 3. Pore size distribution in the TiO2 bead

The experimental layout utilized for cross-sensitivity examinations between atmospheric humidity and ethanol contamination is schematically shown in Figure 4.The temperature at the outer chamber is controlled using a system comprising a PT100 temperature sensor and a PID controller connected to a distributed fine wire heating element on the chamber, which stabilizes the temperature in the chamber at 298±0.2 K.

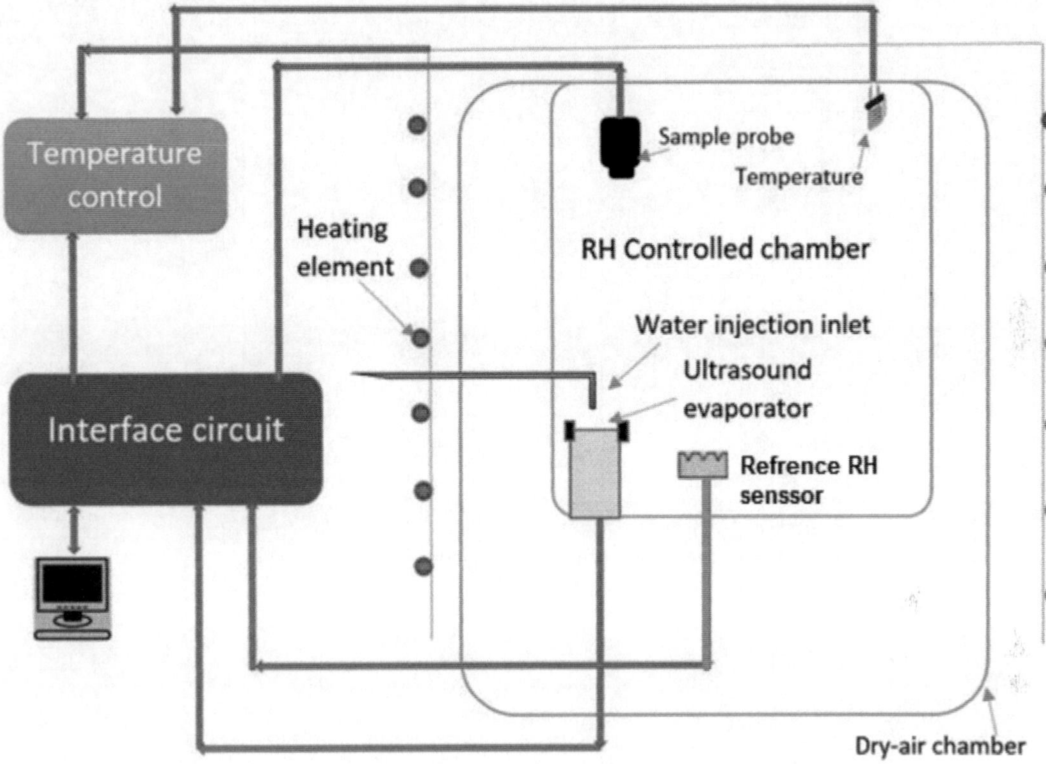

Figure 4. The schematic diagram of the experimental setup.

To minimize the electrode-ceramic connection deterioration due to the electrochemical corrosion, conductance measurements are carried out at 17 Hz. At this frequency, the capacitive admittance of the sample and parasitic parallel to the sample are insignificant compared to its conductance. Closed chambers of different RH levels, in the range of 2-95%, are obtained by injecting predetermined volumes of distilled water into the dry air (RH=2%) chamber using a micro-sampler. The injected water is evaporated to obtain the specified humidity level inside the chamber. The produced stable RH levels are monitored by a commercial humidity sensor (HIH 4000) prior to and during each measurement. The output of the reference sensor is calibrated in closed chambers of standard RH levels obtained using different saturated salt solutions (see Table-1).To examine the effect of ethanol on the humidity sensing process, ethanol is introduced to the chamber air at predetermined conditions; the temporal changes in resistivity are continuously recorded.

IOP Conf. Series: Materials Science and Engineering **108** (2016) 012039 doi:10.1088/1757-899X/108/1/012039

3. Results and discussion:

The variations of sample conductance vs. the ambient relative humidity are given in Figure 5.It is clear from this diagram that the examined sample is more appropriate for the lower relative humidity range as the slope of the characteristic diagram is higher at RH<25%. At this humidity range, water vapor condensation mainly occurs in the smaller pores of the bead [2]. This is consistent with the result showing over 80% of the pore volume to be less than 70 nm in diameter (see Figure 3). Further increase in the humidity level saturates the nano sized pores, while it still can condense in the micron sized pores leading to a considerable decrease in slope of the characteristic curve of the sensor.

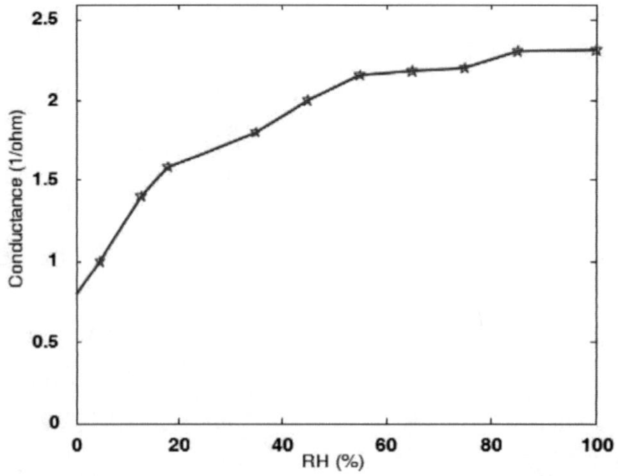

Figure 5. Variations of bead conductance with the relative humidity in the surrounding atmosphere.

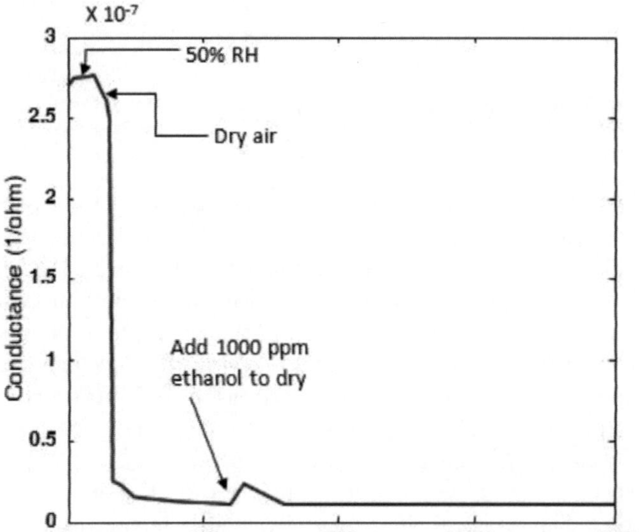

Figure 6. Temporal variation of the bead conductance; the sensor is exposed to dry air (RH<2%) at t=10; significant level of ethanol

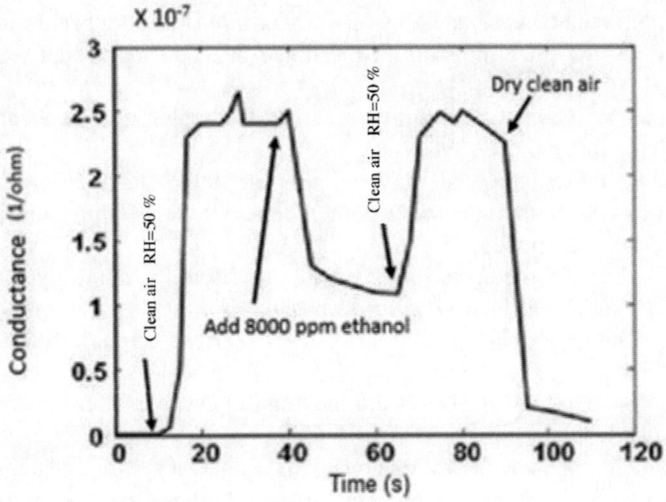

Figure 7. Response of the sensor to ethanol contamination in the wet chamber.

At room temperature, no chemical reaction occurs between the effective surface of the sensor and the adsorbed ethanol molecules, neither do the surface adsorbed ethanol molecules promote surface conductivity. This is clear from the experimental results illustrated in Figure 6, demonstrating that the sensor does not responds to the addition of substantial level of ethanol vapor to the atmosphere. The simultaneous effect of humidity and ethanol is investigated by repeating similar experiment in humid air chamber. Based on the results shown in Figure 7, the conductance of the sample at equilibrium in RH=50% air drops significantly when the atmosphere is contaminated with ethanol vapor. Since ethanol does not influence the resistivity of the device (see Figure 6), the decrease in the bead conductance is attributed to the effective competition of the ethanol molecules with water molecules in the process of surface adsorption. The results of this experiment clarify the immensity of the errors caused by the presence of organic vapor contaminants in humidity sensing.

4. Conclusion

The fabrication of a TiO2-TiO2 composite resistive humidity sensor was reported. The sensor performs best in the RH< 25%. The nanoporous structure necessary for low level humidity detection was created by sol impregnation followed by thermal annealing in air. The sensor is insensitive to the presence of ethanol in dry air, but it was shown that ethanol contamination decreases conductance of the sensor in a humid atmosphere. A model was proposed for this negative cross sensitivity.

5. References

[1] Chen, Z. and C. Lu, Humidity Sensors: A Review of Materials and Mechanisms. Sensor Letters,

2005. 3(4): p. 274-295.

[2] Hossein-Babaei, F. and S. Rahbarpour, Alteration of pore size distribution by sol–gel impregnation for dynamic range and sensitivity adjustment in Kelvin condensation-based humidity sensors. Sensors and Actuators B: Chemical, 2014. 191: p. 572-578.

[3] Hossein-Babaei, F. and P. Shabani, A gold/organic semiconductor diode for ppm-level humidity sensing. Sensors and Actuators B: Chemical, 2014. 205: p. 143-150.

[4] Islam, T., L. Kumar, and S.A. Khan, A novel sol–gel thin film porous alumina based capacitive sensor for measuring trace moisture in the range of 2.5–25ppm. Sensors and Actuators B: Chemical, 2012. 173: p. 377-384.

[5] Yamazoe, N. and Y. Shimizu, Humidity sensors: Principles and applications. Sensors and Actuators, 1986. 10(3): p. 379-398.

[6] Hossein-Babaei, F., P. Shabani, and M. Azadinia, Air-stable electrical conduction in oxidized poly [2-methoxy-5-(2-ethylhexyloxy)-p-phenylene vinylene] thin films. Applied Physics Letters, 2013. 103(22): p. 223303.

[7] Urbiztondo, M., et al., Zeolite-coated interdigital capacitors for humidity sensing. Sensors and Actuators B: Chemical, 2011. 157(2): p. 450-459.

[8] Kulwicki, B.M., Humidity Sensors. Journal of the American Ceramic Society, 1991. 74(4): p. 697-708.

[9] Seiyama, T., N. Yamazoe, and H. Arai, Ceramic humidity sensors. Sensors and Actuators, 1983. 4: p. 85-96.

[10] Chen, X., et al., Comparison of Y-doped Bazro3 thin films for high temperature humidity sensors by RF sputtering and pulsed laser deposition. Sensors and Actuators B: Chemical, 2010. 148(1): p. 173-180.

[11] Wang, C.-T., et al., Humidity sensors based on silica nanoparticle aerogel thin films. Sensors and Actuators B: Chemical, 2005. 107(1): p. 402-410.

[12] Traversa, E., Ceramic sensors for humidity detection: the state-of-the-art and future developments. Sensors and Actuators B: Chemical, 1995. 23(2–3): p. 135-156.

[13] Cosentino, I.C., E.N.S. Muccillo, and R. Muccillo, Development of zirconia-titania porous ceramics for humidity sensors. Sensors and Actuators B: Chemical, 2003. 96(3): p. 677-683.

[14] Bayhan, M. and N. Kavasoğlu, A study on the humidity sensing properties of ZnCr2O4–K2CrO4 ionic conductive ceramic sensor. Sensors and Actuators B: Chemical, 2006. 117(1): p. 261-265.

[15] Pingale, S.S., et al., Mechanism of humidity sensing of Ti-doped MgCr2O4 ceramics. Materials Chemistry and Physics, 1996. 46(1): p. 72-76.

[16] Dražič, G. and M. Trontelj, Preparation and properties of ceramic sensor elements based on MgCr2O4. Sensors and Actuators, 1989. 18(3–4): p. 407-414.

[17] Hossein-Babaei, F., S. Moghadam, and S. Masoumi, Forming ohmic Ag/SnO 2 contacts. Materials Letters, 2015. 141: p. 141-144.

[18] Katayama, K., et al., Humidity sensitivity of Nb2O5-doped TiO2 ceramics. Sensors and Actuators A: Physical, 1990. 24(1): p. 55-60.

[19] Hossein-Babaei, F., M.M. Lajvardi, and N. Alaei-Sheini, The energy barrier at noble metal/TiO2 junctions. Applied Physics Letters, 2015. 106(8): p. 083503.

[20] Hossein-Babaei, F. and S. Rahbarpour, Titanium and silver contacts on thermally oxidized titanium chip: electrical and gas sensing properties. Solid-State Electronics, 2011. 56(1): p. 185-190.

[21] Hossein-Babaei, F. and S. Rahbarpour, Separate assessment of chemoresistivity and Schottky-type gas sensitivity in M–metal oxide–M′ structures. Sensors and Actuators B: Chemical, 2011. 160(1): p. 174-180.

[22] Nahar, R.K., V.K. Khanna, and W.S. Khokle, On the origin of the humidity-sensitive electrical properties of porous aluminium oxide (sensor application). Journal of Physics D: Applied Physics, 1984. 17(10): p. 2087.

IOP Conf. Series: Materials Science and Engineering **108** (2016) 012039 doi:10.1088/1757-899X/108/1/012039

[23] Islam, T., et al., A micro interdigitated thin film metal oxide capacitive sensor for measuring moisture in the range of 175–625 ppm. Sensors and Actuators B: Chemical, 2015. 221: p. 357-364.

[24] Mistry, K.K., D. Saha, and K. Sengupta, Sol–gel processed Al2O3 thick film template as sensitive capacitive trace moisture sensor. Sensors and Actuators B: Chemical, 2005. 106(1): p. 258-262.

[25] Wilson, A.D., Diverse Applications of Electronic-Nose Technologies in Agriculture and Forestry. Sensors (Basel, Switzerland), 2013. 13(2): p. 2295-2348.

[26] Hossein-Babaei, F. and V. Ghafarinia, Compensation for the drift-like terms caused by environmental fluctuations in the responses of chemoresistive gas sensors. Sensors and Actuators B: Chemical, 2010. 143(2): p. 641-648.

First-principles analysis on Seebeck coefficient in zinc oxide nanowires for thermoelectric devices

K Nakamura[1,2]

[1] Center for the Promotion of Interdisciplinary Education and Research, Kyoto University, Kyoto 615-8540, Japan
[2] Department of Materials Science and Engineering, Egypt-Japan University of Science and Technology, New Borg El-Arab, Alexandria 21934, Egypt

E-mail: koichi@cpier.kyoto-u.ac.jp

Abstract. The Seebeck coefficient of ZnO⟨0001⟩ nanowires was simulated on the basis of first-principles calculation, to discuss the potential for future application to thermoelectric devices. Simulation procedure by means of the electronic band structure with one-dimensional periodic boundary condition was presented, and dependences of the Seebeck coefficient on temperature and carrier concentration have been investigated for many kinds of n- or p-doped ZnO⟨0001⟩ nanowire models with 1.00–2.65 nm diameter. For the direct band-gap semiconducting models, a magnitude of the Seebeck coefficient increases gradually as temperature rises in the p-doped state, and a significant effect of miniaturization to nanowire on the Seebeck coefficient has been brought out, such as about 1000 μV/K in the p-doped state and –820 μV/K in the n-doped state for the $(ZnO)_{24}$ nanowire model with 1×10^{17} cm^{-3} carrier concentration at room temperature. Similar characteristics of the Seebeck coefficient were observed for some indirect band-gap semiconducting models. At the end of this paper, the simulation was extended to the no band-gap conducting models with some modification.

1. Introduction

Zinc oxide (ZnO) is an important semiconducting material with excellent performance in optics, electronics, energy science, and so forth [1-7]. Thermoelectric properties of ZnO-based materials have been investigated in terms of thermoelectric power generation at high temperature, by introduction of microstructure, nanovoid or solid solution with other metal oxides to improve thermoelectric performance [1-3]. On the other hand, it has been reported both experimentally and theoretically that miniaturization to lower-dimensional materials at nanoscale is one of promising techniques to enhance energy conversion efficiency [8,9]. From the viewpoint of electronic state, density of state (DOS), $g(\varepsilon)$, of thermoelectric materials is radically affected by the dimension reduction, leading to the improvement of thermoelectric performance.

The detailed electronic states of wurtzite-ZnO nanowires with ⟨0001⟩ orientation have been reported in the previous study [10]; according to the difference in wall structure, ZnO⟨0001⟩ nanowires can be divided into three groups by their conductivities such as no band-gap conducting models, direct band-gap semiconducting models, and indirect band-gap semiconducting models. As expected, different characteristics of thermoelectricity should be observed in response to the classification of conductivity or electronic state in these ZnO nanowires. In this study, the Seebeck coefficient, one of the contributing terms to the power factor in the thermoelectric figure of merit, of ZnO⟨0001⟩ nanowires was simulated on the basis of the first-principles calculation of model structures; particularly, carrier concentration and temperature dependences of the Seebeck coefficient were discussed to estimate the potential for future application to thermoelectric devices.

2. Simulation method

2.1. First-principles calculation

Electronic band structures of the one-dimensional periodic boundary models for wurtzite-ZnO nanowires have been calculated by VASP program package [11,12] based on the density functional theory (DFT) [13]. For the DFT exchange-correlation interaction, the generalized-gradient approximation method was used with the Perdew-Wang (PW91) functional [14,15]. We adopted the three-dimensional supercell approximation technique with ultrasoft pseudopotentials [16]. The cutoff energy for wave functions of electrons with plane-wave expansion was set at 40 Ry (544 eV) for bulk ZnO crystal and 30 Ry (408 eV) for ZnO nanowire models.

ZnO nanowire models have been devised in the previous papers [10,17], as shown in figure 1. The list of wurtzite-ZnO⟨0001⟩ nanowire models is tabulated in table 1. These models were devised by cutting out a fragment with one-dimensional periodic boundary condition from the optimized bulk ZnO. The direction of the fragment with the one-dimensional periodic boundary is set to [0001] and can be defined as the longitudinal direction of ZnO nanowire. A periodic boundary condition along the transverse or perpendicular direction to the nanowire in the three-dimensional hexagonal supercell was given by inserting sufficient space between nanowires with a large cell parameter. On the condition that sufficient space in the ZnO nanowire model disturbs the interaction between nanowires, the electronic band structure of the ZnO nanowire model can be reduced to be one dimensional, which is dependent on only one reciprocal coordinate, k_z.

The characteristics of electronic state for each nanowire group listed in table 1 have been discussed in the previous study [10]. We can find some common features of band structure among the nanowire models belonging to the same group. Figure 2 shows examples of band structure for each nanowire group.

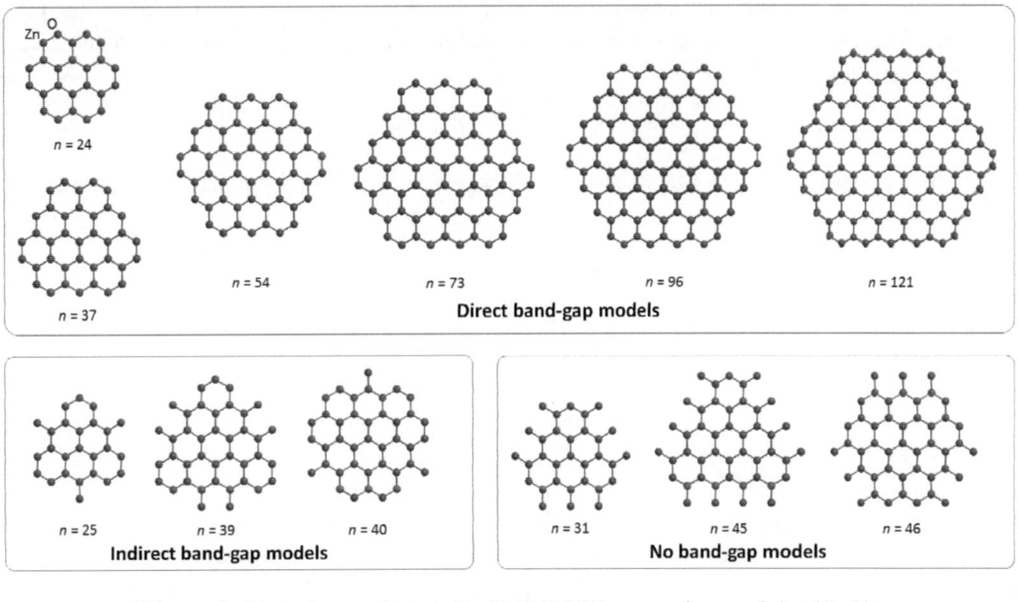

Figure 1. Top views of wurtzite-ZnO⟨0001⟩ nanowire models, $(ZnO)_n$.

IOP Conf. Series: Materials Science and Engineering **108** (2016) 012040 doi:10.1088/1757-899X/108/1/012040

Table 1. Structural parameters of wurtzite-ZnO⟨0001⟩ nanowire models and their hexagonal unit cells

Model	Wire radius R (Å)	Wire volume per unit cell V (Å^3)	Cell parameter (Å)	
			$L_{//}$	L_z
Direct band-gap				
$(ZnO)_{24}$	5.019	394.3	22.000	5.280
$(ZnO)_{37}$	6.383	665.3	25.000	5.280
$(ZnO)_{54}$	8.266	1035.0	27.000	5.280
$(ZnO)_{73}$	10.034	1453.9	30.000	5.280
$(ZnO)_{96}$	11.534	1971.4	33.000	5.280
$(ZnO)_{121}$	13.272	2538.1	37.000	5.280
Indirect band-gap				
$(ZnO)_{25}$	5.690	443.6	22.00	5.280
$(ZnO)_{39}$	7.586	763.9	25.00	5.280
$(ZnO)_{40}$	7.586	763.9	25.00	5.280
No band-gap				
$(ZnO)_{31}$	6.570	591.4	22.00	5.280
$(ZnO)_{45}$	8.266	911.8	25.00	5.280
$(ZnO)_{46}$	8.266	911.8	25.00	5.280

2.2. Seebeck coefficient

The Seebeeck coefficient S is given by the ratio of electromotive force $\vec{E}$ to temperature gradient $\vec{\nabla}_{\vec{r}}T$ incorporated implicitly in the Boltzmann transport equation [18,19]. In the stationary state, velocity distribution function of carriers f in semiconductor should be derived from the Boltzmann transport equation by use of relaxation time τ as

$$f(\varepsilon)-f_0(\varepsilon)=\tau(\varepsilon)\vec{v}(\varepsilon)\left(df_0/d\varepsilon\right)\cdot\left\{e\vec{E}+[(\varepsilon-\varepsilon_{\rm F})/T]\vec{\nabla}_{\vec{r}}T\right\}, \tag{1}$$

where e is the elementary charge, $\vec{v}$ is carrier velocity, ε_F is the Fermi energy, and f_0 is the Fermi-Dirac distribution defined by

$$f_0(\varepsilon)=\{\exp[(\varepsilon-\varepsilon_{\rm F})/k_{\rm B}T]+1\}^{-1} \tag{2}$$

with the Boltzmann constant k_B. Then, S can be evaluated from the information from band structure of the material in terms of integrals including DOS with respect to energy ε by

$$S=-(1/eT)\left\{\left[\int g(\varepsilon)\tau(\varepsilon)\varepsilon v^2(\varepsilon)(df_0/d\varepsilon)d\varepsilon\Big/\int g(\varepsilon)\tau(\varepsilon)v^2(\varepsilon)(df_0/d\varepsilon)d\varepsilon\right]-\varepsilon_{\rm F}\right\}, \tag{3}$$

where interval of the integrals corresponds to the region of carrier occupation. All physical properties in (3) except for the relaxation time can be easily determined from the first-principles electronic state, because the amount of carrier velocity v can be evaluated from the slope of each band energy curve. At the case that the relaxation time is constant regardless of ε, the effect of τ on S is canceled.

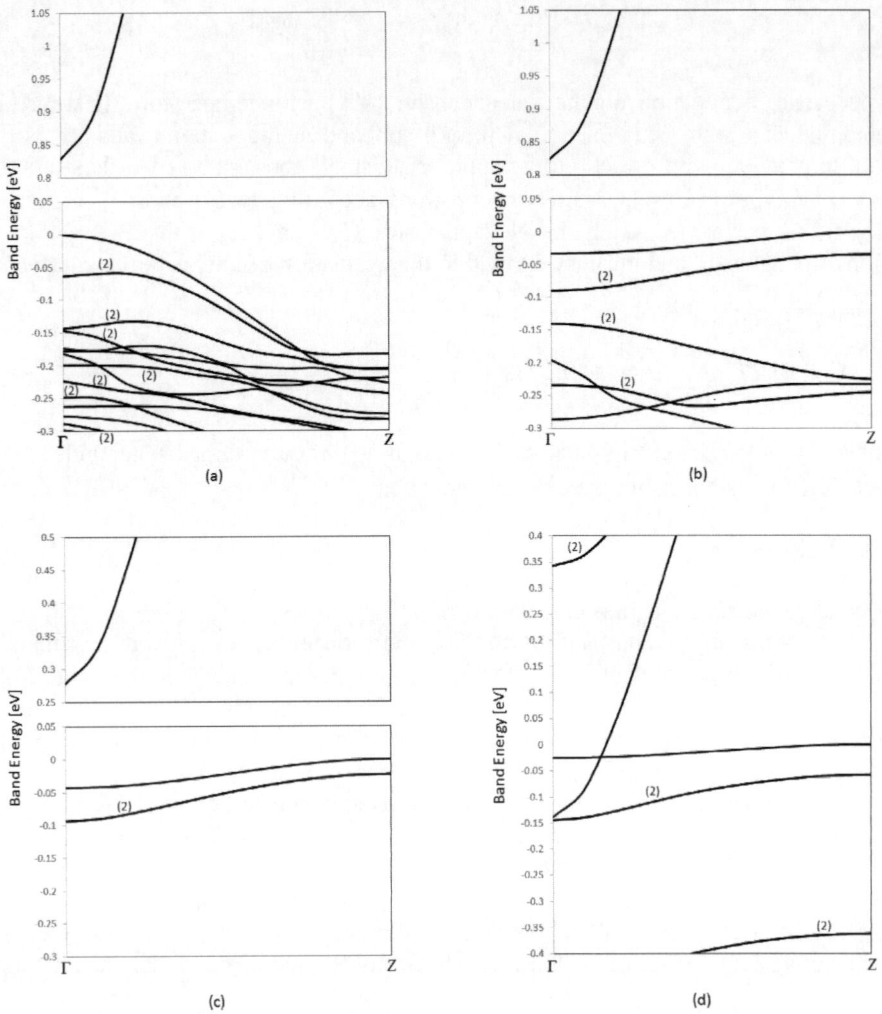

Figure 2. Band diagrams of (a) $(ZnO)_{121}$, (b) $(ZnO)_{25}$, (c) $(ZnO)_{39}$, and (d) $(ZnO)_{31}$ models in the vicinities of conduction-band bottom and valence-band top. The energy zero is defined respectively at the valence-band top or the maximum of occupied subband, and mark (2) indicates degenerate subbands..

Carrier doping should be considered for the application to devices, and we have presented the original procedure to represent the n- or p-doped state in semiconductors [20,21]. An excess or lack of electrons by as much as one particle per unit cell in order to carry out regular first-principles calculations for the n- or p-type semiconductor state should lead to an enormous overestimation of carrier concentration as much as the 10^{21} cm^{-3} order for our ZnO nanowire models. In practical n- or p-type ZnO nanowires, the total number of carriers per unit cell, $\delta = nV$, must be a few orders less than 1, where n is the carrier concentration and V is the volume of ZnO nanowire in the unit cell. Under the condition that a small amount of carrier occupation does not cause significant change in the band structure, δ can be given by an appropriate shift of ε_F as

$$\delta = \int_{CB} 2g(\varepsilon) f_0(\varepsilon) d\varepsilon = 2 \sum_{j \in CB} \sum_{k_z} w_{k_z} f_0(\varepsilon_{j,k_z}) \tag{4}$$

in the n-type carrier occupation on the conduction band (CB) or

$$\delta = \int_{\mathrm{VB}} 2g(\varepsilon)[1 - f_0(\varepsilon)]d\varepsilon = 2\sum_{j\in\mathrm{VB}}\sum_{k_z} w_{k_z}[1 - f_0(\varepsilon_{j,k_z})] \quad (5)$$

in the p-type carrier occupation on the valence band (VB) with temperature T, where $\varepsilon_{j,kz}$ is the intrinsic-semiconductor-state band energy of the jth subband at the k_z point, and w_{kz} is the k-point weight for k_z. In practice, we first set δ to an appropriate small constant a few orders less than 1, and then solve ε_F in n-type and p-type carrier occupations according to (4) and (5), respectively. The sampling for w_{kz} is performed using the Newton-Cotes formula [22] with 29 k_z points along the reciprocal longitudinal path, and it can be applied to the evaluation of the Seebeeck coefficient S in (3) as follows,

$$S = -(1/eT)\left\{\left[\sum_j\sum_{k_z} w_{k_z} v^2(\varepsilon_{j,k_z})\varepsilon_{j,k_z}\frac{df_0(\varepsilon_{j,k_z})}{d\varepsilon}\Big/\sum_j\sum_{k_z} w_{k_z} v^2(\varepsilon_{j,k_z})\frac{df_0(\varepsilon_{j,k_z})}{d\varepsilon}\right] - \varepsilon_{\mathrm{F}}\right\}, \quad (6)$$

where jth subbands belong to CB in the n-doped state or VB in the p-doped state, under the condition that the relaxation time is constant regardless of energy ε.

3. Results and discussion

3.1. Direct band-gap semiconducting ZnO nanowires

Figure 3 shows temperature and carrier concentration dependences of the calculated Seebeck coefficients for the $(ZnO)_{24}$ and $(ZnO)_{121}$ direct band-gap nanowire models, in the range from –200 to 300 °C. The results satisfy the well-known fundamental features of Seebeck coefficient for semiconductors [18]: (i) S should be positive in the p-doped state and negative in the n-doped state, and (ii) the absolute value of S is reduced as the carrier concentration increases.

Regarding to temperature dependence for each carrier concentration, S increases gradually as temperature rises in the p-doped state, whereas S almost keeps a fixed value regardless of temperature more than –100 °C in the n-doped state. Actually, even at high temperature, most all of carrier electrons should be distributed only on the lowest CB subband, which is located far from other higher CB subbands as shown in figure 2(a). It is considered that no redistribution of carrier electrons would suppress temperature dependence of the Seebeck coefficient for one-dimensional nanomaterials.

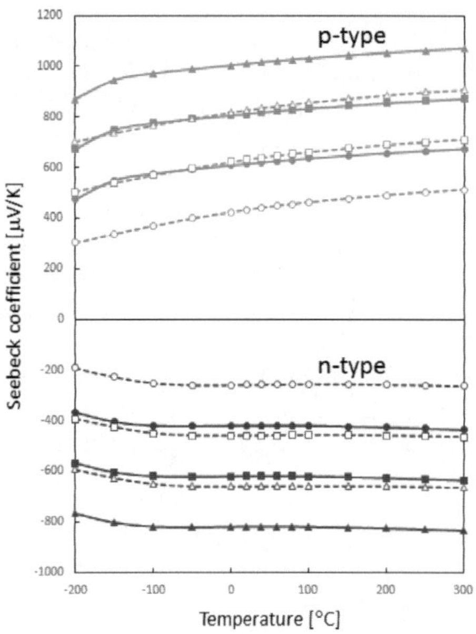

Figure 3. Temperature and carrier concentration dependences of the calculated Seebeck coefficients for the direct band-gap models. Solid and dashed lines denote the $(ZnO)_{24}$ and $(ZnO)_{121}$ models, respectively. Circle, square, and triangle plots respectively mean the carrier concentration of 1×10^{19}, 1×10^{18}, and 1×10^{17} cm^{-3}.

Table 2. Calculated Seebeck coefficients (μV/K) for the direct band-gap models with the carrier concentration of 1 × 10^{17} cm^{-3}.

Model	Temperature			
	−50 °C	20 °C	100 °C	200 °C
p-type				
$(ZnO)_{24}$	989	1010	1031	1053
$(ZnO)_{37}$	1005	1033	1050	1064
$(ZnO)_{54}$	889	927	958	985
$(ZnO)_{73}$	834	869	903	935
$(ZnO)_{96}$	806	841	875	909
$(ZnO)_{121}$	794	828	858	886
n-type				
$(ZnO)_{24}$	−820	−819	−819	−825
$(ZnO)_{37}$	−776	−774	−774	−778
$(ZnO)_{54}$	−737	−736	−735	−739
$(ZnO)_{73}$	−708	−707	−706	−708
$(ZnO)_{96}$	−681	−681	−680	−681
$(ZnO)_{121}$	−659	−660	−658	−659

The same trend in the behavior of temperature dependence was observed for all direct band-gap models. As tabulated in table 2, it has been also clarified that the absolute values of S are raised as the wire diameter decreases in both of n- and p-doped states. For the $(ZnO)_{24}$ model, S expands to as much as 1000 μV/K in the p-doped state and −820 μV/K in the n-doped state, respectively, with the carrier concentration of 1 × 10^{17} cm^{-3} at room temperature. Thus, significant effect of miniaturization to nanowire on the Seebeck coefficient can be displayed.

3.2. Indirect band-gap semiconducting ZnO nanowires

Temperature and carrier concentration dependences of the calculated Seebeck coefficients for the $(ZnO)_{25}$ and $(ZnO)_{39}$ indirect band-gap nanowire models are shown in figure 4. For the $(ZnO)_{25}$ and $(ZnO)_{40}$ models shown in table 3, similar characteristics of the Seebeck coefficient as the direct band-gap nanowire models were observed; namely, S increases gradually as temperature rises in the p-doped state and does not change regardless of temperature more than −100 °C in the n-doped state, respectively. On the other hand, for the $(ZnO)_{39}$ model, S is independent of temperature more than 0 °C in the p-doped state as well as the n-doped state.

Difference in the electronic structure between the $(ZnO)_{25}$ and $(ZnO)_{39}$ models can be clearly found in figures 2(b) and 2(c). For the $(ZnO)_{39}$ model, the highest VB subband and the second highest VB subbands in double degeneracy are completely isolated from other lower VB subbands, and therefore, the energy range of hole redistribution due to temperature change should be restricted to these three highest VB subbands. The behavior of the Seebeck coefficient for the p-doped $(ZnO)_{39}$ model strongly supports the consideration discussed above that temperature dependence of the Seebeck coefficient for one-dimensional nanomaterials would not be clearly observed unless the carrier redistribution occurs adequately.

IOP Conf. Series: Materials Science and Engineering **108** (2016) 012040 doi:10.1088/1757-899X/108/1/012040

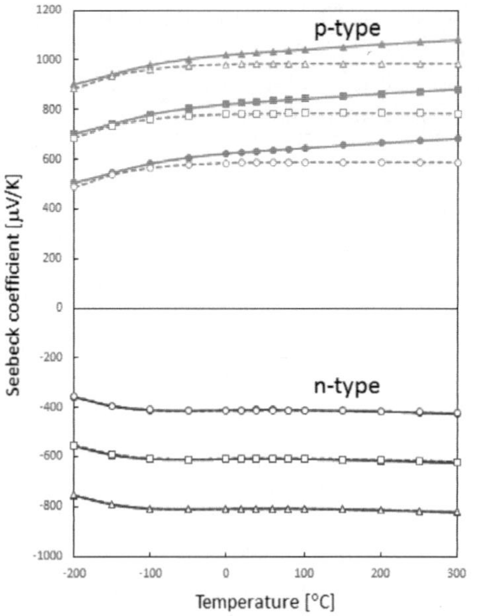

Figure 4. Temperature and carrier concentration dependences of the calculated Seebeck coefficients for the indirect band-gap models. Solid and dashed lines denote the $(ZnO)_{25}$ and $(ZnO)_{39}$ models, respectively. Circle, square, and triangle plots respectively mean the carrier concentration of 1×10^{19}, 1×10^{18}, and 1×10^{17} cm^{-3}.

Table 3. Calculated Seebeck coefficients (μV/K) for the indirect band-gap models with the carrier concentration of 1×10^{17} cm^{-3}.

Model	Temperature			
	–50 °C	20 °C	100 °C	200 °C
p-type				
$(ZnO)_{25}$	1004	1024	1043	1063
$(ZnO)_{39}$	974	982	985	985
$(ZnO)_{40}$	972	979	994	1021
n-type				
$(ZnO)_{25}$	–810	–809	–809	–815
$(ZnO)_{39}$	–811	–809	–809	–813
$(ZnO)_{40}$	–764	–762	–762	–766

3.3. No band-gap conducting ZnO nanowires

To apply the simulation method in this paper to conducting systems, (4) and (5) can be modified by using a general form of electron distribution [23] as follows,

$$\delta = \pm\left[\int 2g(\varepsilon)f_0(\varepsilon)d\varepsilon - N\right] = \pm\left[2\sum_{\text{All } j}\sum_{k_z} w_{k_z} f_0(\varepsilon_{j,k_z}) - N\right], \tag{7}$$

where N is the total number of electrons in the system, and plus-minus sign means plus for n-doped state or minus for p-doped state. In this study, the Fermi energy ε_F was determined according to (7) for each condition, and then the value of S was evaluated by using (6) under the condition that the relaxation time is constant. Practically, the effect of relaxation time on the Seebeck coefficient is relatively small for the semiconducting material, because the effective $\tau(\varepsilon)$ has a small overlap with the derivative of the Fermi-Dirac distribution $df_0/d\varepsilon$. On the contrary, the energy dependence of $\tau(\varepsilon)$ in the vicinity of the Fermi energy will affect the Seebeck coefficient significantly in the case of metallic conducting material. Therefore, the relaxation time should be taken into consideration for the quantitative analysis on the Seebeck coefficient for the conducting material, but the simulation by using (6) is available for easy evaluation.

Figure 5 shows temperature dependence of the calculated Seebeck coefficients for the $(ZnO)_{31}$, $(ZnO)_{45}$, and $(ZnO)_{46}$ no band-gap nanowire models. A magnitude of S for the conducting material is much smaller than that for the semiconducting one as is well-known, and plus/minus in S does not correspond to p-/n-doping. In contrast to the semiconducting material, a large amount of carriers in the conducting one exists in the quite vicinity of the Fermi energy. Accordingly, the characteristics of the Seebeck coefficient for the conducting material is very sensitive according to the small difference in the electronic structure.

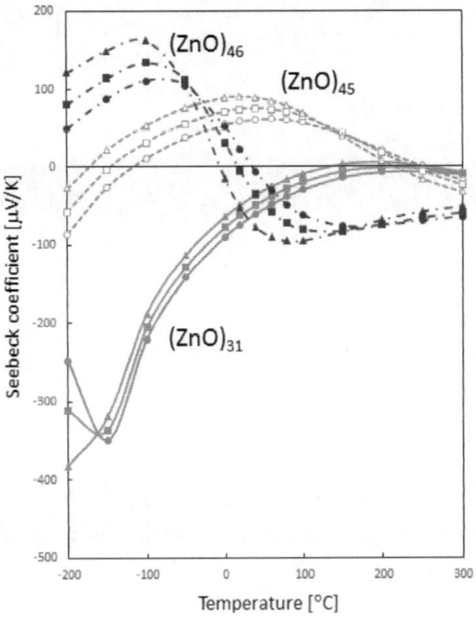

Figure 5. Temperature and doped state dependences of the calculated Seebeck coefficients for the no band-gap models. Solid, dashed, and chain lines denote the $(ZnO)_{31}$, $(ZnO)_{45}$, and $(ZnO)_{46}$ models, respectively. Square, circle, and triangle plots respectively mean the intrinsic state, p-doped state, and n-doped state. The carrier concentration in the doped states is set to 1×10^{20} cm^{-3}.

4. Summery

In this paper, the Seebeck coefficient of ZnO⟨0001⟩ nanowires was simulated on the basis of first-principles calculation. Simulation procedure by means of the electronic band structure with one-dimensional periodic boundary condition was presented, and dependences of the Seebeck coefficient on temperature and carrier concentration have been demonstrated for many kinds of n- or p-doped ZnO⟨0001⟩ nanowire models with 1.00–2.65 nm diameter [10,17]. For the direct band-gap semiconducting models, a magnitude of the Seebeck coefficient increases gradually as temperature rises in the p-doped state, and a significant effect of miniaturization to nanowire on the Seebeck coefficient has been brought out, such as about 1000 μV/K in the p-doped state and –820 μV/K in the n-doped state for the $(ZnO)_{24}$ nanowire model with 1×10^{17} cm^{-3} carrier concentration at room temperature. Similar characteristics of the Seebeck coefficient were observed for some indirect band-gap semiconducting models. The simulation was extended to the no band-gap conducting models with some modification for determination of the Fermi energy [23], and much smaller values of the Seebeck coefficient were obtained, as compared with the semiconducting models.

Acknowledgments
This work was supported by JSPS KAKENHI Grant Number 26420665, for which the author expresses his gratitude.

References

[1] Hossein-Babaei F and Masoumi S 2014 *Key Eng. Mater.* **605** 185
[2] Tsubota T, Ohtaki M, Eguchi K and Arai H 1998 *J. Mater. Chem.* **8** 409
[3] Ohtaki M, Shige S and Maehara S 2004 *Trans. Mater. Res. Soc. Jpn.* **29** 2727
[4] Özgür Ü, Aivov Y I, Liu C, Teke A, Reshchikov M A, Doğan S, Avrutin V, Cho S-J and Morkoç H 2005 *J. Appl. Phys.* **98** 041301
[5] Kamiya T and Kawasaki M 2008 *MRS Bull.* **33** 1061
[6] Wang Z L 2010 *Nano Today* **5** 540
[7] Wang Z L 2012 *Adv. Mater.* **24** 4632
[8] Pichanusakorn P and Bandaru P 2010 *Mater. Sci. Eng. R* **67** 19
[9] Cornett J E and Rabbin O 2011 *Appl. Phys. Lett.* **98** 182104
[10] Nakamura K 2015 *Key Eng. Mater.* **644** 16
[11] Kresse G and Hafner J 1993 *Phys. Rev. B* **47** R558
[12] Kresse G and Furthmüller J 1996 *Phys. Rev. B* **54** 11169
[13] Hohenberg P and Kohn W 1964 *Phys. Rev.* **136** B864
[14] Perdew J P and Wang Y 1992 *Phys. Rev. B* **45** 13244
[15] Perdew J P, Chevary J A, Vosko S H, Jackson K A, Pederson M R, Singh D J and Fiolhais C 1992 *Phys. Rev. B* **46** 6671
[16] Vanderbilt D 1990 *Phys. Rev. B* **41** 7892
[17] Nakamura K 2015 *Jpn. J. Appl. Phys.* **54** 06FJ11
[18] Ioffe A F 1957 *Semiconductor Thermoelements and Thermoelectric Cooling* (London: Infosearch)
[19] Lundstrom M 2000 *Fundamentals of Carrier Transport* (Cambridge: Cambridge Press)
[20] Nakamura K, Isono Y and Toriyama T 2008 *Jpn. J. Appl. Phys.* **47** 5132
[21] Nakamura K, Isono Y, Toriyama T and Sugiyama S 2009 *Phys. Rev. B* **80**, 045205
[22] Hildebrand F B 1956 *Introduction to Numerical Analysis* (New York: McGraw-Hill) p 160
[23] Gamil M, Nakamura K, Fath El-Bab A M R, Tabata O and Abd El-Moneim A 2013 *Model. Numer. Simul. Mater. Sci.* **3** 117

Using granular Co-Al_2O_3 spacer for optimization of functional parameters of the FeMn/$Fe_{20}Ni_{80}$ magnetoresistive films

A N Gorkovenko, V N Lepalovskij, O A Adanakova and V O Vas'kovskiy

Department of Magnetism and Magnetic Nanomaterials, Ural Federal University, Ekaterinburg 620002, Russia

E-mail: a.n.gorkovenko@urfu.ru

Abstract. In this paper we studied the possibility of tailoring the functional properties of the multilayer magnetoresistive medium with unidirectional anisotropy and the anisotropic magnetoresistance effect (AMR). Objects of the research were composite Co-Al_2O_3 films and Ta/$Fe_{20}Ni_{80}$/$Fe_{50}Mn_{50}$/$Fe_{20}Ni_{80}$/Co-Al_2O_3/$Fe_{20}Ni_{80}$/Ta multilayers structures obtained by magnetron sputtering and selectively subjected vacuum annealing. Structure, magnetic and magnetoresistive properties of the films in the temperature range 77÷440 K were investigated.

1. Introduction

Magnetic films containing exchange-coupled antiferromagnetic and ferromagnetic layers are widely used in magnetic field sensors based on the effect of giant magnetoresistance. However, in some cases, the value of the output signal is not as important as the functionality of the magnetic sensors, for example, odd conversion function [1]. For such purpose, it is more appropriate to use medium which has both the exchange bias and the anisotropic magnetoresistance (AMR). The main objective of this work was to determine the optimal structural characteristics of this medium in the classical ferro-antiferromagnetic FeMn/$Fe_{20}Ni_{80}$ system. As was shown earlier [2] its practical potential is realized in a SiO_2/Ta(5)/$Fe_{20}Ni_{80}$(5)/$Fe_{50}Mn_{50}$(20)/$Fe_{20}Ni_{80}$(45)/Ta(5) multilayer. The presence of two layers of permalloy (seed layer with thickness of 5 nm and thicker layer is responsible for AMR) allows realization of the exchange bias and ensure the thermal stability of the exchange bias field H_e. At the same time within this structure variation of H_e is complicated which means it is difficult to regulate intentionally the sensitivity of the magnetic field sensors. Efficiency of the interlayer exchange coupling can be changed by introduction of ultra-thin non-magnetic spacer (for example, Al_2O_3) between the interacting layers. However, direct separation of $Fe_{50}Mn_{50}$ and $Fe_{20}Ni_{80}$(45) layers, even with a very thin spacer (thickness L is no more than 0.2 nm), leads to the absence of the magnetic exchange bias. Therefore, the spacer introduced into the permalloy layer, which is responsible for AMR, divides it into two unequal parts. The smaller one with thickness of 5 nm is directly adjacent and exchange coupled to the antiferromagnetic layer and larger one with thickness of 40 nm is providing AMR. Furthermore, in order to impart additional features of the spacer for variation of the interlayer interaction the spacer was enriched with Co which is localized in $Co_x(Al_2O_3)_{100-x}$ with x<50% in form of nanoscale granules [3]. Thus, the films which were investigated in this work had the following general structural formula SiO_2/Ta(5)/$Fe_{20}Ni_{80}$(5)/$Fe_{50}Mn_{50}$/$Fe_{20}Ni_{80}$(5)/Co-Al_2O_3(L)/$Fe_{20}Ni_{80}$(40)/Ta(5).

2. Samples and techniques

Composite Co-Al_2O_3 films with thickness of 100 nm and multilayers with variable thickness of Co-Al_2O_3 spacer were prepared by magnetron sputtering (Orion sputtering system, AJA International, Inc.) onto Corning glass substrates. The composite layers were formed by co-sputtering of Co and Al_2O_3 targets. Other layers were deposited using homogeneous targets of appropriate compositions. Films were deposited at presence of the uniform magnetic field which was used to ensure unidirectional anisotropy. A series of samples was annealed in vacuum at 350 ° C for one hour in order to increase AMR effect. Structure and elemental composition of the films were investigated by Carl Zeiss AURIGA CrossBeam scanning electron microscope and high-resolution 30 CM Super Twin transmission electron microscope. Magnetic properties and magnetoresistance of multilayers were determined using the LakeShore 7407VSM measurement system at temperatures ranging from 77 to 440 K.

3. Results and discussion

As was mentioned above in composites such as Co-Al_2O_3 cobalt can be in granular form. Moreover, the granules at room temperature due to the small granule size demonstrate superparamagnetic behavior. However, being a part of the ultra-thin spacer, which separate magnetically ordered layers, the granules can contact them, become ferromagnetically ordered and serve as a bridges for the implementation of the interlayer exchange coupling at a greater distance. In this context, important factors determining the efficiency of the exchange separation of the main layers are composition and thickness of the spacer.

In a preliminary experiment carried out on single-layer films, influence of the composition on the structure and chemical state of the $Co_x(Al_2O_3)_{100-x}$ composite was evaluated. The composition was determined by the ratio of partial deposition rates of Co и Al_2O_3. Figure 1 shows a typical picture of the secondary X-ray radiation of Co obtained by a scanning electron microscope from the film with $x = 40$. In fact, it can be concluded that the metal macroscopically is distributed uniformly over the area of the sample. And if the metal is localized in the granules, their dimensions are beyond the resolution limit of this technique which is less than 10 nm. So we used indirect method for evaluation of the granules size. It was based on the analysis of magnetization curves considering compositional film as an ensemble of noninteracting superparamagnetic particles of the same size [4]. In our case, the prerequisites for choosing such model were virtually anhysteretic magnetization curves and their adequate approximation using Langevin function. This approach allowed us to obtain the dependence of average grain size *d* on the composition (for $35<x<45$ composition range see figure 1, *b*).

Calculations confirmed granules size being in the nanometer range, and the increase of Co concentration entails significant and almost linear growth of *d*. Granular structure was directly observed using transmission electron microscopy on a special $Co_{50}(Al_2O_3)_{50}$ sample, which was deposited onto NaCl crystal and subsequently separated from the substrate (figure 1, *c*). In photography provided numerous dark areas can be identified as Co granules. They are uniform in size (~ 5 nm) and separated by lighter streaks ~ 1 nm thick, which represent the dielectric matrix.

For further experiments on optimization of multilayers properties we have chosen $x = 40$ composition. If we assume that the quantitative results presented above are applicable to ultra-thin composite layers (spacers), the average size of Co grain should exceed 3 nm. In addition, such compositions are characterized by high resistivity values (3.76 Ohm*cm). This suggests good localization of Co granules and can be useful for minimizing the shunting of the responsible for AMR effect permalloy layer.

Figure 2, *a-c* shows examples of hysteresis loops of as-deposited $SiO_2/Ta(5)/Fe_{20}Ni_{80}(5)/Fe_{50}Mn_{50}/Fe_{20}Ni_{80}(5)/Co_{40}(Al_2O_3)_{60}(L)/Fe_{20}Ni_{80}(40)/Ta(5)$ films. They are measured along the axis characterized by a maximum exchange bias. It can be seen that increase of the thickness *L* of the granular spacer modifies significantly shape of the hysteresis loops. Without spacer two-stage hysteresis loop is observed (figure 2, *a*). It describes the magnetization reversal of both main (in the magnetic field H_{e1}) and auxiliary (in the magnetic field H_{e2}) permalloy layers. Moreover,

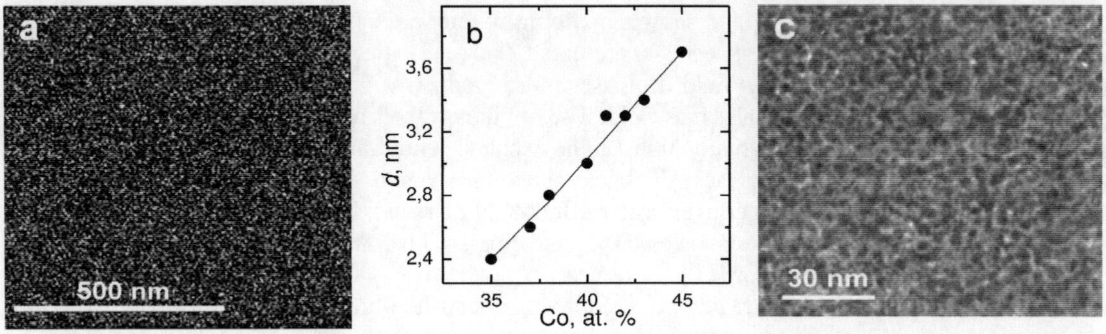

Figure 1. *a* – SEM scan of $Co_{40}(Al_2O_3)_{60}$ film in elemental analysis of Co mode; *b* – dependence of the average grain size on Co concentration; *c* – TEM image of $Co_{50}(Al_2O_3)_{50}$ film obtained by the transmission electron microscope.

both minor loops are shifted along the magnetic field axis which indicates the presence of an interlayer exchange coupling. Introduction of the spacer leads to the three-stage hysteresis loop corresponding to the magnetization reversal of the main permalloy layer and two auxiliary layers of smaller thicknesses (in magnetic fields H_{e2} and H_{e3}). The H_{e1} value of the main layer decreases monotonically with *L* increasing, and when L reaches 5 nm, it becomes almost zero. At the same time, exchange bias fields of auxiliary layers remain and become almost the same (figure 2, *c*).

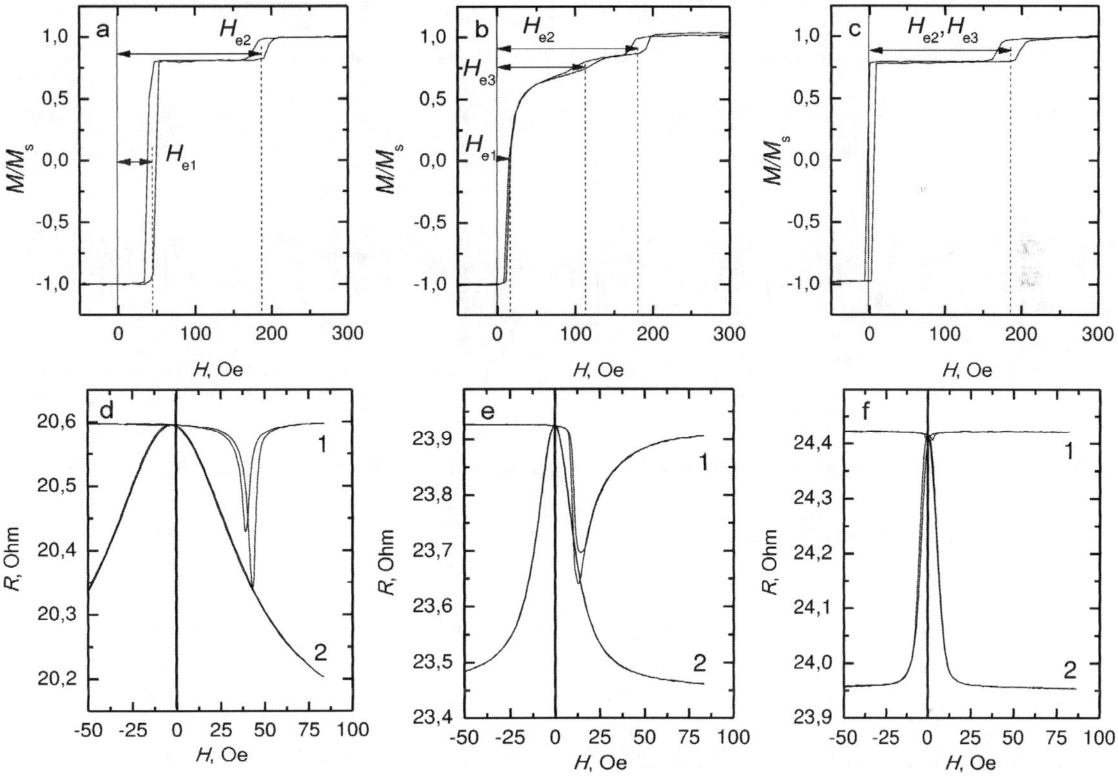

Figure 2. Magnetometric (*a-c*) and magnetoresistive (*d-f*) hysteresis loops measured along (curve 1) and perpendicular (curve 2) to electric current lines on as-deposited Ta(5)/$Fe_{20}Ni_{80}$(5)/$Fe_{50}Mn_{50}$(20)/$Fe_{20}Ni_{80}$(5)/$Co_{40}(Al_2O_3)_{60}$(*L*)/$Fe_{20}Ni_{80}$(40)/Ta(5) multilayers with different thickness *L* of the granular spacer: *a,d* – 0 nm; *b, e* – 2 nm; *c, f* – 5 nm.

In general, dependence $H_{e1}(L)$ is shown in figure 3 (curve 1). For comparison, similar $H_{e1}(L)$ dependence obtained for Al_2O_3 spacer is presented (curve 2). It can be seen that in case of the granular spacer the exchange bias field decreases more gradually. This feature can be useful for the intentional tailoring of the exchange bias field. The obvious reason for that is its Co granules, which provide the inter-exchange at relatively high *L*. The exchange coupling disappearance for L> 5 can be associated with excessing of the spacer thickness above the average size of Co grain, which leads to the disappearance of magnetic interlayer bridges. It should be noted, that $H_{e1}(L)$ dependence obtained for as-deposited films and samples subjected to annealing are qualitatively the same. In the last case, however, somewhat higher level of H_{e1} is observed.

Figure 2, *d-f* shows the magnetoresistive loops of as-deposited multilayers with different *L*. They are measured by application of the external magnetic field along (longitudinal loop) and perpendicular (lateral loop) to the electric current. Probing current in both cases was applied along the induced easy magnetization axis. The amplitude of the applied magnetic field did not exceed H_{e2} and H_{e3} values. Thus, the data obtained reflect hysteresis loops devoted to only main permalloy layer. It can be seen that the longitudinal loops are biased along the magnetic field axes, and the H_e value is close to the value of the exchange bias obtained from magnetometric loops measured on the corresponding samples. Transverse magnetoresistance loops are free from anomalies and have a typical form for magnetouniaxial films. As can be expected, exchange bias is decreasing for large L and resistivity on the transversal loops changes more rapidly. Thus, the introduction of spacer allows to increase sensitivity, but reduces the dynamic range of the medium considered as a material for magnetoresistive sensors. The advantage of the composite spacer is the possibility of a relatively smooth adjustment of these characteristics.

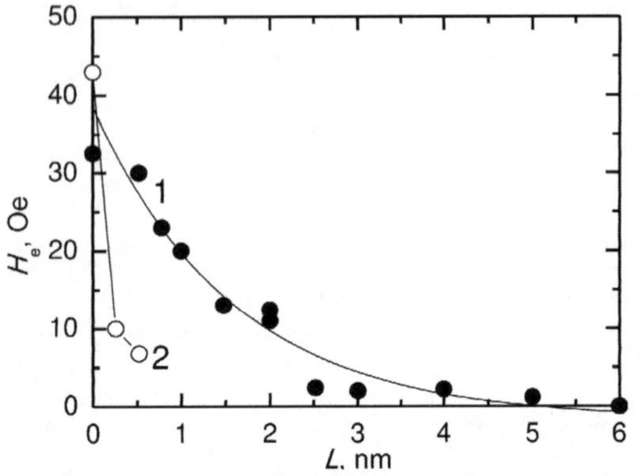

Figure 3. Dependence of the exchange bias field on the thicknesses of $Co_{40}(Al_2O_3)_{60}$ (curve 1) and Al_2O_3 (curve 2) spacers.

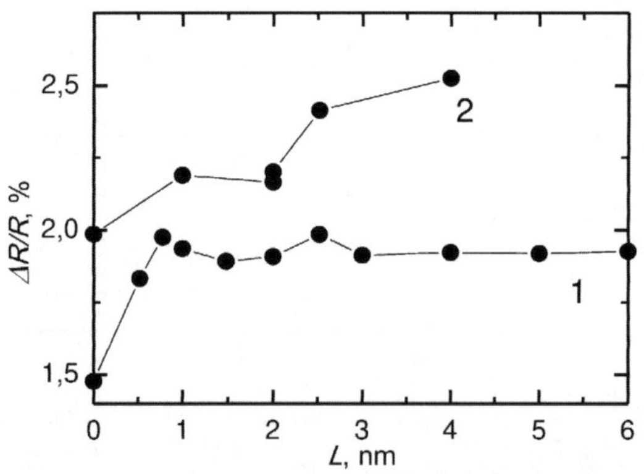

Figure 4. Dependencies of AMR effect on the thickness of the granular spacer of the as-deposited films (curve 1), and annealed films (curve 2).

Figure 4 shows dependences of anisotropic magnetoresistance effect $\Delta R/R$ measured in magnetic field of 60 Oe on thickness of granular spacer on the two types of films. As-deposited samples showed an increase of $\Delta R/R$ with *L* increasing to ~ 1 nm. For larger thicknesses of spacer magnetoresistance remained almost unchanged and had a value of about 2%. For the annealed films, ΔR/R grows throughout the considered range and reaches 2.5%. Since measurements of AMR effect were held in the fixed magnetic field, the observed increase of $\Delta R/R$ reflects the weakening of the interlayer coupling and reduction of unidirectional anisotropy caused by introduction of spacers. Annealed films demonstrate stronger exchange interaction, which might lead to the change of $\Delta R/R$ in a larger range of *L*.

Temperature measurements of the magnetic and magnetoresistive properties were performed on annealed $Ta/Fe_{20}Ni_{80}/Fe_{50}Mn_{50}/Fe_{20}Ni_{80}/Co_{40}(Al_2O_3)_{60}/Fe_{20}Ni_{80}/Ta$ film with L = 2.5 nm. Figure 5 shows the temperature dependence of reduced values of the exchange bias field δH_{e1} and the magnetoresistive effect δR. As normalizing values H_{e1} and $\Delta R/R$ at room temperature were used . As can be seen, with temperature increasing in the range from 77 to 320 K, δH_{e1} decreases slightly. The sharp fall of the exchange bias field is observed at higher temperatures. The blocking temperature of this sample was about 400 K. The described behaviour of the exchange bias field is mostly due to the influence of temperature on the magnetic ordering in the antiferromagnetic layer. Besides that transition of Co granules from superparamagnetic to ferromagnetic state can take place with temperature decreasing. The AMR value decreases almost linearly with temperature increasing, which corresponds to the typical temperature behaviour of the anisotropic magnetoresistance effect value due to temperature changes of the mean free path of conduction electrons.

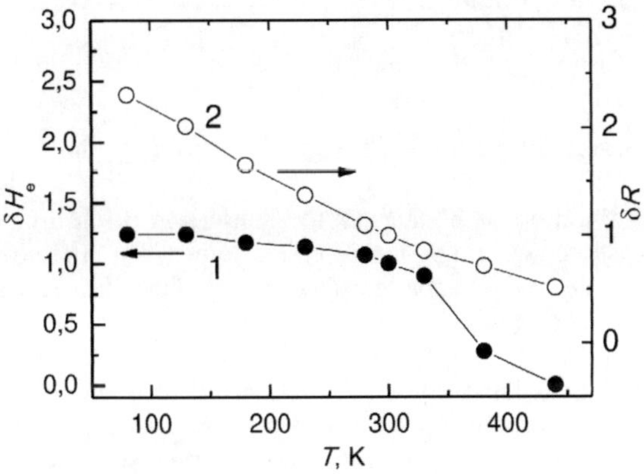

Figure 5. Dependences of reduced values of exchange bias field (curve 1) and the AMR effect (curve 2) on the temperature of the sample with 2.5 nm granular spacer.

4. Conclusion

In the paper we studied the magnetic and magnetoresistive properties of multilayers based on $FeMn/Fe_{20}Ni_{80}$ with exchange bias, which can be used as a magnetoresistive medium for magnetic field sensors. It was demonstrated that introduction of the composite spacer in the layered structure is an effective way of tailoring its hysteretic properties. The presence of the fine Co granules in the composite allows to extend interlayer interaction for a wide range of the spacer thicknesses. This makes possible the intentional influence on sensitivity of such media. For films with selected layered structure dependences of the exchange bias field and the value of AMR effect on the thickness of the spacer, annealing and temperature are defined.

Acknowledgments

This work was supported by The Ministry of Education and Science of the Russian Federation, project № 2582.

References

[1] Nogues J, Sort J, Langlais V, Skumryev V, Surinach S, Munoz J S and Baró M D 2005 *Phys. Rep.* **422** 65
[2] Vas'kovskiy V O, Lepalovskij V N, Gorkovenko A N, Savin P A, Kulesh N A and Shchegoleva N N 2014 *IEEE Transactions on Magnetics* **50** 4800504
[3] Vas'kovskiy V O, Gorkovenko A N, Lepalovsky V N, Yuvchenko A A and Shchegoleva N N 2011 *Russian Physics Journal* **54** 773
[4] Yakushiji K, Mitani S, Takanashi K, Ha J-G and Fujimori H 2000 *JMMM* **212** 75

IOP Conf. Series: Materials Science and Engineering **108** (2016) 012042 doi:10.1088/1757-899X/108/1/012042

Electrochemical measurements on a droplet using gold microelectrodes

Amin Jenabi[1,a], Asma Souri[1,b], Ali Rastkhadiv[1,c]

[1]Electronic Materials Laboratory, Electrical Engineering Department, K. N. Toosi University of Technology, Tehran 16317-14191, Iran

[a]aminjenabi@yahoo.com, [b]asma_souri@yahoo.com, [c]ali.rastkhadiv@gmail.com

Abstract. Facile methods of ion recognition are important for the fabrication of electronic tongue systems. In this work, we demonstrate performing pulsed conductometry on microliter electrolyte droplets dropped on gold microelectrodes vapor deposited on soda lime glass slides. A droplet is dropped between two microelectrodes when a voltage waveform from a preprogramed power supply is applied on them. The temporal variation of the electric current passing through the droplet is recorded, digitized and stored. The obtained data are compared with the database formed out of the previous experiences for the classification of the sample electrolytes. It is shown that the shape of the voltage waveform is the important parameter of the process. We devised a method for the optimization of the voltage waveform profile for obtaining the maximum of discriminating information from the recorded current variations.

1. Introduction

Electronic tongues are usually formed by arraying a number of miniaturized electrochemical cells [1-3] providing uncorrelated information on an electrolyte [4-8]. The drift of array components, particularly due to the electrochemical corrosion on the electrodes, affects the performance of the system and system responses shift away from the training-based calibrations [9-13]. Moreover, owing to the rapid advancement of the microfluidics, it has become important to develop analytical methods compatible with these techniques and can be performed on microliter volumes of the analyte electrolyte [9, 14].

In the standard conductometry, the conductance of an electrolyte is measured by measuring the electric current passing through it when an AC voltage is applied between two conductive electrodes immersed in the liquid [15-17]. The measurements are carried out at the steady state of the system. The results of these standard tests can hardly be used for the recognition of the ionic species responsible for the conduction. The data provided by the standard conductometry can only be utilized as measures of the overall concentration of a predetermined ionic substance. For identifying unknown ionic species in electrolytic solutions in the standard procedure, a complex combination of the conductometry data and other physicochemical phenomena is used [18-20].

Here, we are reporting conductometry with gold microelectrodes on microliter droplets of different electrolytes. The applied voltage is a time varying pulse train which causes a transient current in the electrolyte with temporal variations depending on the nature of the mobile ions present in the electrolyte. The voltage waveform applied is specifically selected to obtain the maximum of diagnostic information out of the current vs. time diagrams. It is shown that the information extracted from these data is enough to facilitate correct classification of a number of electrolytes containing different concentrations of KF, KCl, and KF. The device and measurement system are cost effective and easy to use.

2. Experimental

Gold electrodes are deposited on soda lime glass substrates by physical vapor deposition in vacuum. As shown in Figure 1, each substrate contains 8 gold electrode couples with varying effective electrode areas. Experiments are carried out on 2 different electrode sizes, but those with smaller electrode area, evidently, rendered better results. The results reported here are related to the smaller area electrodes marked in Figure 1.

Three target electrolytes are prepared by dissolving 800 mg of the analytically pure salts in doubly distilled water. The conductivities of the obtained electrolytes are all adjusted at 2 mS/cm and are kept in a clean glass bottle as backups. Each sample is prepared by mixing 1 cc of the backup electrode with appropriate volume of distilled water. For each experiment 2 microliter from the sample electrolyte is dropped at the line separating two gold electrodes, as shown in Figure 1.

The measurement system is schematically shown in Figure 2. It comprises a programmable pulse generator capable of providing voltage waveforms applied to the electrodes [21] and a data acquisition unit which samples the voltage drop on a resistor connected in series with microcell. The sampling rate is 100 s^{-1}, and the data is recorded by the computer connected to the data acquisition system. Due to the low current levels involved, the measurement system is placed in a Faraday cage for noise reduction. The wire connections from the system to the microelectrodes are shown as an inset in Figure 2. To prevent electrolyte evaporation during the experiment, the experiments were carried out in a closed chamber with 90% humidity level. The temperature in the chamber is 25 +/-1°C.

The voltage waveform applied to the electrodes is shown in Figure 3. Experiments are carried out on KCl, KL and KF samples. After dropping the solution between the electrodes, the voltage pulse train is applied to the electrodes and the voltage on the series resistor shown in Figure 2, which is proportional to electric current passing through the droplet, is measured and its temporal variations are recorded. The recorded temporal profiles obtained for various electrolytes are given in Figure 3, confirming that each solution has its particular response profile (see the inset).

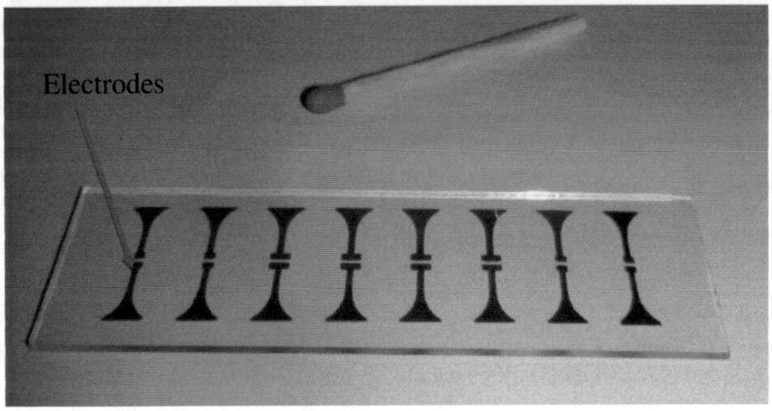

Figure 1. The glass substrate and gold electrodes with different effective surface areas used for the conductometry on electrolyte droplets.

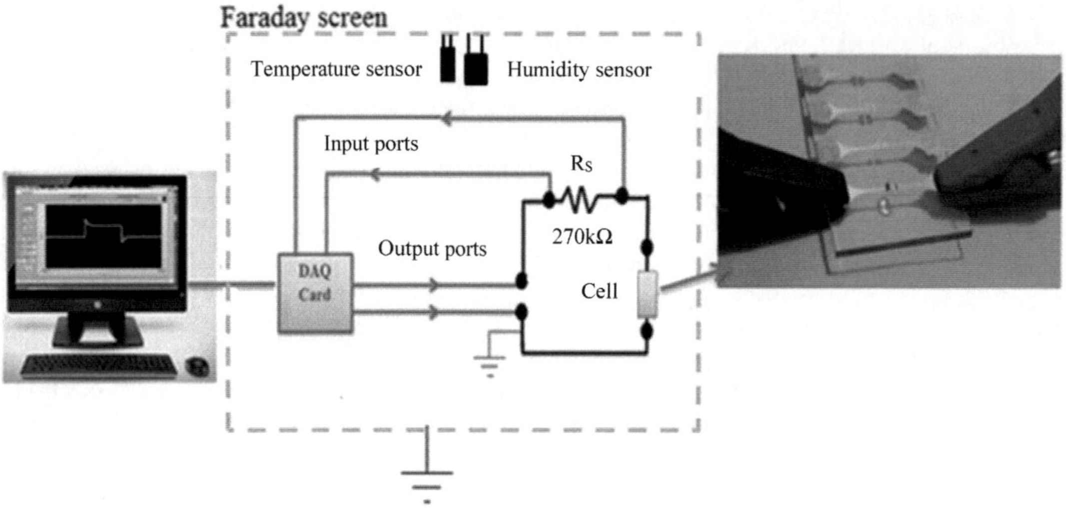

Figure 2. The schematics of the experimental setup. The inset shows the voltage connection to the gold microelectrodes.

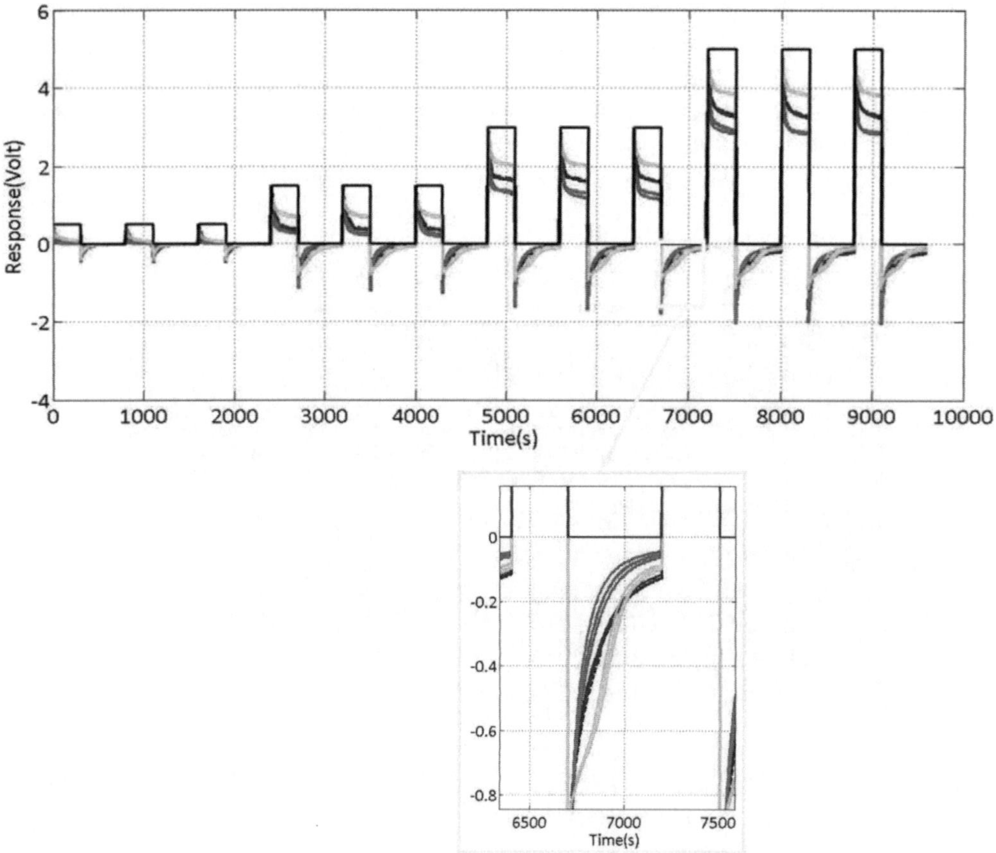

Figure 3. The voltage waveform applied to the electrodes and the temporal responses recorded for different electrolytes each at five different concentrations.

3. Result

According to Figure 3, recorded waveforms for different electrodes are clearly distinct. For distinguishing the recorded waveforms, 48 features are extracted from each response profile. The selection of the features is heuristic and based on the most dissimilar regions of the responses collected. Applying principle component analysis (PCA) reduces the feature space dimensions from 48 to 3 and allows visualization in a 3-dimensional feature space. The heuristic feature extraction and the dimension reduction method for feature space visualization is similar to those utilized in references [21, 22]. This facilitates observation of the different sample classes from each other in the feature space. The results obtained are shown in Figure 4, where unfilled markers are the verification results. Verification experiments are carried out independently in a different work session.

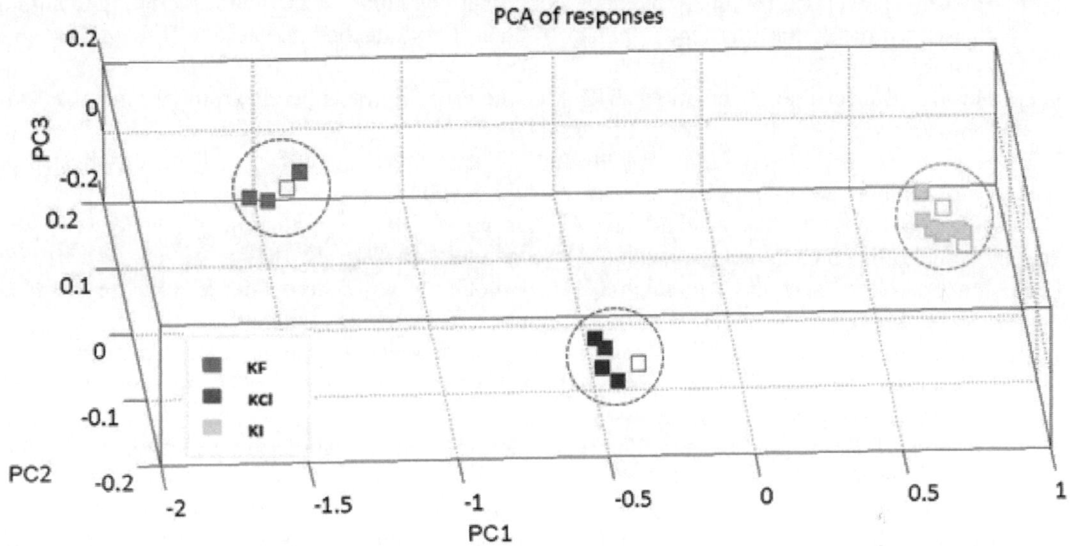

Figure 4: Successful classification of the stated analytes in the 3-dimmensional feature space. The unfilled markers are related to the verification tests.

4. Conclusion

A simple and fast method was presented for performing conductometry on electrolyte droplets. It was shown that the information acquired from a simple conductometry is enough for distinguishing between three different potassium halides. The method uses only 2 µl of the electrolyte and is a good match to the common microfluidic systems and techniques.

References

[1] Ciosek P, Brzózka Z and Wróblewski W 2006 Electronic tongue for flow-through analysis of beverages *Sens. Actuat. B* **118** 454

[2] Ciosek P and Wróblewski W 2007 Sensor arrays for liquid sensing–electronic tongue systems *Analyst* **132** 963

[3] Wilson A D and Baietto M 2009 Applications and advances in electronic-nose technologies *Sensors* **9** 5099

[4] Lacher N A, Garrison K E, Martin R S and Lunte S M 2001 Microchip capillary electrophoresis/electrochemistry *Electrophoresis* **22** 2526

[5] Mathwig K, Mampallil D, Kang S and Lemay S G 2012 Electrical cross-correlation

spectroscopy: Measuring picoliter-per-minute flows in nanochannels *Phys. Rev. lett.* **109** 118302

[6] Scampicchio M, Ballabio D, Arecchi A, Cosio S M and Mannino S 2008 Amperometric electronic tongue for food analysis *Microchim. Act.* **163** 11

[7] Vandaveer W R, Pasas-Farmer S A, Fischer D J, Frankenfeld C N and Lunte S M 2004 Recent developments in electrochemical detection for microchip capillary electrophoresis *Electrophoresis* **25** 3528

[8] Wang J 2002 Electrochemical detection for microscale analytical systems: a review *Talanta* **56** 223

[9] Hossein-Babaei F and Ghafarinia V 2010 *G*as analysis by monitoring molecular diffusion in a microfluidic channel *Anal. Chem.* **8** 8349

[10] Hossein-Babaei F, Hemmati M and Dehmobed M 2005 Gas diagnosis by a quantitative assessment of the transient response of a capillary-attached gas sensor *Sens. Actuat. B* **107** 461

[11] Hossein-Babaei F and Nemati K 2012 A concept of microfluidic electronic tongue *Microfluid. Nanofluid.* **13** 331

[12] Hossein-Babaei F and Orvatinia M 2005 Transient regime of gas diffusion-physisorption through a microporous barrier *Sens. J. IEEE* **5** 1004

[13] Hossein-Babaei F, Paknahad M and Ghafarinia V 2012 A miniature gas analyzer made by integrating a chemoresistor with a microchannel *Lab Chip.* **12** 1874

[14] Jenabi A, Zare A H and Erfantalab S 2014 Voltammetry of Micro-Liter Electrolyte Samples on ITO Microelectrodes for Analyte Recognition *Key Eng. Mat.* **605** 601

[15] Lambrechts M and Sansen W 1992 *Biosensors: microelectrochemical devices* CRC Press

[16] Pungor E, Damokos T and Townshend A 1965 *Oscillometry and conductometry* (London) Pergamon Press

[17] Dasgupta P K and Bao L 1993 Suppressed conductometric capillary electrophoresis separation systems *Anal. Chem.* **65** 1003

[18] Kappes T, Galliker B, Schwarz M A and Hauser P C 2001 Portable capillary electrophoresis instrument with amperometric, potentiometric and conductometric detection *TrAC Trends in Analytical Chemistry* **20** 133

[19] Kovács A, Slezsak I, McLaughlin W L and Miller A 1995 Oscillometric and conductometric analysis of aqueous and organic dosimeter solutions *Rad. Phys. Chem.* **46** 1211

[20] Ronkainen N J, Halsall H B and Heineman W R 2010 Electrochemical biosensors *Chem. Soc. Rev.* **39** 1747

[21] Hossein-Babaei F and Amini A 2014 Recognition of complex odors with a single generic tin oxide gas sensor *Sens. Actuat. B* **194** 156

[22] Hossein-Babaei F and Amini A 2012 A breakthrough in gas diagnosis with a temperature-modulated generic metal oxide gas sensor *Sens. Actuat. B* **166** 419

IOP Conf. Series: Materials Science and Engineering **108** (2016) 012043 doi:10.1088/1757-899X/108/1/012043

Cyclopean gauge factor of the strain-resistance transduction of indium oxide films

J Ivančo[1], Y Halahovets[1], K Vègsö[1], I Klačková[2], M Kotlár[1,3], A Vojtko[1], M Micušík[4], M Jergel[1] and E Majková[1]

[1] Institute of Physics, Slovak Academy of Sciences, Dúbravská cesta 9, SK-845 11 Bratislava, Slovakia

[2] Faculty of Electrical Engineering and Information Technology, Slovak University of Technology, Ilkovičova 3, SK-812 19 Bratislava, Slovakia

[3] STU Centre for Nanodiagnostics, Vazovova 5, SK-812 43 Bratislava, Slovakia

[4] Institute of Polymers, Slovak Academy of Sciences, Dúbravská cesta 9, SK-845 41 Bratislava, Slovakia

E-mail: jan.ivanco@savba.sk

Abstract. The resistance of indium-oxide covered polyethylene terephthalate foils (IO-PET) shows an extreme sensitivity to tensile strain. In terms of the deformation-resistance transduction, the gauge factor as high as about 60 000 was recorded upon the relative elongation up to 1 %. Except the onset of deformation, the nearly exponential dependence of the resistance on strain suggests that the conductivity of the strained films is governed by tunnelling mechanism; this notion is supported by the formation of scattered cracks in the IO-PET film. The cracks are oriented perpendicularly to the strain vector and are characterized by a rather similar and uniform width. Appropriateness of the standard definition of the gauge factor for strain sensors, which are governed by tunnelling conductance, is critically discussed.

1. Introduction

Applications of strain sensors span from traditional pressure and torsion gauges to tactile sensors [1-3]. Their design flexibility reflects employment of various physical and chemical methods, yet simple resistive strain sensors attract a lot of attention. The principal parameter characterizing the strain sensor is the gauge factor, g

$$g = \frac{\frac{R-R_0}{R_0}}{\frac{L-L_0}{L_0}} \qquad (1)$$

where R_0 is the initial resistivity of the pristine sample of the length L_0, R is the resistivity of the sample stretched to the length of L. The denominator $(L-L_0)/L_0$ is termed the strain, which is commonly expressed in per cents.

Reported more than four decades ago, strain gauges based on metal films formed by discrete islands demonstrated $g > 100$ [4], thereby markedly higher compared to the continuous bulk metal films having $g \sim 2$. This was shown to be due to the conductance governed by tunneling current mechanism, which occurs in discontinuous films [5-8]. A recent impetus has been provided by employment of nanostructured films (see *e.g.* [9-16] and references therein). The exponential dependence of the tunneling current on the separation gap between adjacent nanoobjects results in large conductivity changes responsive to fractional deformation; the gauge factors as high as 200 [10], 300 [15], and even 700 [13] were reported for closely-packed nanoparticle (NP) assemblies.

The indium-tin oxide (ITO) films have became standardly used as transparent conductive electrodes in optical planar devices such as solar cells and light emitting diodes. The ITO films supported by either glassy or flexible substrates have been available on commercial basis. Alternatively, indium-oxide (IO) or fluorine-tin-oxide (FTO) films can be considered. The massive usage of ITO in optical devices may overshadow their much less frequent utilization in the role of active sensing films, *e.g.* for piezo [17] and strain [18,19] sensors. In this study, we report on strain-resistivity transduction of the indium-oxide coated polyethylene terephthalate flexible substrates; the gauge factor up to about two orders higher magnitude as compared to up-to-now reported highest values for the nanoparticle-based strain gauges was determined. The nature of the extremely high response is examined.

2. Experimental

The films under study were stretched using the TS 600 device (Anton Paar, GmbH), which controls the tensile strain by axial stretching of the probed sample. The total length of the examined stripes was 37 mm, while the actual length subjected to the strain was 24 mm. The width of the stripes was chosen to be either of 8 or 14 mm. The samples were indium-oxide coated polyethylene terephthalate (IO-PET) foil.

The IO-PET foil with the surface resistance below 10 Ω/□ and the thickness of 0.2 mm was purchased from Sigma-Aldrich. Except the specification "In_2O_3/Au/Ag coated PET film", no further information was ascertainable. Silver was evaporated on a pair of areas separated by 10 mm for resistance characterization and the wired terminals to the resistance meter were glued to these areas by colloidal Ag paste. The film resistance of the stretched film was measured at bias of 1 V. The gradual stretching of the IO-PET foils (typically with a constant ramp rate of $\Delta L/L_0 \approx 0.23$ %/min) was carried out in the direction of the current flow up to a preset maximal strain, *e.g.* 1 %. The reached maximal strain was maintained for 5 minutes followed by a strain release either with the same ramp or abruptly.

The morphology of the IO-PET films was examined by the field-emission gun Scanning electron microscopy (SEM). The SEM images were recorded by means of the FEG-SEM JEOL JSM 7500F facility using combined signals originating from both secondary and back-scattered electrons with accelerating voltage of 10 kV and working distance of 4.1 mm. For the SEM examination of the strained IO-PET films, the deformation of the pre-stretched film was preserved by gluing a rigid solid substrate to the reverse side of the IO-PET sample.

3. Results

The sequential strain cycles were applied to the IO-PET samples. The cycles consisted of a gradual increase of the strain up to the targeted maximal strains of 0.5 or 1.0 %, the retention of the maximum strain for 5 min, and a gradual decrease of the strain with the same ramp. The first cycle up to the strain of 0.5 % showed relatively low resistance change. The next cycle up to 1 % led to the resistance normalized to the initial resistance $R/R_0 > 600$. The following cycle up to 0.5 % showed markedly higher R/R_0 compared to the first 0.5 % cycle. The next cycles either up to 0.5 % or 1.0 % showed R/R_0 values resembling to those of the 2nd or 3rd cycle. The reproducibility of the cycling will be addressed below.

Figure 1 presents one cycle of the R/R_0 *versus* strain dependence for the IO-PET film in the semi-logarithmic plot. The dependence shows a hysteresis presumably due to a large time constant of the supporting foil. The gauge factor of the presented curve calculated at the strain of 1 % is about 60 000. The issue of the evaluation of the gauge factor will be discussed in more detail below. The dependence is nearly linear in the log scale in the range between ca 0.5 and 0.8 % suggesting the exponential evolution of the resistance upon the applied strain.

IOP Conf. Series: Materials Science and Engineering **108** (2016) 012043 doi:10.1088/1757-899X/108/1/012043

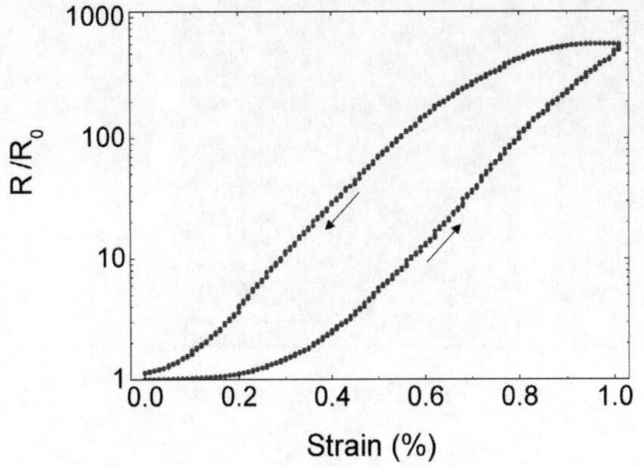

Figure 1. The R/R_0 dependence on the strain for the IO-PET films. Both the increase and decrease of the strain were performed with the ramp of 0.23 %/min.

In the given experimental layout, electrical contacts had to be located within the strained area. Yet, our tests showed that the Ag contacts were not affected by strain and accordingly they do not contribute to the resistance-strain transduction of the probed IO films.

The surface mapping of the IO-PET films by the SEM (figure 2) revealed grains arranged into a network of irregular elongated “chaplets” with their elongation equally oriented.

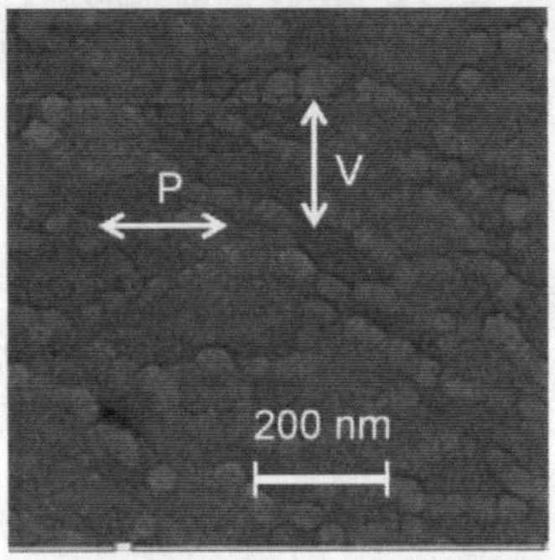

Figure 2. SEM of the pristine IO-PET surfaces. “P” and “V” refer to the directions of the applied strain those are nearly parallel and perpendicular, respectively, to the elongation of “chaplets”.

The same technique was employed for the inspection of the surface morphology of the *strained* IO-PET film. For this purpose, the sample stretched up to 1 % was fixed by gluing its reverse side to a rigid ceramic plate followed by the release of the external force. The strain had partially released in spite of the gluing, yet it sustained the value of about 0.5 %, as deduced from the remaining elevated film resistance still markedly exceeding the initial resistance. The SEM inspection (figure 3) suggests roughly equidistantly parallel-oriented cracks with orientation perpendicular to the deformation vector. Figure 3b details one crack at higher magnification and it reveals fairly uniform width of about 30 nm across the sample. By estimating the crack frequency of about 1 crack per 10 μm (figure 3a), the crack width of ca 30 nm (figure 3b), and considering the stretched length of 24 mm, the sum of crack widths is about 0.072 mm yielding $\Delta L/L_0 \approx 0.3$ %. This is in qualitative agreement with the strain of about 0.5 % deduced from the resistance of the pre-stretched sample and supports the reasoning on the presence of cracks over the entire strain sample length.

We presume that the high resistance-strain transduction is due to the current tunneling over the cracks emerging in the IO films subjected to strain. Their uniform widths over the entire width of the sample are presumably owing to both well-defined cleavage of the polycrystalline IO films and the uniform deformation of the PET substrate.

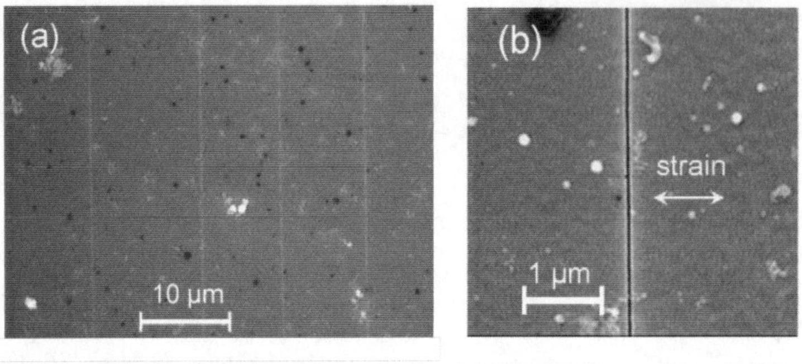

Figure 3. The SEM maps of the strained IO-PET film with various magnifications.

In view of the strain-resistance transduction governed by the formation of cracks, admittedly, a question on the reproducibility of a such response may arise. To illustrate the reproducibility of the transduction for a particular IO-PET sample, five sequential stretching cycles (figure 4a) from unstrained state up to strain $\Delta L/L_0 = 1$ % were measured after several previous identical cycles. In each cycle, the strain increased with a linear ramp of 0.23 %/min up to the maximum strain of 1 % that was preserved for five minutes and followed by an abrupt release of the stress. The strain evolution is indicated by red full line in figure 4a. Obviously, the resistance change is reversible and the almost perfect overlapping of the curves of particular cycles suggests the good reproducibility for the probed strain interval.

The same cycling procedure was applied for several IO-PET samples to estimate the sample-to-sample reproducibility. Here, the strain-resistance transduction depended on the stretching orientation with respect to the orientation of "chaplets", whereas the initial resistance was not sensitive to the orientation of the current flow. Figure 4b presents two triplets of the samples that were cut in mutually perpendicular directions. For each sample, the fifth of the five sequential cycles is shown in figure 4b. The full/open symbols refer to the samples, where the strain was oriented nearly parallel/perpendicular to the longer chaplet axis (the arrows "P" and "V" in figure 2). The latter configuration yields roughly one order higher ratio R/R_0 compared to the former one.

4. Discussion

For further discussion, we mention our recent investigations on the strain-resistance transduction of the one monolayer of ordered thiolalkane-linked Au nanoparticles (NP*s*) with the core diameter of 6-7 nm deposited onto the BoPET foil (Mylar) [16]. The sample structure was the similar to that of the IO-PET samples except the separation of electrodes being of 50 µm. The gauge factor of ca 45 was assessed at the strain $(L-L_0)/L_0 = 5$ %, where L_0 and L are the pristine and strained sample lengths, respectively. As determined by the small-angle x-ray scattering (SAXS) characterization, the nanoparticle distance being of 6.87 nm for the pristine sample changed proportionally with the deformation of the supporting foil; the constant of proportionality was about 0.48 Å per % of strain.

To provide a qualitative explanation of the NP*s*- and cracks-derived transductions differing by 2-3 orders of magnitude, simplified models can be considered. We recall that the tunneling resistance depends approximately exponentially on the interparticle gap [10]

$$R \propto \exp(\beta \Delta l) \qquad (2)$$

where Δl is the interparticle gap change induced by strain. The cracks can be considered as the tunneling gaps too. β is the tunneling decay constant describing the tunneling of electrons. It ranges from 0.8 to 1.18 $Å^{-1}$ [23,26,27] for thiolalkane molecules and is ≈ 2 $Å^{-1}$ [15] for the tunneling in air. When comparing the NP*s*- and IO-derived samples exposed to the same strain, much higher difference of exponent–and accordingly the resistance change–is brought by the distinct width change of the tunneling gap. This is owing to the high disproportion in their numbers; given the geometries of the respective samples, a rough estimate yields about 2400 cracks and 7000 gaps across the active length of 24 mm and 50 μm, respectively. Consequently, the crack width is about three orders higher than the interparticle distance change to conform the same strain.

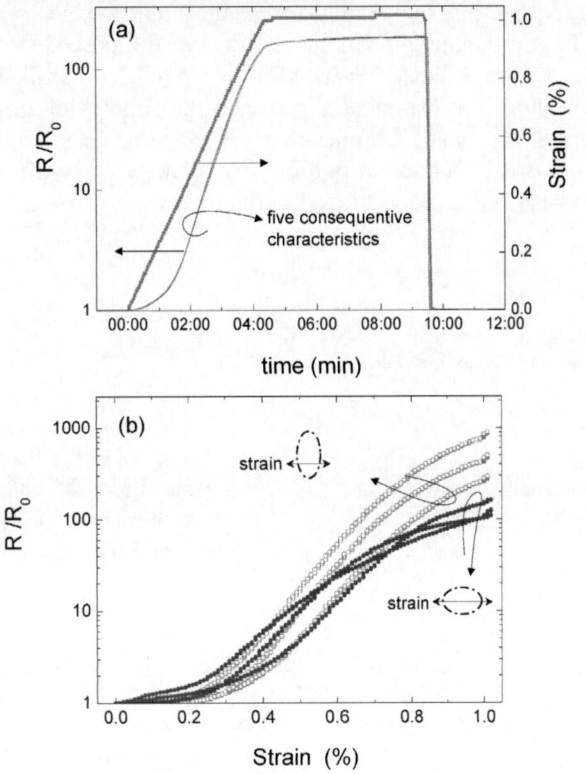

Figure 4. The strain-induced resistance normalized by the initial resistance R/R_0 for (a) the five sequential stretching cycles from unstrained state up to strain $\Delta L/L_0 = 1$ % measured for a particular IO-PET film. The strain evolution, which is indicated by the red full line, increased with the ramp of 0.23 %/min and was released abruptly. (b) The fifth of the five sequential cycles measured for two triplets of IO-PET samples cycled under the same conditions as in (a). The strain orientation with respect to the "chaplets" is indicated next to the triplets. The leading edges are shown only.

The tunneling mechanism of the charge transport seems to be a key factor for getting the high variance of the film conductivity in dependence on the strain. In reference to the nanostructured films, the fabrication approach often leads to the frequency of tunneling gaps commensurate to the density of nanoobjects and with the nano-gap widths being comparable with the size of nanoobjects themselves. Such structures tend to display lower gauge factor due to the deformation distributed to exiguous changes of the interparticle gaps. Conversely, the decreasing number of gaps would increases the width accruement of gaps upon the unit strain, which in turn increases the gauge factor exponentially due to the exponential character of the tunneling current.

Assuming the crack width of ca 30 nm–as visualized by the SEM (figure 3b) for the strain of ca 0.5 %–one would expect open circuit behavior. Admittedly, $R/R_0 \approx 100$ only. Such a resistance increase would correspond to the much thinner tunneling gap increase by about 2.3 Å. Therefore, we speculate that whereas the number of tunneling gaps in the IO-PET samples corresponds to the number of cracks visualized by SEM, yet the tunneling gaps are effectively much narrower–likely at the IO/PET interface–than the crack width suggested by SEM. Such behavior could be due to the multilayer structure of the IO-PET films: the depth profiling by the x-ray photoemission spectroscopy (XPS) revealed that the indium oxide film is, in fact, composed of two approximately equally thick indium oxide layers sandwiching a thin Au/Ag layer (not shown here).

Assuming the linear transduction, *i.e.* the linear resistance dependence on strain, the gauge factor evaluated by equation (1) is insensitive to the upper limit of the probed strain range. However, the gauge factors of the nanostructured films are very high owing to their non-linear responses. Typically, the resistance depends on the strain exponentially due to the conduction governed by the tunneling mechanism. The exponential dependence implies that the gauge factor strongly depends on the upper attainable limit of strain. This is illustrated in figure 5 where the empty symbols show the evolution of the gauge factor for the IO-PET sample introduced in figure 1, which was calculated by means of (1).

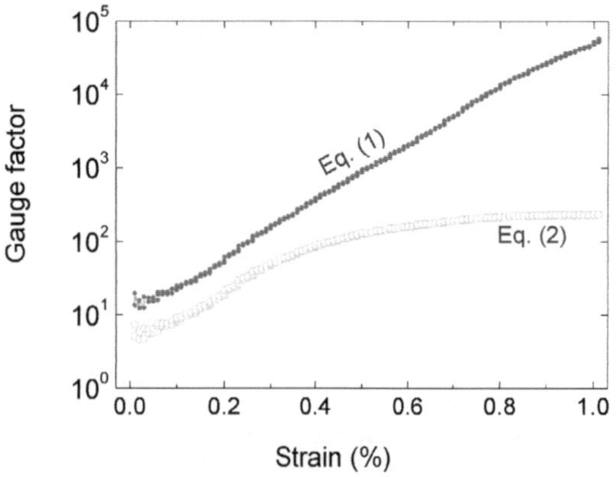

Figure 12. The comparison of gauge factors of the IO-PET film presented in figure 1 which were calculated either according to equation (1) (top curve) or (3) (bottom curve).

The IO-PET samples get broken at the strain well exceeding about 1.5 %; obviously, a much higher gauge factor is attainable than the one calculated here according to equation (1).

It follows that the gauge factor determined by (1) lacks information on the strain-gauge sensitivity and the probed strain interval. A modified formula, presumably more appropriate for evaluation of the transductions governed by the exponentially-dependent mechanisms, reads

$$g = \frac{\frac{log(R/R_0)}{logR_0}}{\frac{L-L_0}{L_0}}. \quad (3)$$

The gauge factor defined by (3) reflects also the signal conversion efficiency and it is much less dependent on the measured strain interval provided an ideal mechanism characterized by the exponential strain-to-resistance transduction.

5. Conclusions

The strain-resistivity transduction of the indium-oxide coated polyethylene terephthalate (IO–PET) flexible substrates reaches the gauge factor of tens of thousands, *i.e.*, exceeds by about two orders of magnitude the up-to-now reported highest values for the nanostructured strain gauges. The extremely high response is due to the transduction mechanism governed by the tunneling conductivity across the cracks formed in the strained films. A redefined gauge factor for the transductions based on exponentially dependent mechanism instead of that for the linear transduction was proposed.

Acknowledgment

The work was supported by Scientific Grant Agency (project No. VEGA 2/0010/15) and Slovak Research and Development Agency (project No. APVV-14-0891).

References

[1] Mannsfeld S C B, Tee B C K, Stoltenberg R M, Chen C V H H, Barman S, Muir B V O, Sokolov A N, Reese C and Bao Z 2010 *Nat. Mater.* **9/10** 859-64

[2] Pang C, Lee G-Y, Kim T-i, Kim S M, Kim H N, Ahn S-H, Suh K-Y 2012 *Nat. Mater.* **11** 795-801

[3] Alvares D, Wieczorek L, Raguse B, Ladouceur F and Lovell N H 2013 *Sens. & Actuat. A: Phys.* **196** 38-47

[4] Neuman M R and Sutton W G 1969 *J. Vac. Sci. & Technol.* **6** 710-13

[5] Neugebauer C A and Webb M B 1962 *J. Appl. Phys.* **33** 74-82

[6] Hartman T E 1963 *J. Appl. Phys.* **34** 943-47

[7] Campbell D S and Morley A R 1971 *Rep. Prog. Phys.* **34** 283-368

[8] Morris J E and Coutts T J 1977 *Thin Solid Films* **47** 3-65

[9] Brust M, Schiffrin D J, Bethell D and Kiely C J 1995 *Adv. Mater.* **7** 795-97

[10] Herrmann J, Müller K-H, Reda T, Baxter G R, Raguse B, Groot G J J B d, Chai R., Roberts M and Wieczorek L 2007 *Appl. Phys. Lett.* **91** 183105-3

[11] Vossmeyer T, Stolte C, Ijeh M, Kornowski A and Weller H 2008 *Adv. Funct. Mater.* **18** 1611-16

[12] Farcau C, Moreira H, Viallet B, Grisolia J, Ciuculescu-Pradines D, Amiens C and Ressier L 2011 *J. Phys. Chem. C* **115** 14494-99

[14] Tanner J L, Mousadakos D, Giannakopoulos K, Skotadis E and Tsoukalas D 2012 *Nanotechnology* **23** 285501-6

[15] Burzhuev S, Dâna A and Ortaç B 2013 *Sens. & Actuat. A: Phys.* **203** 131-36

[16] Ivančo J, Végsö K, Šiffalovič P, Kostiuk D, Halahovets Y, Klačková I, Kotlár M, Luby Š and Majková E 2015 *Key Eng. Mater.* **644** 31-34

[17] Gregory O J, Chen X and Crisman E E 2010 *Thin Solid Films* **518** 5622-25

[18] Gregory O J, You T and Crisman E E 2005 *Thin Solid Films* **476** 344-51

[19] Miller T M, Fang H, Magruder III R H and Weller R A 2003 *Sens. & Actuat. A: Phys.* **104** 162-70

[20] Terrill R H, Postlethwaite T A, Chen C-h, Poon C-D, Terzis A, Chen A, Hutchison J E, Clark M R and Wignall G 1995 *J. Am. Chem. Soc.* **117** 12537-48

[21] Wuelfing W P and Murray R W 2002 *J. Phys. Chem. B* **106** 3139-45

[22] Holmlin R E, Haag R, Chabinyc M L, Ismagilov R F, Cohen A E, Terfort A, Rampi M A and Whitesides G M 2001 *J. Am. Chem. Soc.* **123** 5075-85

IOP Conf. Series: Materials Science and Engineering **108** (2016) 012044 doi:10.1088/1757-899X/108/1/012044

Open-circuit sensitivity model based on empirical parameters for a capacitive-type MEMS acoustic sensor

Jaewoo Lee[1,2], J.H. Jeon[1], C.H. Je[1], S.Q. Lee[1], W.S. Yang[1], and S.-G. Lee[2]
[1]Nano Sensor Lab, Electronics and Telecommunications Research Institute (ETRI), Republic of Korea
[2]Department of Electrical Engineering, Korea Advanced Institute of Science and Technology (KAIST), Republic of Korea

E-mail: jaewoo@etri.re.kr and jaewoolee@kaist.ac.kr

Abstract. An empirical-based open-circuit sensitivity model for a capacitive-type MEMS acoustic sensor is presented. To intuitively evaluate the characteristic of the open-circuit sensitivity, the empirical-based model is proposed and analysed by using a lumped spring-mass model and a pad test sample without a parallel plate capacitor for the parasitic capacitance. The model is composed of three different parameter groups: empirical, theoretical, and mixed data. The empirical residual stress from the measured pull-in voltage of 16.7 V and the measured surface topology of the diaphragm were extracted as +13 MPa, resulting in the effective spring constant of 110.9 N/m. The parasitic capacitance for two probing pads including the substrate part was 0.25 pF. Furthermore, to verify the proposed model, the modelled open-circuit sensitivity was compared with the measured value. The MEMS acoustic sensor had an open-circuit sensitivity of -43.0 dBV/Pa at 1 kHz with a bias of 10 V, while the modelled open-circuit sensitivity was -42.9 dBV/Pa, which showed good agreement in the range from 100 Hz to 18 kHz. This validates the empirical-based open-circuit sensitivity model for designing capacitive-type MEMS acoustic sensors.

1. Introduction

MEMS microphones based on capacitive-type acoustic sensors compatible with CMOS processes have been rapidly replacing electret condenser microphones (ECMs) in the market of communication systems, especially smart phones, due to their competitive cost and technology. Recently, their application range has been expanded to a variety of monolithic integrated circuits, such as digital MEMS microphone modules and beam-forming modules, because they have monolithic integrated capability with ROIC. As a result, various design issues and fabrication methods related to capacitive-type MEMS acoustic sensors have been reported based on surface and bulk micromachining [1]-[8]. On the other hand, to investigate the characteristics of the MEMS acoustic sensors, an analogous equivalent circuit model has been applied in sensor modules [2], [6], [9]-[12]. Generally, the behaviour of acoustic-electro transducers has been modelled with lumped parameters. However, in spite of theoretically complicated analysis or finite-element method (FEM) simulation, it may be difficult to model some parameters, such as the effective residual spring constant in a diaphragm and parasitic capacitance for a MEMS sensor chip, owing to their inherently non-separability from an intrinsic part. In this paper, an empirical-based model of open-circuit sensitivity at a low frequency for a capacitive-type MEMS acoustic sensor is proposed and investigated. Each parameter was modelled through an extracted method based on empirical measurement, a theoretical method, or a mixed method of the two. In particular, both the residual stress and the parasitic capacitance are critical for determining the open-circuit sensitivity. To verify the proposed model, the modelled open-circuit sensitivity was compared with the measured sensitivity using an evaluation board for a packaged MEMS acoustic sensor.

2. Empirical based open-circuit sensitivity model

2.1. Lumped spring-mass model

Figure 1 shows a schematic diagram of a conventional lumped spring-mass model for a capacitive-type acoustic sensor with a circular diaphragm. In this model, the movable plate is assumed to act like a piston without any difference between the centre point and an edge of the moving plate, when a uniform force is applied to the diaphragm.

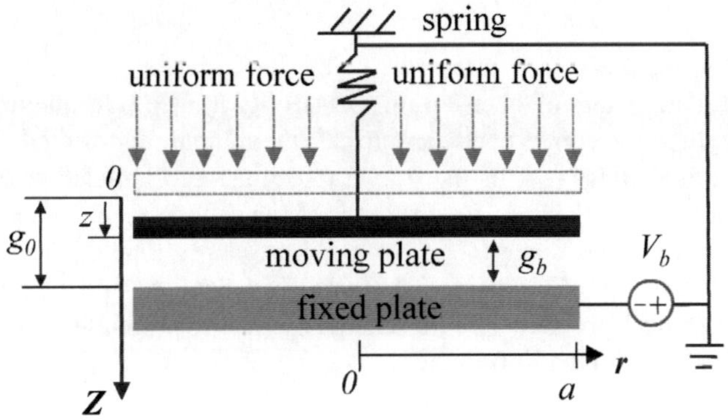

Figure 1. Schematic diaphragm of a conventional lumped spring-mass model for a capacitive-type acoustic sensor with a circular diaphragm

Generally, from the large deflection theory, linearized restoring force (F_{res}) from a uniform force is expressed by

$$F_{res} \approx \left(\frac{64\pi D}{a^2} + 4\pi\sigma_0 h\right) \cdot z \quad , \tag{1}$$

where D is the flexural rigidity, a is the diaphragm radius, σ_0 is the residual stress, h is the diaphragm thickness, and z is the axis direction in a cylindrical coordinate. In addition, the electrical force (F_{ele}) between two electrodes is described as

$$F_{ele} = -\frac{{V_b}^2}{2}\frac{\varepsilon_0 \pi a^2}{(g_0 - z)^2} \quad , \tag{2}$$

where g_0 is the height of the air gap, V_b is the bias voltage, and the permittivity of vacuum. Thus, at an equilibrium condition equal to

$$F_{res} = F_{ele} \; at \; z = g_0/3 \quad , \tag{3}$$

the pull-in voltage can be determined by

$$V_P = \sqrt{\frac{8}{27}\left(\frac{64\pi D}{a^2} + 4\pi\sigma_{eff}h\right)\frac{{g_0}^3}{\varepsilon_0 \pi a^2}} \quad . \tag{4}$$

From equation (4), it is obvious that the effective residual stress (σ_{eff}) can be extracted from both the pull-in voltage and the air gap between two electrodes. If its value is more than +5 MPa, the effective residual stress plays an essential role on the spring constant of a diaphragm, resulting in dominant impact on open-circuit sensitivity.

2.2. Each parameter extraction

As shown in figure 2, parameters of open-circuit sensitivity at a low frequency for a capacitive-type acoustic sensor are divided into three parts: empirical, theoretical, and mixed data group. Among the empirical data, the residual stress (σ_0) of the diaphragm must be extracted from the measured values due to its strong dependence on the specific structure of the sensor itself. In addition, it is not a simple task to extract the parasitic capacitance (C_{par}) from the measured total capacitance owing to the non-linear substrate effect. Thus, on the basis of the empirical data, σ_0 and C_{par} must be determined to reflect some practical impact on open-circuit sensitivity at a low frequency. The modelling process based on empirical data is described in figure 3.

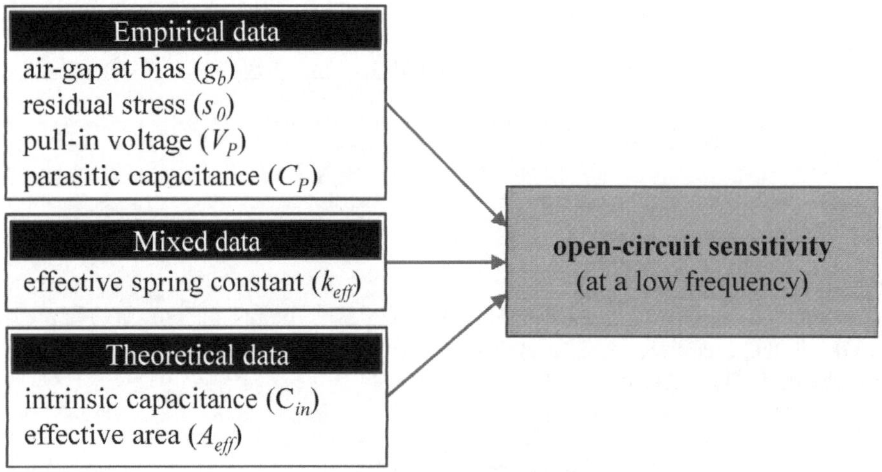

Figure 2. Three parameter groups of open-circuit sensitivity at a low frequency for a capacitive-type MEMS acoustic sensor

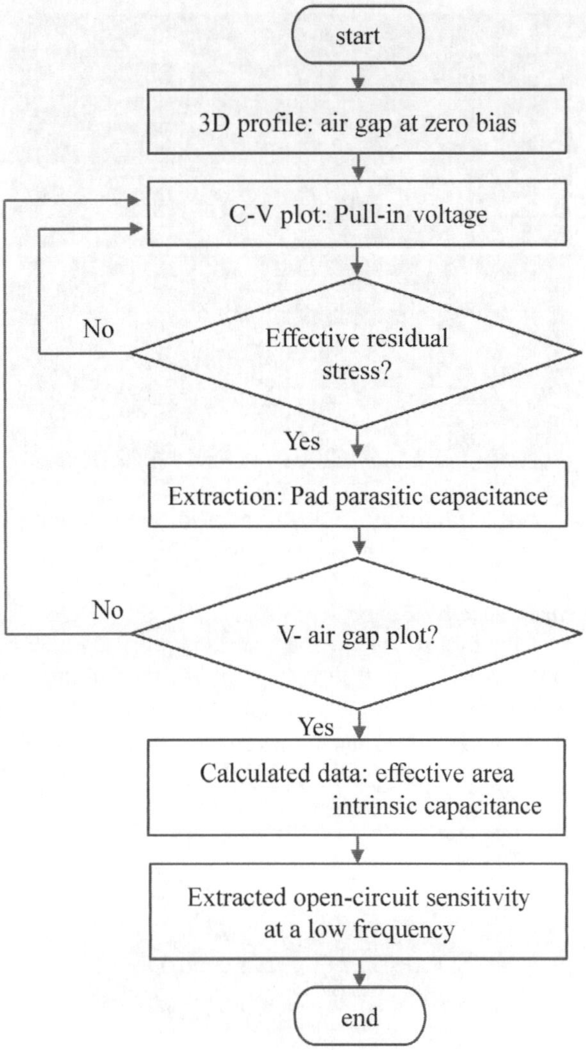

Figure 3. Modelling process of open-circuit sensitivity at a low frequency for a capacitive-type MEMS acoustic sensor

2.3. MEMS acoustic sensor

For a capacitive-type MEMS acoustic sensor with a Al/PECVD-Si_3N_4/Al diaphragm, each step is carried out with a CMOS compatible process only, except a DRIE process related to back chamber etching, having 7 photo masks. A 2.0 μm low-stress LPCVD Si_3N_4 was implemented as a rigid back plate with a back plate electrode of 0.5 μm Al. As a sacrificial layer, 2.6 μm of polyimide was adapted for wide material selectivity. Also, the membrane consisted of a 100 nm Al electrode, a 400 nm PECVD Si_3N_4 membrane, and a 100 nm Al passivation layer, which formed a multilayer sandwich stack. The average mean stress of the diaphragm was measured to be +20 MPa by a curvature method. After the back chamber was etched by DRIE, the sacrificial layer was easily released using O_2 gas. Figure 4 shows the fabrication process for the MEMS acoustic sensor.

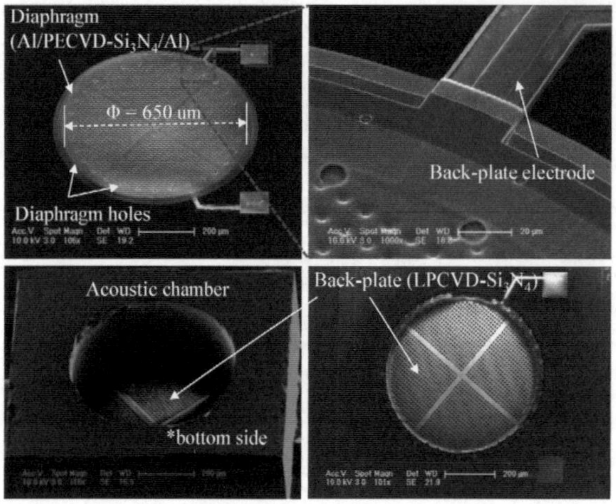

Figure 4. SEM images for the fabricated MEMS acoustic sensor with an Al/PECVD-Si_3N_4/Al diaphragm

2.4. Residual stress and Parasitic capacitance

To extract the residual stress, the pull-in volatage and surface profile of the diaphragm were measured as shown in figure 5. From the measured pull-in voltage of 16.7 V and the height of air gap of 2.6 um, the effective residual stress was extracted by using equation (4). Here, σ_{eff} was determined to be +13 MPa, showing a difference of +7 MPa from the average mean stress. Thus, it is noted that σ_{eff} must be evaluated by using the practical value rather than the mean stress by the curvature method.

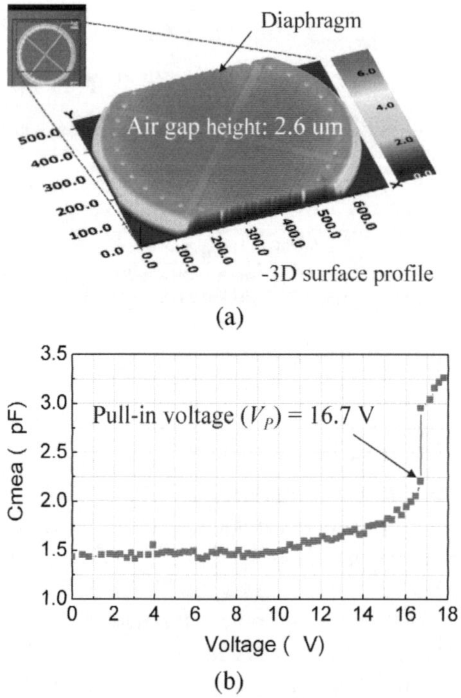

Figure 5. Surface topology of the diaphragm (a) and pull-in characteristic (b) for the MEMS acoustic sensor

On the other hand, the relation of capacitance as a function of bias voltage can be applied in the extraction process of C_{par} with an additional test sample: only a two-pad chip without an active parallel-plate capacitor. The extracted C_{par} from the test sample was 0.25 pF. Also, the measured total capacitance (C_{mea}) at a zero bais and the cable paracitic capacitance (C_{cab}) were 1.43 pF and 0.05 pF, respectively. Thus, the intrincic capacitance (C_{in}) can be extracted as 1.13 pF by using

$$C_{in} = C_{mea} - C_{par} - C_{cab} \quad . \tag{5}$$

In addtion to C_{par}, the model process of separating C_{in} from C_{mea} is critical because both C_{in} and C_{par} are applied to model the open-circuit sensitivity.

2.5. *Open-circuit sensitivity*

When a bias voltage (V_b) is forced in the sensor, the open-circuit sensitivity at a low frequency can be expressed as

$$S_o = \frac{V_b\, A_{eff}}{g_b\, k_{eff}} \left(\frac{C_b}{C_b + C_{par}} \right) \quad , \tag{6}$$

where g_b is the height of the air gap at the bias voltage (V_b), C_b is the intrinsic capacitance at the bias voltage, A_{eff} is the effective capacitor area excluding the etching hole area, and k_{eff} is the effective spring constant containing the effective resisual stress. Here, A_{eff} and k_{eff} were 2.33 x 10^{-7} m^2 and 110.9 N/m, respectively. As a result, when V_b was 10 V, S_0 was modelled as -42.9 dBV/Pa at 1 kHz, where g_b, C_b, C_{par} were 2.45 um, 1.2 pF, and 0.25 pF. The design and modelled parameters are shown in Table. 1.

Table 1. Design and modelled parameters.

Parameters	Values
Diaphragm radius (r)	325 μm
Diaphragm thickness (h)	0.6 μm
Flexural rigidity (D)	4.78 x 10^{-9} N-m
Effective residual stress (σ_{eff})	+13 MPa
Pull-in voltage (V_p)	16.7 V
Air-gap height (g_o)	2.6 μm
Effective spring constant (k_{eff})	110.9 N/m
Effective area (A_{eff})	2.33 x 10^{-7} m^2

3. Results and Discussion

To analyze the frequency characterisitic of an acoustic-electro transducer, an equivalent circuit model reflecting the influence of all of the acoustic, mechanical, and electrical signal domains [10] was used for open-circuit sensitivity (S_o). Figure 6 shows the structure-based model and its equivalent circuit for the dependency of open-circuit sensitivity on the frequency. In the steady-state condition, S_o is defined as the transfer function of the output voltage as a function of the input pressure. Thus, it can be expressed [10] as

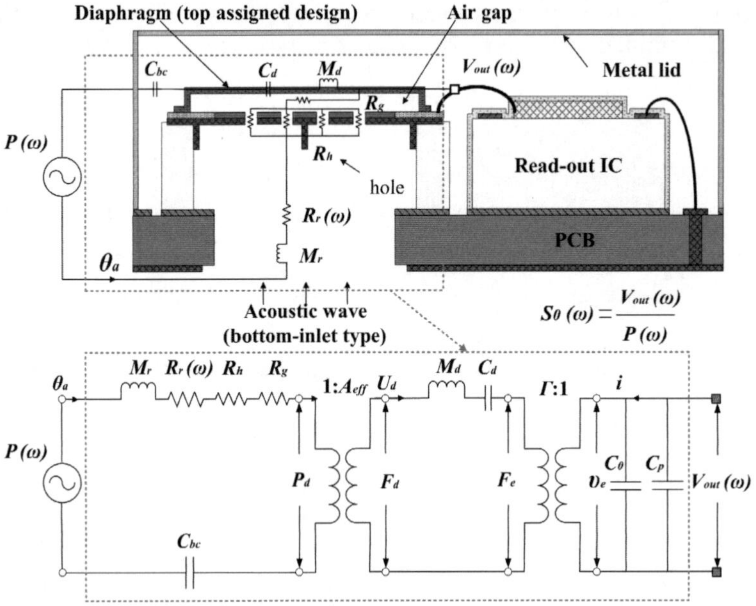

Figure 6. Schematic view for the MEMS acoustic sensor and an equivalent circuit for the open-circuit sensitivity

$$S_o = \frac{v_{out}}{P} = \frac{1}{A_{eff}K_1(\omega) + \left(\frac{1}{A_{eff}}\right)K_2(\omega)} \cdot \frac{\Gamma}{j\omega(C_0 + C_p)} \quad , \tag{7}$$

where $K_1(w)$ and $K_2(w)$ are given by

$$K_1(\omega) = R_r(\omega) + R_h + R_g + j\omega M_r + \frac{1}{j\omega C_{bc}} \quad , \tag{8}$$

$$K_2(\omega) = j\omega M_d + \frac{1}{j\omega C_d} + \frac{\Gamma^2}{j\omega(C_0 + C_p)} \quad . \tag{9}$$

As S_o is determined by the complicated interconnection among elements in three signal domains, the proposed equivalent circuit is suitable to accurately design and evaluate microphone performance.

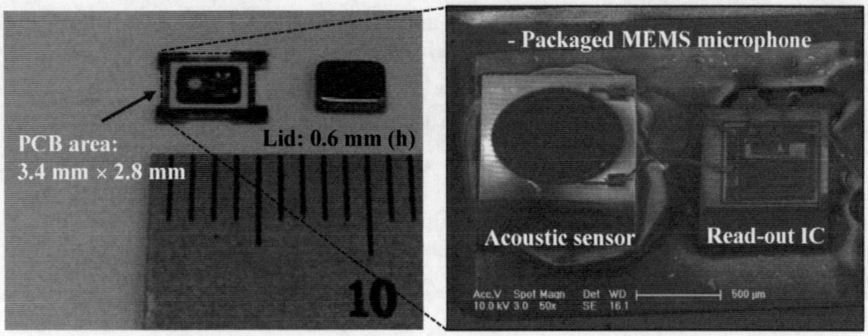

Figure 7. Images of the packaged MEMS acoustic sensor with ROIC

Based on the equivalent circuit model shown in figure 6, each parameter was calculated or extracted by either theoretical calculation or empirical measurement. The modelled S_o was -42.9 dBV/Pa at 1 kHz at the bias voltage of 10 V, and the first resonance frequency was 67 kHz. Furthermore, to compare the modelled values with the measured values, the MEMS acoustic sensor was packaged with a read-out IC (ROIC). Figure 7 shows an integrated MEMS microphone module, where the PCB had an area of 3.4 mm × 2.8 mm, and the metal lid had the height of 0.6 mm. The measurement was performed in an 8.0 m × 8.0 m × 7.0 m anechoic chamber with a commercial analysis set (B&K) in the sweep range of 100 Hz to 18 kHz. Accordingly, the total measured sensitivity (S_{mea}) of the packaged acoustic sensor was -42.0 dBV/Pa at 1 kHz (0 dB = 1 V/Pa), where the ROIC had 0.7 pF input capacitance, 10.0 V output bias, and 6 dB gain. Subsequently, S_o was extracted from S_{mea} using

$$S_{mea} = S_o \frac{C_b}{C_b + C_{par} + C_{in}} G \quad , \tag{10}$$

where G is the gain of the ROIC, and C_{in} is the input capacitance of the ROIC.

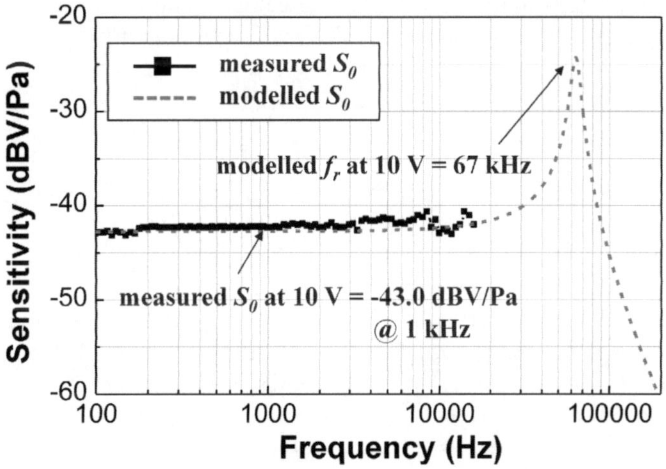

Figure 8. Modelled and measured S_0 as a function of frequency for the MEMS acoustic sensor

From (10), S_o was extracted to be -43.0 dBV/Pa at 1 kHz with the bias point of 10.0 V, while the modelled open-circuit sensitivity was -42.9 dBV/Pa, as shown in figure 8. This validates the structure-based equivalent circuit modelling and the open-circuit sensitivity at a low frequency presented in the previous section. Table II summarizes the measured and modelled data of the proposed MEMS acoustic sensor.

Table 2. Measured and modelled data of the proposed MEMS acoustic sensor

Parameters	**Values**
Bias voltage (V_b)	10.0 V
Aig-gab height at V_b (g_b)	2.45 μm
Intrinsic capacitance at V_b (C_b)	1.2 pF
Parasitic capacitance (C_p)	0.25 pF
Modelled open-circuit sensitivity at V_b ($S_{o\text{-}mod}$)	-42.9 dBV/Pa
Measured open-circuit sensitivity at V_b ($S_{o\text{-}med}$)	-43.0 dBV/Pa
ROIC input capacitance (C_{in})	0.7 pF
ROIC gain (G)	6 dB
Measured total sensitivity at V_b (S_{mea})	-42.0 dBV/Pa

4. Conclusions

An empirical-based model of open-circuit sensitivity at a low frequency for a capacitive-type MEMS acoustic sensor was proposed. To model the open-circuit sensitivity, both the residual stress and the parasitic capacitance were extracted from empirical data because it is difficult to analytically obtain both of them due to their structure dependence. The residual stress was modelled from both the pull-in voltage and surface topology of the diaphragm, whereas the parasitic capacitance was determined from an additional test pattern. To evaluate the proposed empirical model, the modelled open-circuit sensitivity was compared with the measured sensitivity by using a packaged MEMS acoustic sensor. The modelled open-circuit sensitivity was in good agreement with the measured value in the range from DC to higher frequency, demonstrating the validity of the modelling.

5. References

[1] J. Bergqvist and F. Rudolf, "A silicon condenser microphone using bond and etch-back technology," *Sensor and Actuators A*, vol. 45, pp. 115-124, 1994
[2] P. R. Scheeper, A. G. H. var der Donk, W. Olthuis, and P. Bergveld, "A review of silicon microphones," *Sensor and Actuators A*, vol. 44, pp. 1-11, 1994
[3] Q. Zou, Z. Li, and L. Liu, "Design and fabrication of silicon condenser microphone using corrugated diaphragm technique," *IEEE J. Microelectromech. Syst.*, vol. 5, pp. 197-204, 1996
[4] A. Torkkeli, O. Rusanen, J. Saarilahti, H. Seppa, H. Sipola, and J. Hietanen, "Capacitive microphone with low-stress polysilicon membrane and high-stress polysilicon backplate," *Sensor and Actuators A*, vol. 85, pp. 116-123, 2000.
[5] J. W. Weigold, T. J. Brosnihan, J. Bergeron, and X. Zhang, "A MEMS condenser microphone for consumer applications," *IEEE MEMS 2006 Conf.*, Istanbul, Turkey, pp. 86-89, Jan. 2006

[6] M. Gato, Y. Iguchi, K. Ono, A. Ando, F. Takeshi, S. Matsunaga, Y. Ya- suno, K. Tanioka, and T. Tajima, “High-performance condenser microphone with single-crystalline silicon diaphragm and backplate,” *IEEE Sens. J.*, vol. 7, pp. 4-10, Jan. 2007

[7] C.-K. Chan, W-C. Lai, M. Wu, M.-Y. Wang, and W. Fang, “Design and implementation of a capacitive-type microphone with rigid diaphragm and flexible spring using the two poly silicon micromachining processes,” *IEEE Sens. J.*, vol. 11, pp. 2365-2371, Oct. 2011

[8] J. Lee, C. H. Je, J. H. Jeon, W. S. Yang, and J. Kim, “A surface micromachined MEMS acoustic sensor with back-plate anchors of 100 μm depth,” *IEEE Sensor 2011 Conf.*, pp. 1978-1981, Oct. 2011

[9] J. Bergqvist, “Finite-element modelling and characterization of a silicon condenser microphone with a highly perforated backplate,” *Sensor and Actuators A*, vol. 39, pp. 191-200, 1993

[10] H. A C Tilmans, “Equivalent circuit representation of electromechanical transducers: I. lumped-parameter systems,” *J. Micromech. Microeng.*, vol. 9, pp. 157-176. Mar. 1996

[11] W. J. Wang, R. M. Lin, Q. B. Zou, and X. X. Li, “Modeling and characterization of a silicon condenser microphone,” *J. Micromech. Microeng.*, vol. 14, pp. 403-409. 2004

[12] M. Fuldner, A. Dehe, and R. Lerch, “Analytical analysis and finite element simulation of advanced membrane for silicon microphones,” *IEEE Sens. J., Journal,* vol. 5, pp. 857-863, Oct. 2005.

Acknowledgments
This work was supported by Electronics and Telecommunications Research Institute (ETRI) grant funded by the Korea government (15ZB1500, Development of Environment & User Adaptable MEMS Microphone Solution).

A new expression of N_S versus E_F to an accurate control charge model for AlGaAs/GaAs

I. Bouneb[1], F. Kerrour [1]

Laboratory of renewable energy devices modeling and nanoscale- MODERNA-
University of Constantine 1, Algeria
E-mail: bounebilhem@yahoo.fr
E-mail:f_kerrour@yahoo.fr

Abstract. Semi-conductor components become the privileged support of information and communication, particularly appreciation to the development of the internet. Today, MOS transistors on silicon dominate largely the semi-conductors market, however the diminution of transistors grid length is not enough to enhance the performances and respect Moore law. Particularly, for broadband telecommunications systems, where faster components are required. For this reason, alternative structures proposed like hetero structures IV-IV or III-V [1] have been.The most effective components in this area (High Electron Mobility Transistor: HEMT) on IIIV substrate. This work investigates an approach for contributing to the development of a numerical model based on physical and numerical modelling of the potential at heterostructure in AlGaAs/GaAs interface. We have developed calculation using projective methods allowed the Hamiltonian integration using Green functions in Schrödinger equation, for a rigorous resolution "self coherent" with Poisson equation . A simple analytical approach for charge-control in quantum well region of an AlGaAs/GaAs HEMT structure was presented. A charge-control equation, accounting for a variable average distance of the 2-DEG from the interface was introduced. Our approach which have aim to obtain ns-Vg characteristics is mainly based on: A new linear expression of Fermi-level variation with two-dimensional electron gas density in high electron mobility and also is mainly based on the notion of effective doping and a new expression of ΔEc

1. Introduction

The IV-IV or III-V hetero structures [1] use semi-conductors in which, the electrons effective mass is less than the massive Si mass, thus enhancing their mobility. In this domain The most efficient devices are HEMTS on III-V substrate.

The development of semi-conductor materials hetero structures gave clear illustrations on some quantum mechanics concepts as confinement or quantification of energy levels. A new electronic devices generation especially the HEMT have been developed benefitting from these quantum effects. The HEMT) or TEGFET (Two Dimensional Electron Gas FET) is essentially made of one hetero junction GaAsAl/GaAs, which is considered as abrupt of references because both materials have the same crystal structure and the physical parameters of these materials are very close (parameters of meshes very close to the two semiconductors) [2].

The study of this hetero junction (abrupt) inevitably passes by the resolution of the Schrödinger-Poisson equation system due to its quantum effects occurring at the interface level [3, 4].

Different models of TEGFET have been suggested, but in the literature the modeling of TEC is abundant, the amount of works about the modeling of TEGFET is limited.

IOP Conf. Series: Materials Science and Engineering **108** (2016) 012045 doi:10.1088/1757-899X/108/1/012045

However, two types of models have been suggested: analytic models and the two dimensional classical models or Monte-Carlo type [5].
The first models are based on controlling the accumulated charge suggested by Dr. Delagebeaudeuf [5]. The potential well combined with the hetero junction is supposed triangular, giving a simple analytic expression of the energies from different sub-bands.
The second models based on the two -dimensional analysis of devices, in this case transport phenomenon is governed by standard electro kinetics equations or of Monte-Carlo type.
Therefore, it seems interesting to us to precisely study this component from a theoretical point view in order to assess its performances.
This work is dedicated to nanometric modeling of HEMT in a rigorous and complete manner. That we started with a preliminary and precise study of the hetero junction, then, the exact numerical resolution of globally non linear system Schrödinger-Poisson, in an self coherent way, was carried out[6, 7]. Thus allowing to have the repartition the potential model of GaAsAl/GaAs hetero junction interface.
This allowed to us to deduce the influence of the different physical and geometric parameters (as doping and the rate of aluminum) on characterizing parameters of quantum wells (such as the density of charge, ΔE_c) [6, 7], Which led us to a study of the accumulated charge and its command, to highlight its importance in the comprehension of the principle functioning of HEMTs.

2. Theory

Generally, high gap materials are preferred for manufacturing high power components; many of them have used these materials especially, the HEMT. The development of Semiconductors materials hetero structures led to clear illustrations of some quantum mechanics concepts as the confinement and/or energy levels quantification.
One of quantum hetero structures applications is the HEMT transistor. Witch is made of e GaAs substrate, a low gap buffer layer , 2D GAZ will be formed in this material at the upper defining the canal which is the confinement zone of carriers; and then, the non doped AlGaAs layer it allows to reduce interactions between donor's electrons and donor layer which provide electrons to the canal [3].

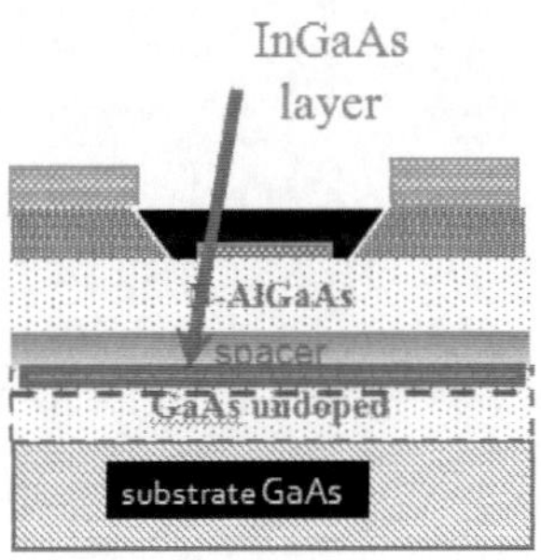

Figure 1. HEMT structure.

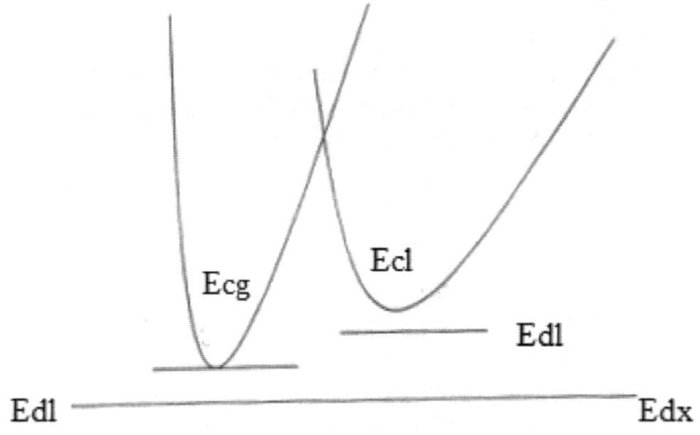

Figure 2. Energy diagram.

The electrons provided by donor layer will pass from high gap material to low gap material. These electrons fill the lower states of GaAs side, as a result, energy bands bend, the curvature of B_c conduction band become a potential well in which electrons are confined. Energies are always measured in electron volts and dentified relative to the lower side of the conduction band of L valley) [2, 3, 7]. The high donor level E_{dl}, is linked to the low side of conduction band, therefore it is wise to suppose that all electrons belonging to this level participate in conduction (there is a total ionization of donor atoms). In the opposite, the deep level E_{dx} is linked to the bottom of L valley, thus an electron trapped on this level must pass through a potential barrier of the order of 140 mev to reach the L.

Therefore we can suppose that the electrons of deep level E_{dx} are trapped forever and can never participate in conduction.

With:

$$\begin{cases} E_{cg} = 0 \\ E_{dl} = 0.006 \\ E_{dx} = 0.17 - 0.58x - 0.31x^2 \\ E_{cl} = 0.31 - 0.58x - O.31x^2 \end{cases} \tag{1}$$

The values of the bottom of the Γ and L conduction band, respectively denoted E_{cg} and E_{CL}, where x represents the rate of Aluminum GaAsAl.
These hypotheses will allow us to calculate the effective total concentration N_{eff} in the GaAsAl according to the concentration of Si doping N_D and the aluminum rate.
Namely N_D the doping in Si donor atoms, we call N_D, and N_{dx} the respective concentrations in light and deep donor atoms, in an evident manner, we have (figure 2):

$$N_D = N_{dl} + N_{dx} \tag{2}$$

According to the statistic of Fermi-Dirac, N_{dx}, N_{dx} are given by the following formulas [8]:

IOP Conf. Series: Materials Science and Engineering **108** (2016) 012045 doi:10.1088/1757-899X/108/1/012045

$$N_{dl} = \frac{N}{1 + \frac{1}{2} e^{(\frac{E_{dl} - E_F}{KT})}} \quad (3)$$

$$N_{dx} = \frac{N}{1 + \frac{1}{2} e^{(\frac{Edx - E_F}{KT})}} \quad (4)$$

Where E_F is the level of Fermi and N is a constant of the problem.

A simple calculation allows us to access the value N_D, that is, in our hypotheses simply effective doping N_{eff} [8]:

$$N_{eff} = \frac{N_d}{1 + (1 + \frac{1}{2} e^{(\frac{E_{dl} - E_F}{KT})}) / (1 + \exp(\frac{E_{dx} - E_F}{KT}))} \quad (5)$$

In our recent works [6, 7], we have determined the parameters characterizing the potential quantum well like as: potential energy , spatial dispersion, wave function and surfaced density , these parameters can be used to determine the behavior dynamics of the HEMT structure . In this work we have proposed a tow new expressions one for the offset ΔE_C and the second is for N_S versus E_F or while also studying the influence of physic parameterizes (Al rate) and the geometric parameters of this heterojunction.

In our model, this new linear expression of Fermi level varied with two-dimensional electron gas (2DEG) density in high electron mobility is proposed, we have obtained this expression by exploiting the numerical calculation of the reference [6,7] in the static case by numerical interpolation, it is given by the following formula:

$$n_s = a + B \times E_F \quad (6)$$

Where $a=2.43.10^{12}/m^2$, and $B=0.05ev/m^2$

The surface density is given by [9]:

$$n_s = \frac{\varepsilon}{qd}(V_G - V_{off} - E_F) \quad (7)$$

ε: is the permittivity,

d: is the total thickness of the AlGaAs layer, V_{off}: is the threshold voltage of the HEMT, given by:

$$V_{off} = \varphi_B - \Delta E_C - \frac{qn_D d_d^2}{2\varepsilon} - \frac{d}{\varepsilon}\sigma) \tag{8}$$

φ_B: is the barrier of the Schottky gate,

ΔEc: is the conduction band discontinuity at the hetero junction,

ND: is the doping concentration, "dd": is the thickness of doped AlGaAs layer

σ: is the polarization sheet charge density of the heterojunction AlGaAs/GaAs.

In our structures, we use a Ni- Schottky barrier contact at the surface; φ_B and ΔE_C are considered as the function of N_D, N_{eff} and Al rate and we use the following formulations for the physical properties of AIGaAs in our calculations, Ni Schottky barrier [9]:

$$\varphi_B = 1.38x + 0.84 \tag{9}$$

We also suggested a new formula at V_{off} level where we have introduced a new expression (our formula) of ΔE_c given by the following expression [10]:

$$\Delta E_C = 0.023N_D^4 - 0.18N_D^3 + 0.53N_D^2 - 0.66N_D + 0.34 \tag{10}$$

With [9]:

$$\sigma(x) = \left| 2\frac{a(0) - a(x)}{a(x)} \left[e_{31}(x) - e_{33}(x)\frac{C_{31}(x)}{C_{33}(x)} \right] \times P_{SP}(x) - P_{SP}(x) \right| \tag{11}$$

Where the lattice constant: a= 5.6533 + 0.0078x A°

The elastic constants: C_{31} = 5.x+103; C_{33} = -32x+405 C/m^2

The piezoelectric constants: e_{31}= -0.11x -0.49; e_{33} = 0.73.x+0.73 C/m^2;

The spontaneous polarization: p_{sp} = (-0.034-0.124.x) 10^{-10} C/m^2;

3. Results and Discussion

3.1. Variation of sheet carrier concentration (n_s) with applied gate voltage (v_{gs}-v_{off})

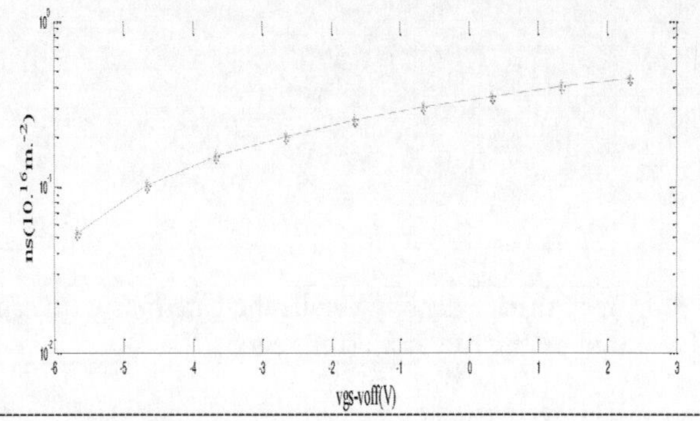

Figure 3. The Variation of sheet carrier concentration (n_s) with applied gate voltage (v_{gs}-v_{off}) à T=300K°.

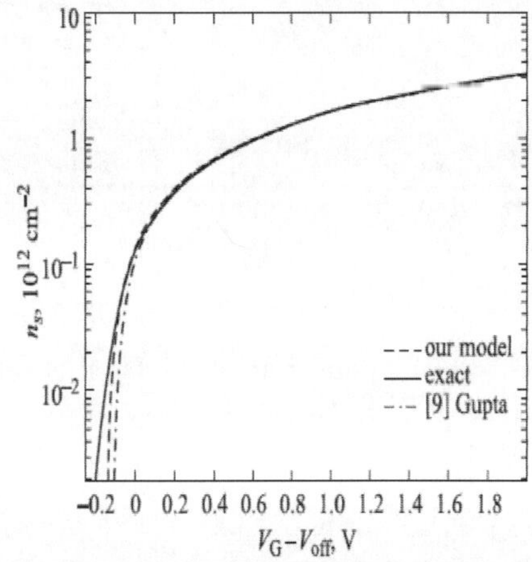

Figure 4. Variation of sheet carrier concentration (n_s) with applied gate voltage (v_{gs}-v_{off}) à T=300K° Results

Figure 3 shows the surface density variation versus to the the grid tension (V_{gs}-V_{off}). This is perfectly agreed with the results obtained in the literature [9] (figure4). This figure shows an increases followed by nearly a saturation of ns when V_{gs}-V_{off} increases , however this limit value differs from the reference [9] value (figure4) because our result is simulated for AlGaAs, one band state, a fixed x (25%) and N_D= 10^{24} /m^3additionally we used a linear formula for E_f=f (n_s) and a polynomial formula of order 4 for ΔE_c but the result of the reference [9] was simulated for NGaAs, two band states, with a non linear formula of ΔE_c.

3.2. sheet carrier density versus the gate bias

Figure 5 shows that ns increases linearly with V_{gs}, we can see that our result is different than the reference [9] because literature value of V_{off} is fixed (it is a constant) while we have calculated V_{off} which is a function of Al rate "x", therefore the AlGAs stays the basic material for manufacturing HEMT transistor.

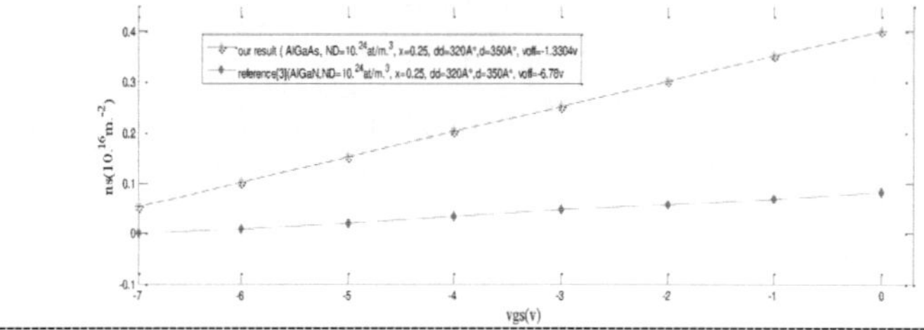

Figure 5. Insheet carrier density versus the gate bias with regard to Results reported by the literature (reference [9]).

3.3. Influence of Aluminum rate

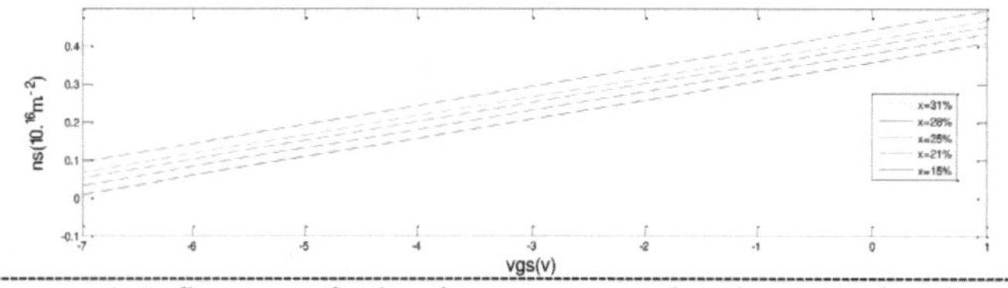

Figure 6. Influence of Aluminum rate on the sheet carrier density versus the gate bias.

The figure 6 shows that the decrease of x causes an increase of ns. This result is reasonable because the decrease of x causes an increase of effective N_d which causes an increase of ns.

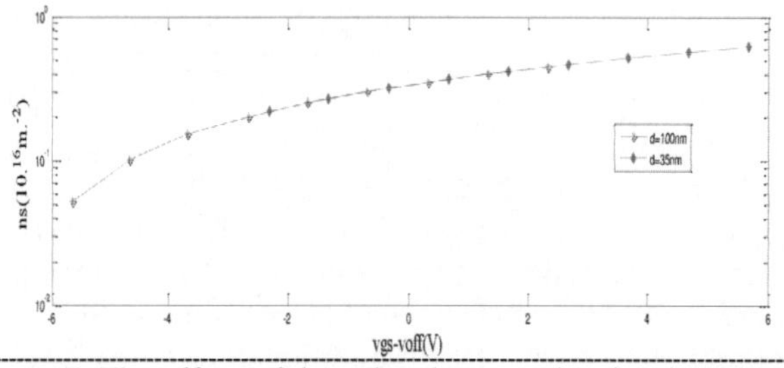

Figure 7. The effect of d on the sheet carrier density versus with applied gate voltage.

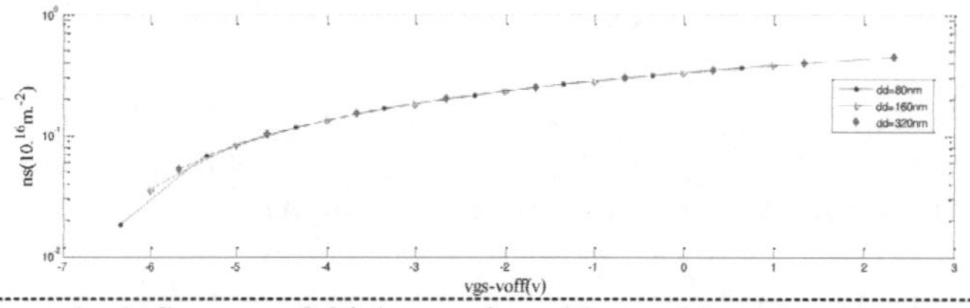

Figure 8. Influence of thickness doped layer dd on the sheet carrier density versus with applied gate voltage

The increase of the positive values of (V_{gs} -V_{off}) where d = 35 the curve tends to a higher value where d = 100nm.
For negative value of (V_{gs} -$V_{off)}$ where d = 35 nm: the curve where d = 100 nm shifts to the right which shows the directly influence of the thickness on the interval of (V_{gs} -$V_{off)}$.
On the other side the increase of the "dd" (figure8) causes a shifted of curve to the right (for an increase of positive tension V_{gs}-V_{off}) with more elevated ns.

4. Conclusion

This work is based on an approach for contributing to the development of a numerical model based on physical and numerical modelling potential at the interface of a hetero structure in GaAsAl / GaAs .
We have elaborate a computation , using projective methods permit the Hamiltonian integration using Green functions in the equation of Schrödinger, for a rigorous resolution auto coherent with the equations of Poisson.
A serious approximation for Fermi level with sheet carrier density was made and some expressions were proposed. These expressions were important in an analytic model development for the HEMT, and it participates to of charge control model research.
A simple analytical approach for charge-control in quantum well region of a AlGaAs/GaAs HEMT structure was presented. A charge-control equation, accounting for a variable average distance of the 2-DEG from the top interface was introduced. Our approach to obtain ns-Vg characteristics is mainly based on: A new linear expression of Fermi-level variation with two-dimensional electron gas density in a high electron mobility and also is mainly based on the notion of effective doping and a new expression of ΔE_c.

References

[1] R.Castagné, J.P.Duchemin, M.Gloanecet Ch.Rumelhard, « circuits intégrés en Arséniure de galilum », physique, et regles de conception, Masson, 1989
[2] G.Bastart, « Wave Mechanics applied to semiconductors heterostructures », les éditions de physiques, 1988.
[3] P.Godts, D.Depreeuw, E.Constant et J.Zmmermann, «Methodes generales de modélisation du transistor à effet de champ à heterojonction », Revue Phys.Appl.24 , 151-170, 1989.
[4] M .Benabbas, B.Marir , D.Bajon ,H.Baudrand, « exact resolution of coupled Schrodinger Poisson equation : application to accurate determination of potential profile in HEMTS » *electronic letter* **,vol 27** , 1848 -1850, 1991
[5] A.Cappy, « propriétés physiques et performances potentielles des composants submicroniques à effet de champ .structure conventionnel à gaz bidimensionnel », thèse de doctorat en science physique, université des sciences et techniques de LILLE Flandres-Artrois , 1986
[6] I.BOUNEB , M .MARIR BENABBAS ,« modelisation of accumulation barrier of nanostructure AlGaAs/GaAs», « 2nd wseas international conference on Nanotechnology » university of Combridge, UK , fubruary 20-22,2010
[7] I.BOUNEB, .M .MARIR BENABBAS,N.RAVEU,H.BAUDRAND, «influence of different technological parameters on the accumulation barrier of nanostructure AlGaAs/GaAs» *International Review of PHYSICS* - April 2011 (**Vol. 5** N. 2).
[8] P TEMPLE-Boyer «Elaboration d'un modèle Pseudo - Bidimensionnel d'un transistor pseudomorphique GaAlAs/GaInAs/GaAs » laboratoire d'électronique Philips, Juin 1990
[9] Pu Jinrong, Sun Jiuxun, Zhang Da «An accurate polynomial-based analytical charge control modelfor AlGaN/GaN HEMT » issn 1063-7826 , semiconductor ,2011,**vol.45**,No.9, pp.1205-1210.
[10] I.BOUNEB,F.KERROUR ; « Nanometric Modelisation to characterize dynamics carriers in a HEMT heterostructure (AlGaAs/GaAs) using an effectif doping», *Key Engineering Materials* **Vol 644** (2015) pp 26-30

Performance analysis of resistive switching devices based on $BaTiO_3$ thin films

Natasa Samardzic[1*], Tijana Kojic[1], Jelena Vukmirovic[2], Djordjije Tripkovic[2], Branimir Bajac[2], Vladimir Srdic[2] and Goran Stojanovic[1]

[1]Department of Microelectronics, Faculty of Technical Sciences, University of Novi Sad, Novi Sad, Serbia
[2]Department of Materials Engineering, Faculty of Technology, University of Novi Sad, Novi Sad, Serbia

[*]E-mail: nsamardzic@uns.ac.rs

Abstract. Resitive switching devices, memristors, have recenty attracted much attention due to promising performances and potential applications in the field of logic and memory devices. Here, we present thin film $BaTiO_3$ based memristor fabricated using ink-jet printing technique. Active material is a single layer barium titanate film with thickness of ~100 nm, sandwitched between metal electodes. Printing parameters were optimized aiming to achieve stable drop flow and uniform printed layer. Current-voltage characteristics show typical memristive behavior with pinched hysteresis loop crossed at the origin, with marked differences between High Resistive State (HRS) and Low Resistive State (LRS). Obtained resistive states are stable during numerous switching processes. The device also shows unipolar switching effect for negative voltage impulses. Variable voltage impulse amplitudes leads to the shifting of the energy levels of electode contacts resulting in changing of the overall current through the device. Structural charcterization have been performed using XRD analysis and SEM micrography. High-temperature current-voltage measurements combined with transport parameter analysis using Hall efect measurement system (HMS 3000) and Impedance Analyzer AC measurements allows deeper insigth into conduction mechanism of ferroelectric memristors.

1. Introduction

Expanding interest for the devices with memristive behaviour is caused by favourable future applications such as in: new generation of memory device, artificial synapses, unconventional computing systems and overcoming of the Moore's law. Various materials and mechanism have gained memristive effect, however one of the most promising types are ferroelectric junctions, due robustness during read-out process, long retention time, possibilities for scaling down to few unit cells, chemical stability and high R_{OFF}/R_{ON} ratio [1].

Widely used ferroelectric memristive structures are based on $BaTiO_3$ thin films [2], while underlying switching mechanisms is explained as field induced charge redistribution at the interface metal/ferroelectric, which lowers the ferroelectric barrier height and enables conduction. Accumulation of the charge carriers at the interface contributes to the stability of the memorized states, wherein interface property, i.e. selection of the electrode material is critical issue for the current control [1].

In this paper, we present detailed performance analysis of Au/$BaTiO_3$/Ag memristive device. Ink-jet printing fabrication technique was chosen for deposition of active $BaTiO_3$ layer with advantages of additive, controlled and low-cost deposition. Memristive effect was identified through current-voltage measurement, while further electrical and structural analysis bring contribution in still open issues of underlying resistive switching effects mechanism.

2. Materials and Methods

Active material is $BaTiO_3$ thin film deposited using Dimatix DMP 3000 ink-jet printer on Si wafer (SG 108, CrysTec) substrate on which was previously evaporated Au layer. Details about $BaTiO_3$ ink synthesis and properties have been reported elsewhere [3].

Stable drop flow during from 10 pl nozzles was obtained with trapezoidal actuation voltage waveform with amplitude of 26V and 1 kHz printing frequency. Deposition was performed in the clean room environment; class ISO 7 at room temperature. Drop spacing was set to 35 μm, matching the half diameter of the printed drop on Si wafer substrate. Square-shaped patterns of $BaTiO_3$ one-layer film were well defined and reproducible. Post-processing step involves sintering at temperature of 700°C for 1 hour. Top electrode of the device is Ag two-component conductive paste (*CircuitWorks® Conductive Epoxy*), cured for 24 hours at room temperature.

3. Results and discussion

3.1. Structural and morphology analysis

Structural XRD analysis (Rigaku XRD Miniflex 600), resolved characteristics peaks for $BaTiO_3$ crystalline phase (Figure 1a), with the presence of Si peaks from the substrate. SEM micrograph (JEOL JSM 6460 LV), Figure 1b, showed uniform layer of $BaTiO_3$ thin film with thickness ~100 nm.

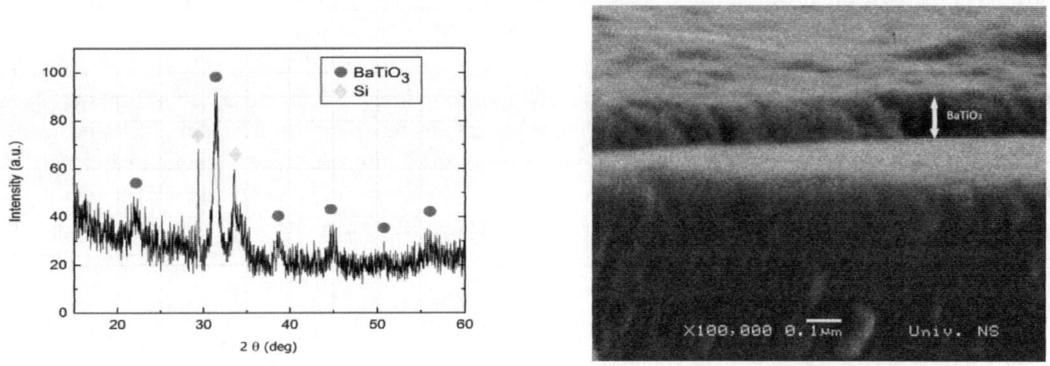

(a) (b)

Figure 1. Structural analysis of Au/$BaTiO_3$/Ag resistive switching device a) XRD patterns b) SEM micrograph

3.2. Electrical analysis

The current-voltage characteristics were measured using Keithley 2410 High Voltage Source Meter in voltage source mode with 0.1V per second step voltage sweep. Bottom electrode Au was grounded during measurement, while voltages pulses are applied to the top (Ag) electrode with amplitude of +/-5V. Pinched hysteresis loop crossed at origin, the fingerprint of memristive effect, with bipolar switching behavior shown in Figure 2. Dissymmetry of the characteristics originates from Joule heating on the contact regions, as well as differences between barrier heights and metal work functions at Au/$BaTiO_3$ and Ag/$BaTiO_3$ interfaces. Exponential behavior of current-voltage relation is most likely caused by dominant Fowler-Nordheim tunneling mechanism above critical electric field [4], while low voltage switching is controlled by barrier height modulation due to charge accumulation and dissipation at the interface. Namely, in the ON state barrier height is reduced, which lowers the resistance values, in contrary in the OFF state resistance increases as a consequence of charge dissipation.

Unipolar negative voltage pulses were applied from -1V to -5V, however only above 3V amplitude hysteretic I-V curve is visible, Figure 2 (inset). The relation between R_{OFF} and R_{ON} state, decreases as the voltage amplitude decrease, which is the resulted of less attracted carriers included in the current flow. Considering that $BaTiO_3$ acts as n-type of semiconductor, at room temperature [5] it is expected that main accumulated charge carriers included in the transport are electrons.

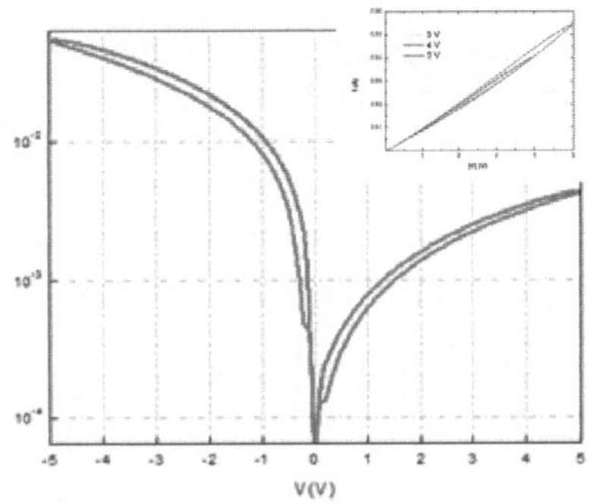

Figure 2. The current-voltage characteristics of Au/$BaTiO_3$/Ag printed memristor

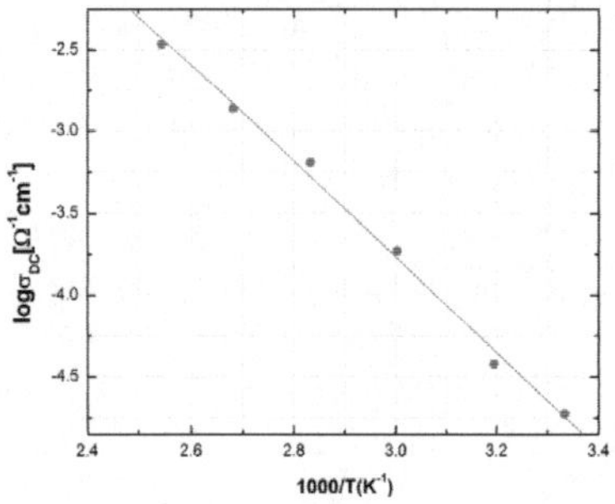

Figure 3. Temperature dependence of DC conductivity for $BaTiO_3$ memristive device

Specific conductivity of the Au/$BaTiO_3$/Ag sample was measured in temperature range from 300K-400K, results are shown in Figure 3. Results are in accordance with typical semiconductor behavior described with Arrhenius relation $\sigma_{DC}(T) = \sigma_0 \exp(-\Delta E/kT)$, where ΔE is activation energy, k-Boltzmann's constant, T-absolute temperature and σ_0 pre-exponential factor $[\Omega\ \mathrm{cm}]^{-1}$. According to measured results and fitted curve, activation energy is estimated $\Delta E = 0.25\ eV$, corresponding to previously obtained value for $BaTiO_3$ thin films [6] . Activation energy indicates that film has shallow traps near conduction band edge, which could origin from the vacancies introduced during high temperature treatment or impurities [6].

Further electrical analysis was performed by means of impedance spectroscopy (Impedance Analyzer HP4194A), resulting in Cole-Cole diagram shown in Figure 4a, for frequency range from 100 Hz- 40 MHz.

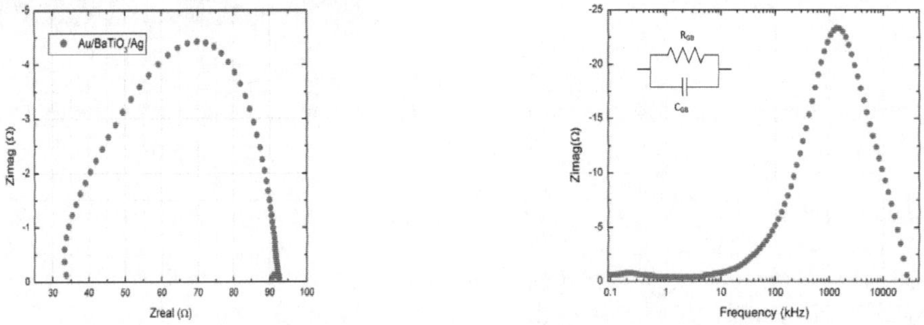

(a) (b)

Figure 4. Cole-Cole diagram (a) and Zimag=f(frequancy) (b)

As it is known from the literature [7], the semicircular in the Cole-Cole diagram formed from resistance and capacitance of the grain boundaries are much larger to then the resistance and capacitance of the bulk grains. Accordingly, dominant middle-frequency semicircular, Figure 4a, corresponds to grain boundaries contribution to the total conductivity of the $BaTiO_3$ sample.

Grain resistivity can be estimated according to graphical impedance spectroscopy analysis approach proposed in [8], as $R_\infty \approx 35\ \Omega$, which is equal to the dislocation of the semicircle from the origin, while parallel resistance and capacitance of the grain boundary arc is estimated to $R_{GB} \approx 110\ \Omega$ and $C_{GB} \approx 1$nF, respectively. Maximum frequency of the complete semicircular is evident in Figure 4b, is ω~1.53 MHz. Accordingly grain boundary relaxation time amounts $\tau \approx \frac{1}{\omega} = R_{GB}C_{GB} = 1.0\ 10^{-7} s$.

Additionally, we performed Hall effect measurement (HMS 3000) of one layer $BaTiO_3$ film deposited on Si wafer substrate connected in four points with the sample holder. Measurement results showed negative Hall coefficient $R_H = -5.91 \times 10^3$ [cm^3/C], which confirms n-type conductivity of the sample, bulk concentration $N_b = 1.05 \times 10^{15}$ [1/cm^3] and mobility $\mu_n = 0.1 \times 10^{-2}$ [cm^2/Vs] in good agreement with results from literature [6].

4. Conclusion

In this paper we originally presented Au/$BaTiO_3$/Ag thin film memristor fabricated using ink-jet printing deposition technique. Current-voltage characteristics is in the form of pinched hysteresis, for both bipolar and unipolar impulse. Ratio of R_{OFF}/R_{ON} state could be improved by increasing the contribution of ferroelectric phase in the $BaTiO_3$. High temperature DC conductivity measurement, revealed typical semiconductor behavior, with 0.25 eV shallow trap near the conductive band gap, probably due to introduced vacancies upon high-temperature treatment. Transport parameter measurement, proved that the sample is n-type semiconductor, thus electrons present the main charge carriers. Impedance spectroscopy graphical analysis revealed separate influence of the grain and grain boundaries in the overall conduction.

Acknowledgements

This work was supported by the Ministry of Education and Science, Republic of Serbia under the Project III-45021.

References

[1] D. J. Kim et al., "Ferroelectric Tunnel Memristor," *Nano Lett.*, vol. 12, pp. 5697-5702, 2012.

[2] X. Chen, C. H. Jia, Y. H. Chen, G. Yang and W. F. Zhang, "Ferroelectric memristive effect in BaTiO3 epitaxial thin films," *Journal of Physics D: Applied Physics*, vol. 47, p. 365102, 2014.

[3] Vukmirovic J, Tripkovic Dj, Bajac B, Kojic S, Stojanovic G and Srdic V, "Comparison of characteristics of barium titanate thin films prepared by inkjet printing and spin coating," *Processing and Application of Ceramics*, 2015.

[4] V. Garcia and M. Bibes, "Ferroelectric tunnel junctions for information," *Nature Communications*, vol. 5, p. 4289, 2014.

[5] J. Nowotny and M. Sloma, "Surface electrical properties of BaTiO3 at elevated temperatures," *Solid State lonics* , vol. 49, pp. 129-133, 1991.

[6] Z. Xiong et al., "Room-temperature resonant quantum tunneling," *Nanoscale*, vol. 6, p. 13876, 2014.

[7] Y.K. Vayunandana Reddy and D.Mergel, "Frequency and temperature-dependent dielectric properties of BaTiO3," *Physica B*, vol. 391, pp. 212-221, 2007.

[8] E. Barsoukov and J. Ross Macdonald, *Impedance Spectroscopy*. Hoboken, New Jersey.: John Wiley & Sons, 2005.

Preparation and Characterization of a PEDOT-Manganese Oxide Composite, and Its Application to Electrochemical Sensing

A. Arena
Dipartimento di Ingegneria Elettronica, Chimica e Ingegneria Industriale
Università di Messina, Contrada di Dio, 98166, Messina, Italy

E-mail: arenaa@unime.it

Abstract. Stable and transparent aqueous dispersions of a hybrid organic-inorganic composite, are prepared by electrochemically doping Manganese Oxide into Polyethylendioxythiophene (PEDOT). Films deposited from the PEDOT-MnOx dispersions, are characterized by means of electrical and optical measurements, and by means of Atomic Force Microscopy (AFM) investigations. The PEDOT-MnOx composite is then used to modify one of the gold electrodes of a simple electrochemical cell, in which Nafion is used as a solid electrolyte. The cell is characterized using time domain electrical measurements. It is found that distinguishable redox peaks arise in the current-voltage loops of the cell, as nanomolar amounts of either acetic acid and ammonia, are added to the deionized water into which the cell is immersed. The intensity of such current peaks, is linearly related to the concentration of the analytes, in the nanomolar range of concentrations.

1. Introduction

The development of hybrid composites combining the attractive properties of organic polymers with the peculiar behaviour of metal oxides, represents one of the most relevant contributions that materials science has given to the advancements in a variety of applied research fields, including energy storage [1, 2], photovoltaic energy generation [3, 4], sensing [5, 6], and catalysis [7]. As far as it concerns sensing, for instance, it is well known that the performance of chemical sensors can be substantially improved by replacing the metal oxides on which commercially available sensors are often based on, with organic-inorganic composites. Hybrid materials, in fact, can be tailored to combine the advantages of doped conjugated polymers, namely the low working temperature and the good mechanical and electrical transport properties, with the sensitivity, and often the selectivity of metal oxides. In such a framework, in 2012 Wu et al. [5] demonstrated the electrochemical sensing properties of a hybrid composite consisting of PEDOT and Manganese Oxide, synthesised by electropolymerization of the 3,4-ethylenedioxythiophene monomer in the presence of Manganese Acetate. Here a similar composite, PEDOT-MnOx, is prepared by electrochemical doping Manganese Oxide into Polyethylendioxythiophene. Electrical measurements performed in the frequency and time domains show that films deposited from the PEDOT-MnOx dispersions have good charge transport properties, that make them a suitable choice for use as electrodes in electrochemical devices. Sensing tests performed in deionized water, reveal that that simple electrochemical cells having one of the electrodes modified with PEDOT-MnOx, reversibly respond to the presence of nanomolar amounts of acetic acid and of ammonia.

2. Experimentals

Water dispersions of highly conducting Polyethilendioxythiophene and Manganese were purchased from Aldrich. Alcoholic dispersions of Nafion 117 were purchased from Ion Power. The PEDOT-MnOx composite was obtained as a fine dispersion by applying an electric field of about 10V/cm between a couple of manganese electrodes placed inside of a beaker containing PEDOT in a water/ethanol mixture. Immediately, as the manganese cathode started to dissolve, the blue transparent colour of the PEDOT dispersion started to change, reaching the dark brown. Thin films deposited from the developed brown dispersions, were characterized by means of spectrophotometric measurements in the VIS-NIR range performed using an Ocean Optics microspectrophotometer HR4000, and by means of AFM morphological analysis performed in the noncontact mode, using a Nanosurf FlexAFM equipped with a C3000 controller. The electrical properties of the PEDOT-MnOx films were characterized by means of impedance measurements, performed in air using an Agilent 4284A LCR meter, in the frequency range between 20 Hz and 1MHz, with a 100 mV amplitude sinusoidal input.
Simple electrochemical cells based on PEDOT-MnOx, were developed using as substrates copier grade transparency sheets, onto the surface of which rectangular parallel gold electrodes, spaced by half a millimetre, were thermally evaporated in vacuum. The PEDOT-MnOx composite was applied onto the top of one of the gold electrodes. The device was then accomplished by depositing a thin Nafion layer onto the insulating gap between the gold electrode and the Au/PEDOT-MnOx electrode, to provide a proton conducting electrical bridge between the electrodes. Time domain electrical characterization of the Au/Nafion/MWNT-Nafion:Cu/Au cells was performed using the 2400 Keithley source meter, by measuring the current flowing through the device, in response to zero average triangular voltage inputs. Measurements were carried out over symmetric voltage windows. Sensing tests were performed with the cells immersed in deionized water. Diluted acetic acid and ammonia were injected into the solution containing the sensor under test by using a micro syringe, in order to increase the analytes concentration in steps of $2 \cdot 10^{-9}$ M.

3. Results and Discussion

Manganese cathodes, immersed into water/ethanol polyethilendioxythiophen mixtures, dissolve in the presence of externally applied electric fields, due to the formation of hydroxides species at the metal surface. After a few minutes of electrochemical doping under 10V/cm electric field, the polyethilendioxythiophen mixtures turn their colour from the original transparent blue to the brown, due to the dispersions of the dissolution products originating form the cathode. Such colour changes

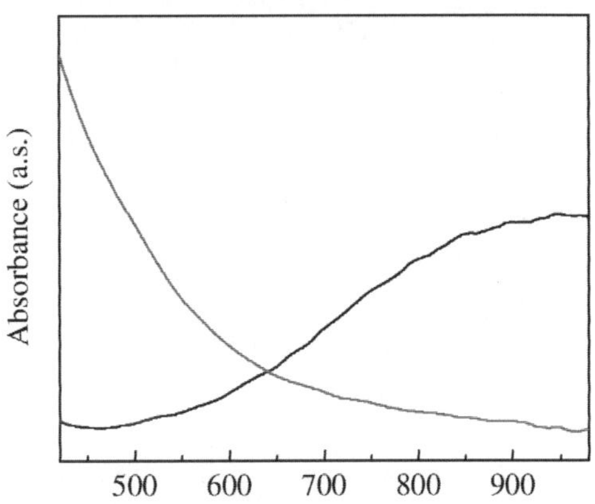

Figure 1. Absorption spectrum of a PEDOT thin film (black) and of PEDOT-MnOx thin film (red).

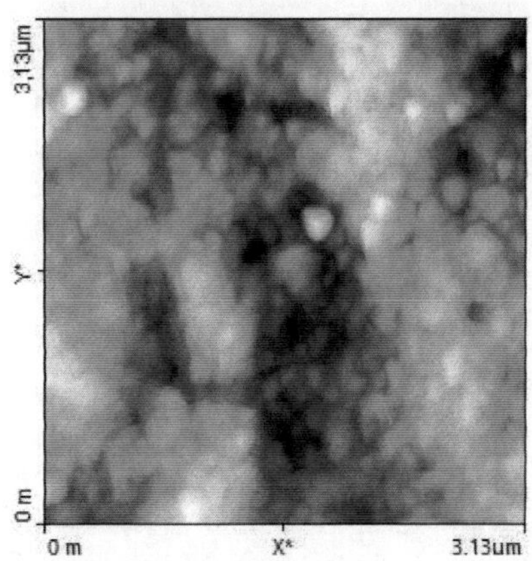

Figure 2. AFM micrography of a film deposited from an aqueous dispersion of PEDOT-MnOx.

are evidenced in Fig.1, where the optical absorption spectrum of PEDOT, nearly transparent in the visible range, and slightly absorbing in the near infrared, is compared with the absorption of PEDOT-MnOx. This latter is found to increase smoothly, as the wavelength decreases.

Figure 2 shows the AFM image, collected in the non contact mode, of a thin film deposited onto a silicon substrate from the electrochemically prepared PEDOT-MnOx aqueous suspensions. The film has a granular morphology, and seems to consist of densely packed particles, having hundreds nanometers average size, randomly distributed into the polymer host. The ratio between the organic and the inorganic counterpart, can be adjusted by acting on the electrochemical doping time: the higher is the doping time, the higher is the nanoparticles concentration.

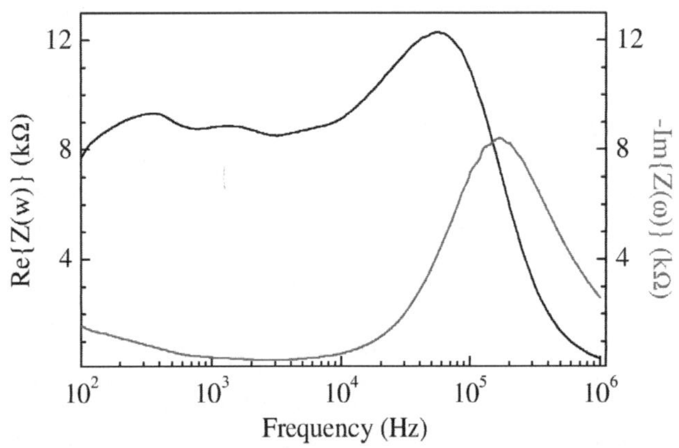

Figure 3. Frequency dependence of the real part (black) and of the opposite of the imaginary part (red) of the complex impedance of a PEDOT-MnOx film.

From the electrical point of view, the electrochemical doping of Manganese Oxide into PEDOT has remarkable effects. The data shown in figure 3 represent the frequency dependence of the complex impedance of a thin PEDOT-MnOx film, deposited onto the gap between a couple of symmetric gold electrodes spaced by about half a mm. Compared with the predominantly resistive impedance of highly conducting pure PEDOT, the composite film has one order of magnitude higher resistive part of the impedance, and six order of magnitude higher modulus of the reactive part.

Time domain electrical characterization of the Au/Nafion/PEDOT-MnOx/Au cells is performed by measuring the current that flows through the cell in response to triangular voltage inputs, as it is sketched in figure 4.

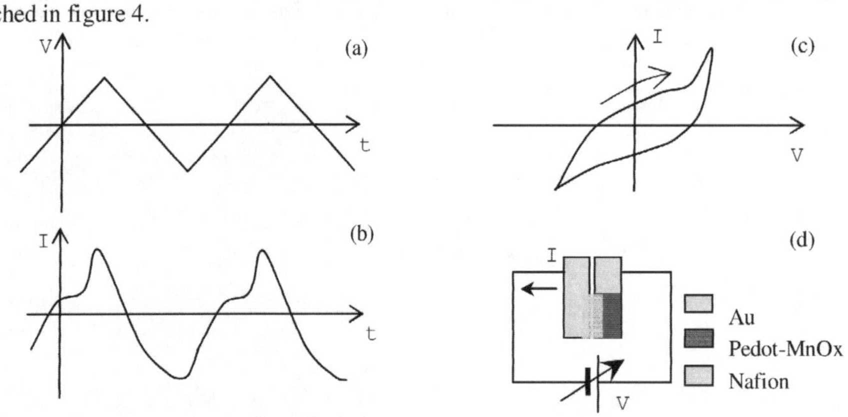

Figure 4. Sketch of the time dependence of the triangular voltage input V (a) and typical shape of the corresponding current output I (b). With increasing time, the current I plotted against voltage V describes loops as the one sketched in (c), advancing in sense of the arrow as the time increases. A schematic view of the cell and the voltage polarity are shown in (d).

It is found that the current measured through a typical Au/Nafion/PEDOT-MnOx/Au cell in deionized water, after a short transient phase, forms stable loops when plotted against voltage. Loops form because the current is affected not only by the applied voltage, but it depends also on the time rate at which the voltage changes, therefore the direct current (i.e. the current measured during the half wave of increasing input voltage) does not coincide with the reverse current, measured while the input voltage has a negative time derivative. This is found to be the case whenever the device under test has a capacitive-resistive behaviour, as it likely happens in the Au/Nafion/PEDOT-MnOx/Au cells, where capacitive effects may arise both from faradaic effects, and from the double layer capacitance that forms at the interface between the electrodes and the ion conducting NAFION layer.

Figures 5a and 5b show examples of steady state current-voltage loops of a typical Au/Nafion/PEDOT-MnOx/Au cell, measured in response to the same voltage inputs (zero average triangular voltage pulse, with 1 V amplitude and 40 s period). The loops differ one from each other by the environment in which the measurements are carried out. In particular, the inner cycle of both figures 5a and 5b (black line), is measured in pure deionized water. The outer cycles represent the steady state current measured respectively when acetic acid (Fig. 5a), and when ammonia (Fig. 5b), are added to the pure water, in subsequent steps of 2 nM.

It can be noticed that the area subtended by the current-voltage cycles shown in Figure 5a, enlarges with the increasing acetic acid concentration. In addition, as the acetic acid content is increased, it is found that a couple of forward current peaks arises, centred at about 100 mV and 700 mV (marked by the arrows, and indicated as A and B in the Figure 5a), while a reverse current peak (marked as C) seems to appear at -30 mV approximately.

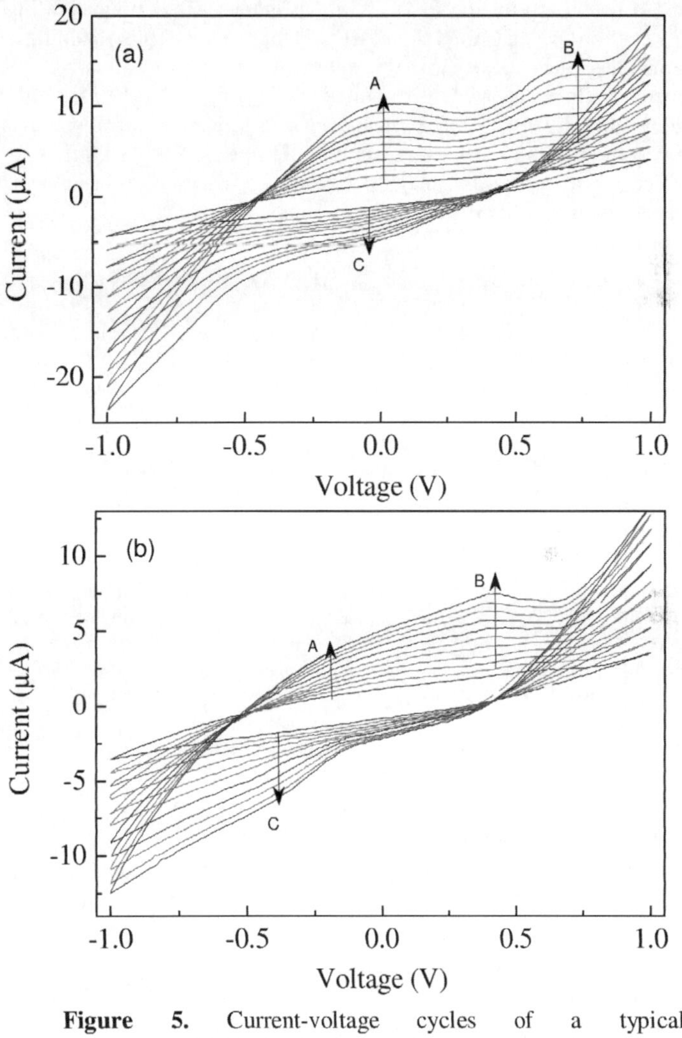

Figure 5. Current-voltage cycles of a typical Au/Nafion/PEDOT-MnOx/Au device, measured in distilled deionized water (black line), and in the presence of acetic acid (a) and of ammonia (b) respectively. The analyte concentration increases going from the inner to the outer loops, in the direction marked by the arrows.

Similar results, with current peaks distinguishable from the former case, are obtained when the cell having a gold electrode modified with PEDOT-MnOx, is exposed to the presence of ammonia. As the NH_3 content increases, in steps of 2 nM going from the inner to the outer current loops of Figure 5b, a forward current peak (marked as B) grows at about 420 mV, and a couple of forward and reverse broad structures seem to arise (marked by A and C in figure 5b), with shoulders positioned at about -200 mV and -400 mV respectively.
Based on the results of figures 5a and 5b, it can be argued that the Au/Nafion/PEDOT-MnOx/Au electrochemical cells are able to detect the presence of weak acids as acetic acid is, and of bases like ammonia, at concentration in the nanomolar range. To demonstrate that the currents peaks of figures 5 are a peculiarity of the PEDOT-MnOx modified electrode, sensing test have been performed using Au/Nafion/Au, and Au/Nafion/PEDOT/Au cells, without obtaining any meaningful response. It can be concluded that the sensitivity towards acetic acid and ammonia is ascribable to the presence of the metal oxide nanoparticles. Responsible of the current peaks shown in figures 5a and 5b may be redox transitions involving the Manganese Oxides, similar to those described by Latif et al. [8], in their review concerning the development of pH sensors [8]. As Latif et al. suggest, in alkaline environment, electrochemical sensing may arise from the reversible transition between the Manganese Oxide and the Manganese Oxyhydroxide redox couple (MnO_2/MnOOH):

$$MnO_2+H^++e^- \rightleftarrows MnOOH.$$

In acidic medium, and in the presence of Nafion [8], the redox equilibrium may instead involve a different couple, consisting of Manganese Oxide and ionized Manganese (MnO_2/Mn^{+2}):

$$MnO_2+4H^++2e^- \rightleftarrows Mn^{+2}+2H_2O.$$

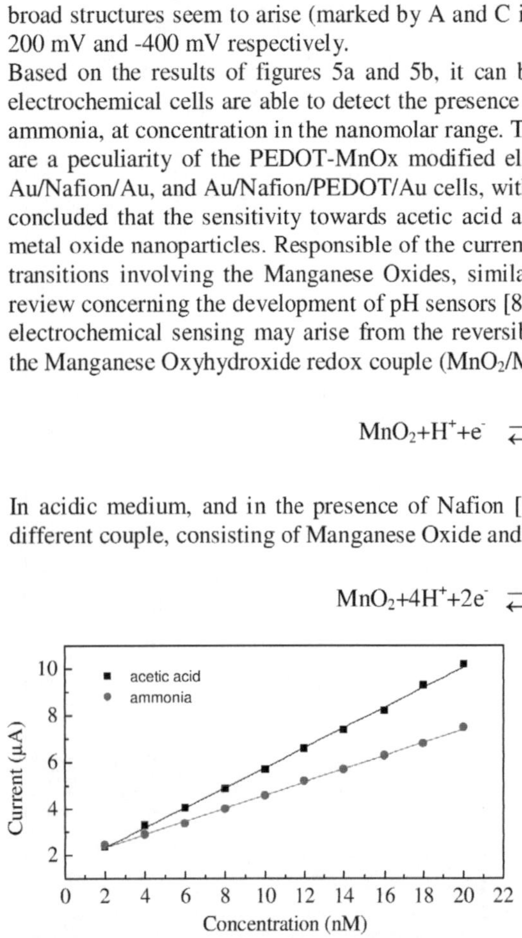

Figure 6. Intensity of the current peak marked as A in figure 5a (squares), and of the current peak marked as B in figure 5b, respectively as a function of the acetic acid and of the ammonia concentration.

Data extracted from figures 5a and 5b, can be used to determine how the intensity of the current peaks due to redox transitions promoted by the analytes, are related to the analytes concentration. Figure 6 shows that the peak marked as A in figure 5a, and the peak marked as B in figure 5b, have intensity that linearly grows, respectively as the acetic acid and as the ammonia concentration grows.

4. Conclusions

A conducting organic-inorganic composite of Manganese Oxide nanoparticles dispersed in PEDOT, has been prepared using a simple electrochemical method, and characterized by means of electrical and optical measurements. The material, applied on gold, has been used as an electrode in electrochemical cells, having Nafion as solid electrolyte. The current-voltage loops of the cells, measured in deionized water, reversibly modify in the presence of acetic acid and ammonia, showing current peaks originating from redox transitions involving the metal oxide. Since the current peaks that arise in response to acetic acid and ammonia, likely originate from different redox systems, it follows that the presence of a weak acid at nanomolar concentration, can be potentially detected and distinguished from the presence of a base. Furthermore, it is found that the current peaks' intensity is a linear function of the analytes' concentration in the nanomolar range. These results, seem to make the PEDOT-MnOx a promising candidate for electrochemical sensing applications.

References

[1] Luo J, Tung VC, Koltonow AR, Jang HD, and Huang J 2012 *J. Mater. Chem.* 22 12993
[2] Sidhu NK and Rastogi AC 2014 *Nanoscale Res. Lett.* **9** 453
[3] Boucle J, Ravirajanac P and Nelson J 2007 *J. Mater. Chem.* **17** 3141
[4] Skompska M 2010 *Synt. Met.* **160** 1
[5] Wu ZY, Thiagarajan S, Chen SM, and Lin KC 2012 *Int. J. Electrochem. Sci.* **7** 1230
[6] Yinhua L, Gong J, He G and Deng G 2011 *Mat. Chem. Phys.* **129** 477
[7] Zhang L, Jamal R, ZhaoQ, Wang M and Abdiryim T 2015 *Nanoscale Res. Lett.* **10** 148
[8] Latif U. and F.L. Dickert in Environmental analysis by electrochemical sensors and biosensors Vol II Applications ISBN 978-1-4939-1300-8 Springer, edited by L.M. Moretto and K. Kalcher 751-777 2014.

IOP Conf. Series: Materials Science and Engineering **108** (2016) 012048 doi:10.1088/1757-899X/108/1/012048

Deposition of Co-doped TiO_2 Thin Films by sol-gel method

A Boutlala[1], F Bourfaa[1], M Mahtili[1] and A Bouaballou[1]

[1]Laboratory of thin films and interfaces, Faculty of science, University Constantine 1, Algeria

abd.boutlala25@gmail.com

Abstract. Cobalt doped TiO_2 thin films have been prepared by sol-gel method onto glass substrate at room temperature. İn this present work, we are interesting to study the effect of Cobalt doped TiO_2 thin films.the concentration of Co was varied from 0 to 6 %at .The obtained films have been annealed at 500°C for 2 hours. X-ray diffraction patterns showed that Co: TiO_2 films are polycrystalline with a tetragonal anatase and orthorhombic brookite types structures. The surface morphologies of the TiO_2 doped with cobalt thin films were evaluated by Atomic Force Microscopy (AFM). The optical properties were studied by mean of UV–visible and near infrared spectroscopy.The calculated optical band gap decreases from 3.30 to 2.96 eV with increasing Co doping.

1. Introduction

Titanium dioxide, TiO_2, can occur in three main crystal structures: anatase, rutile and brookite. The rutile and anatase forms have been intensively studied and have a significant technological relevance, owing, in large measure, to their optical properties: both are transparent in the visible and absorb in the near ultraviolet [1]. In addition, rutile is the most common crystal phase while brookite is scarce. Rutile is then stable high temperature phase (generally in the 600–1855°C), whereas anatase and brookite are metastable and are readily transformed to rutile when heated [2].Therefore, understanding of the mechanism and the factors affecting phase stability and phase transformation is important. However, all these transformations have remained to this day misunderstood. Therefore, understanding of the mechanism and the factors affecting phase stability and phase transformation is very important for more efficient uses. For instance, metal doping can alter the phase transformation and, consequently, the photocatalytic activity of TiO_2. One doping of interest for photocatalysis and other electromagnetic applications is cobalt [3-6]. Indeed, it was reported that the incorporation of metal ions into the titanium crystal lattice can significantly extend the absorption edge into the visible region [7,8] and anatase structure Co-doped TiO_2 films, at room temperature, have ferromagnetism (FM)[9,10]. Recently, it was discovered that Co-doped TiO2 thin films may have potential application as diluted magnetic semiconductors (DMS). Among the elaboration techniques of TiO_2 thin films, the sol-gel method is emerging as one of the most promising process because it is not only effective in producing thin, transparent, homogeneous, multi-layer oxide components many compositions on various substrates at low cost and allows the choice of the refractive index and thickness of the layer by changing the preparation conditions, but also to easily incorporate and control the cobalt.

In this paper, we report the study of the structural, optical and electrical properties of TiO_2 thin films deposited onto glass by sol–gel dip coating technique as a function of the preparation conditions. Structural evolution with annealing temperature and Cobalt concentration are investigated by Differential Scanning Calorimetry (DSC), Xray Diffraction (XRD) and confirmed by Atomic Force Microscopy (AFM) and Scanning Electron Microscope (SEM) images. Transmittance, gap and refractive index of the films are also studied

2. Experimental

Cobalt doped titanium dioxide ($Co:TiO_2$) films were synthesized on glass substrates by sol-gel technique. The sol-gel solution [11] is performed at room temperature in the following way: 2 cm^3 of titanium isopropoxide (Fluka, 99.9) were considered, to which 0.7 cm^3 of isopropanol were added drop by drop. The solution was left under closed stirring during 10 minutes. Then, 2.2 cm^3 of acetic acid were poured, stirred during 15 minutes. Cobalt (II) acetate tetrahydrate powder wasn dissolved in methanol was added to prepare the Co: TiO2 for various at. % of cobalt concentrations $Ti_{1-x}Co_xO_2$ (x =0.00, 0.02, 0.04, 0.06). Finally, 5.2 cm^3 of methanol were added and stirred during 1 hour. The samples were immersed in the sol-gel solution with speed of 12 cm/min, dried at 100°C during 15 min, and finally annealed at 500°C for 2 h. The spectroscopic measurements were performed on Horiba J. Y. Ellipsometer UVISEL. XRD patterns were recorded on Siemens D8 diffractometer using Copper K_α radiation. Optical transmittance was measured on a UV-VIS spectrophotometer. Atomic force microscopy (AFM) was used for the observation of surface morphology for Co: TiO_2 films deposited onto glass substrates in a region of 4.68×4.68 μm^2. The electrical properties of the layers were determined using the I(V) characteristics by the two probes technique in a coplanar structure with two evaporated gold electrodes. A dip-coating apparatus made in our laboratory was used for the depositions. The substrate was lowered into the coating solution and then withdrawn at a regulated speed of 0.6 cm s-1. After each coating, the films were first dried at 100°C for 15 min. The films were then heat-treated at different temperatures ranging between 400 and 500°C with increasing temperature rate of 5°C min^{-1} for 2h in furnace. The electrical properties of the layers were determined using the I(V) characteristics.

3. Results and discussion

3.1 *Films structure*

Figure1 shows the XRD spectrums of the thin films of titaniume oxide after annealing at 500°C for 0, 2, 4, 6% at of Co and 3 dipping. We observed that the TiO_2 thin films with 0 % Co 3 dipping and annealead at 500°C has an amorphose phase while the 2 and 4 % have an anatase phase with plane of (100). For 6%At. Co, we notice the presence of two concurrent phases with three layers, there is a brookite phase. Figure 1 shows the disappearance of Anatase and the appearance of Brookite and the (121) peak intensity increases with increasing concentration of cobalt for 6% At. Co. there is a mixture of two phases and a shifting of peaks to larger angles.

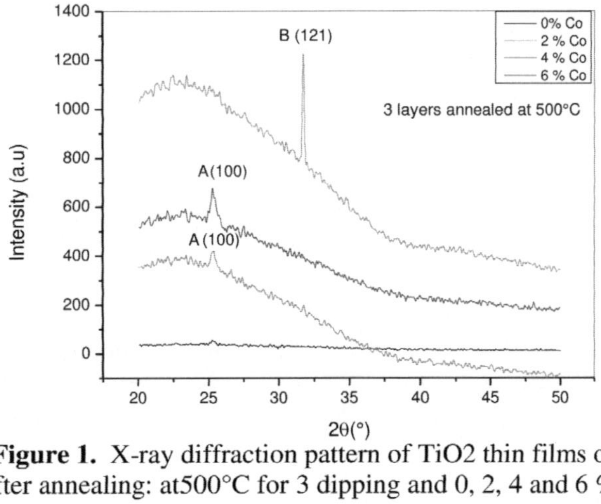

Figure 1. X-ray diffraction pattern of TiO2 thin films obtained after annealing: at500°C for 3 dipping and 0, 2, 4 and 6 %At. Co.

In Figure 2, we present 2D and 3D AFM images of the TiO2 films prepared on glass substrates corresponding to the annealing temperatures of 500°C for 6%at. Co and two layers. This figure shows the surface morphology and indicates a porous and fine structure with small size grains in anatase phase.

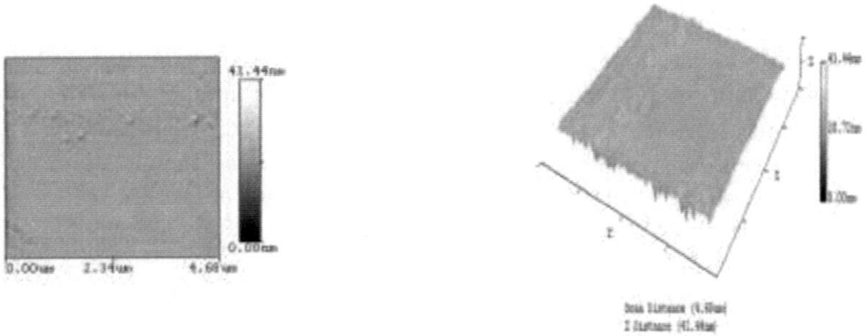

Figure 2. 2D and 3D AFM images of the TiO2 films prepared at 500°C for 6%At. Co

3.2. Optical properties

Figure 3 shows the TiO2 thin film UV-vis spectra for different Co concentrations, in the wavelength range of 300-900 nm. TiO2 thin films annealed at temperatures from 400 to 500°C represent high transparency coefficients in the visible range of 400-800nm. The Figure 5a shows the transmittance increases with increasing in the Co content and the temperature, at the same time, the threshold absorption moves to the long wavelengths. This can be attributed on the one hand to the structural change and the increase in grain size to increase the values of extinction coefficient K with the rise in temperature (500°C, Figure3).

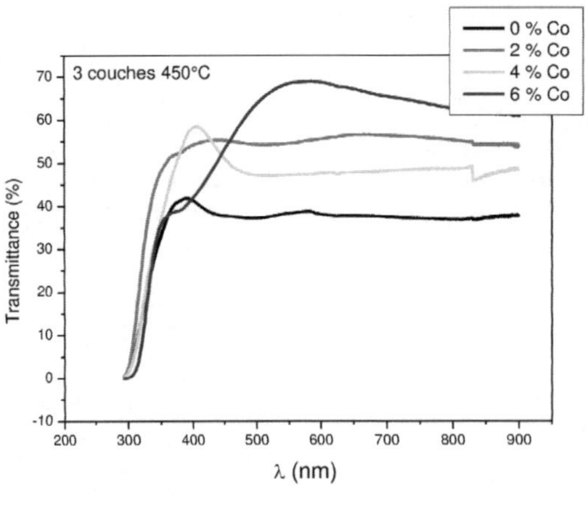

Figure 3. Transmission spectra of the TiO2 thin films obtained after three dipping and annealing at 500°C for 0; 2; 4 and 6% at Co

The optical band gaps of films have been estimated from the plot of their absorption coefficient as a function of photon energy and using Tauc formula for direct band gap semiconductors [12]:

$$(\alpha h\upsilon)^2 = B\,(E_g\text{-}h\upsilon).$$

Were α is a absorption coefficient, B is a constant, h is Planck constant, E_g is the energy band gap and ν is incident photon frequency.

The optical band gap is determinate from the plotting of $(\alpha h\nu)^2$ versus photon energy (eV), the extrapolation intercepting the photon energy axis given the optical band gap Eg. It varied between 3.30to 2.96 eV at 500°C. According to Figure 4, it is easy to notice that the gap value decreases with increasing temperature treatment. At this stage, we can say that the annealing temperature contributes to the reorganization of the structure and impurities that come to their sites more easily occupied when it rises and the decrease of Eg can be correlated with the grain size which increases with temperature.

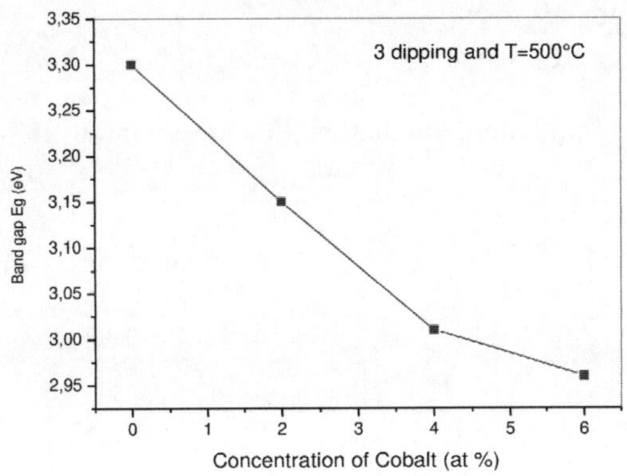

Figure 4. Band gap Eg of TiO_2 thin films deposited after 3 dipping and annealing at 500°C for 0; 2; 4 and 6 % at. Co.

4. Conclusion:

In this article we have shown that cobalt can be easily incorporated by the sol-gel method. The formation of a phase starts at 350 ° C. On aures hand, we verify that the anatase and brookite phases are very unstable. Cobalt can change the absorption edge and the gap of TiO2.

References:

[1] R. Zallem, M. P. Moret, *Solid State Communications* **137**, 154 (2006).

[2] J. Criado, C. Real, *J. Chem. Soc., Faraday Trans.* **1** (79), 2765 (1983).

[3] M. R. Hoffmann, S. T. Martin, W. Choi, and D. W .Bahneman, Environmental applications of semiconductor photocatalysis. *Chem. Rev.* **95**, 69 (1995).

[4] G. H. Li, L. Yang, Y. X. Jin, and L .D .Zhang, Structural and optical properties of TiO2 thin film and TiO2+ 2 wt.% ZnFe2O4 composite film prepared by r.f. sputtering. *Thin Solid Films* **368**, 163 (2000).

[5] V. Mikhelashvili and G. Eisenstein, Optical and electrical characterization of the electron beam gun evaporated TiO2 film. *Microelectr. Reliab.* **41**, 1057 (2001).

[6] V. H. Hsien, C.-F. Chang, Y. H. Chen, and S. Cheng, Photodegradation of aromatic pollutants in water over TiO2 supported on molecular sieves. *Appl. Catal. B: Environ.* **31**, 241 (2001).

[7] K. Wilke and H. D. Breuer , The influence of transition metal doping on the physical and photocatalytic properties of titania. *J. Photochem. Photobiol. A: Chem.* **121**, 49 (1999).

[8] J. M. Herrmann, H. Tahiri, C. Guillar and P. Pichat, Photocatalytic degradation of aqueous hydroxybutandioic acid (malic acid) in contact with powdered and supported titania in water. *Catal., Today* **54**, 131 (1999).
[9] Y. Matsumoto, M. Murakami, T. Shono, T .Hasagawa, T. Fukumura, M. Kawasaki, P. Ahmet, T. Chikyow, S. Koshihara, and H.K oinuma, Room-temperature ferromagnetism in transparent ransition metal-doped titanium dioxide. *Science* **291**, 854 (2001).
[10] S. A. Chambers, S. The vuthasan, R. F C. Farrow, R. F. Marks, J. U. Thiele, L. Folks, M. G. Samant, A. J. Kellock, N. Ruzycki, D. L. Ederer , and U. Diebold, Epitaxial growth and properties of ferromagnetic Co-doped TiO2 anatase. *Appl. Phys. Lett.* **79**, 3467 (2001).
[11] S. Doeuff, M. Henry, C. Sanchez, J. Livage, *J. of Non-Cristalline solids* **89**, 206-216 (1987).
[12] J.J. Tauc, Amorphous and Liquid Semiconductors, Plenum, London, 1974.

Investigation of photocalalytic activity of ZnO prepared by spray pyrolis with various precursors

F Bourfaa[1], M Lamri Zeggar[1], A Adjimi[1], M S Aida[1] and N Attaf[1]

[1]Laboratory of thin films and interfaces, Faculty of science, University Constantine 1, Algeria

fouzia_bourfaa@yahoo.fr

Abstract. Semiconductor photocatalysts such as ZnO has attracted much attention in recent years due to their various applications for the degradation of organic pollutants in water, air and in dye sensitized photovoltaic solar cell. In the present work, ZnO thin films were prepared by ultrasonic spray pyrolysis by using different precursors namely: acetate, chloride and zinc nitrate in order to investigate their influence on ZnO photocatalytic activity. The films crystalline structure was studied by mean of X-ray diffraction measurements (XRD) and the films surface morphology by Scanning Electron Microscopy (SEM). The films optical properties were studied by mean of UV–visible spectroscopy. The prepared films were tested for the degradation of the red reactive dye largely used in textile industry. As a result, we found that the zinc nitrate is the best precursor to prepare ZnO thin films suitable for a good photocatalytic activity.

1. Introduction

Semiconductor-assisted photocatalysis has recently emerged as an efficient method for conversion of photon energy into chemical energy and environmental purification [1, 2]. Among the semiconducting catalysts, zinc oxide (ZnO) has been extensively studied due to its low cost, non-toxicity, outstanding stability, and high efficiency [3-5]. ZnO is an n-type semiconductor with wide band gap (3.3 eV), large exciton binding energy (60 meV). Illuminated ZnO material with ultraviolet (UV) light can generates electron/hole pairs. These electrons and holes can migrate and initiate redox reactions with water and oxygen, by which they degrade organic molecules adsorbed on the surface of a photocatalyst [6, 7]. It is well known that ZnO thin films morphology controls its properties and subsequently their relevant potential applications. ZnO thin films were mainly used in : biomedical [8], piezoelectricity, biosensors, photocatalysis [9-11], biotechnology [12], light emitting diodes (LED's), laser systems [13] transparent electrodes [14], gas sensors [15,16], solar cells [17] and cosmetic products [18]. Various techniques have been used for ZnO thin films deposition namely: magnetron sputtering [19], reactive evaporation [20], chemical vapor deposition (CVD) [21], pulsed laser deposition (PLD) [22] and spray pyrolysis [23]. Among these methods, spray pyrolysis technique has several advantages, such as, simplicity, safety and vacuum less equipments. It is widely used and succeeded in oxide metallic thin films production [24-26].

In the present paper we studied the influence of the solution nature and properties on the characteristics of ZnO thin films on the photocatalytic activity.

2. Experimental

The ZnO films were grown onto glass substrates; using a Spray Pyrolysis Equipments (HOLMARC). Three different solutions, with 0.1 M molarity were prepared by mixing zinc salts with methanol. The used salts are zinc acetate, zinc nitrate and zinc chloride with 99.9995% of purity. ZnO thin films were deposited with the prepared solutions on well cleaned glass substrates with methanol and distilled water. The prepared solutions are then sprayed on the heated glass substrates by a pneumatic pump with flow rate of 100µl/min. During deposition, the substrate temperature is kept at 350°C. Deposition time for each precursor was 5 min. The crystalline structure was studied by X-ray diffraction measurements (XRD) and the films surface morphology was characterized by means of Scanning Electron Microscope (SEM). The UV–Visible transparence of the films is performed by Shimadzu UV–3101 PC spectrophotometer within the wavelength range of 200–800 nm. The films thickness and refractive index were derived from optical transmission measurement. The films electrical conductivity was carried out using D.C electrical measurements in dark and at room temperature.
In the present work, we have studied the photocatalysis application of ZnO thin films prepared by various precursors for study the degradation of the red reactive dye largely used in textile industry. The used solution was prepared from red reactive 184 powder dissolved in distilled water to obtain a solution with concentration of $4*10^{-6}$ mole/l. ZnO samples were placed inside a beaker containing 250 ml of polluted solution in contact with ZnO layer placed in horizontal position. The ZnO sample is exposed to an UV irradiation with a wavelength of 365nm and 2W power.

3. Results and discussion

3.1 *Films structure*

Figure 1 shows XRD spectra of different films deposited with the three studied salts. As can be seen, the ZnO films prepared from zinc nitrate and zinc chloride were polycrystalline with hexagonal wurtzite structure. While, films prepared with zinc acetate exhibits an amorphous phase due to the absence of any relevant peak. The XRD diffraction pattern of both films deposited using nitrate and zinc chloride are composed with several peaks assigned to (101), (002), (101), (110) planes as shown in figure1. We noticed that when we using a chloride as a precursor, diffraction peak, assigned to the plane (002) became the preferential orientation indicating that the film have the Wutrzite hexagonal and the growth is achieved along the axis c normal to the substrate surface. However in the case of zinc nitrate precursor, the deposited film shows that the plane (101) is the most intense peak. The XRD results are in good agreement with Lehraki et al [27] results, they reported that the ZnO thin films obtained by zinc chloride are polycrystalline and the zinc acetate has an amorphous phase.

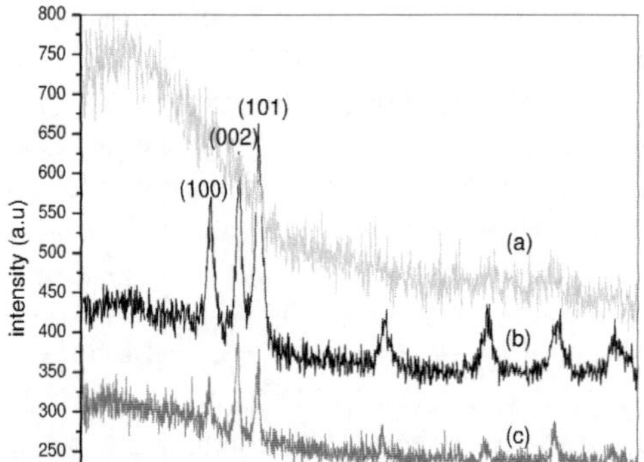

Figure 1. XRD diffraction pattern of ZnO thin films obtained by various precursors: (a) zinc acetate, (b) zinc nitrate and (c) zinc chloride.

Figure 2 shows typical SEM images of ZnO films prepared using the three different precursors. As shown, the films morphology depends strongly on the nature of the used precursor. Film prepared with zinc nitrate as starting solution is porous and exhibits a non continuous network (figure 2.a).While film prepared with zinc chloride exhibits a rough surface morphology and non continuous structure (figure 2.b). However as can be seen in figure 2.c, zinc acetate precursor yields to a dense and continuous structure film structure with a smooth surface. Due to its structure, zinc nitrate films have a larger specific surface then the others films, consequently this precursor will offers larger reactive surface with the environment which is required for gas sensing or for water treatment.

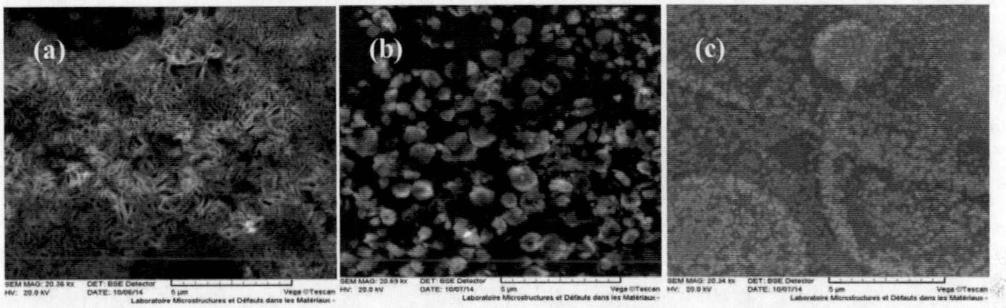

Figure2SEM images of ZnO thin films deposited at substrate temperature equal to 350 °C with different solutions: (a) Zinc nitrate (b) Zinc chloride and (c) Zinc acetate

3.2. Optical properties

In figure 3 we have reported the transmittance spectra of UV-visible range of ZnO films prepared with different precursors. The film deposited with zinc acetate (figure. 3(a)) has the higher transparency of 87% than films deposited with zinc nitride and zinc chloride (curves 3.b and c). This is due to the smooth surface of the former. It is well known that rough surface causes the light scattering resulting in transmittance reduction. This explains the low transmittance measured in films deposited with zinc chloride and nitride despite that they have better crystallinity then the films deposited with zinc acetate. Since these films have rough surfaces as can be seen in SEM images (figure 2).

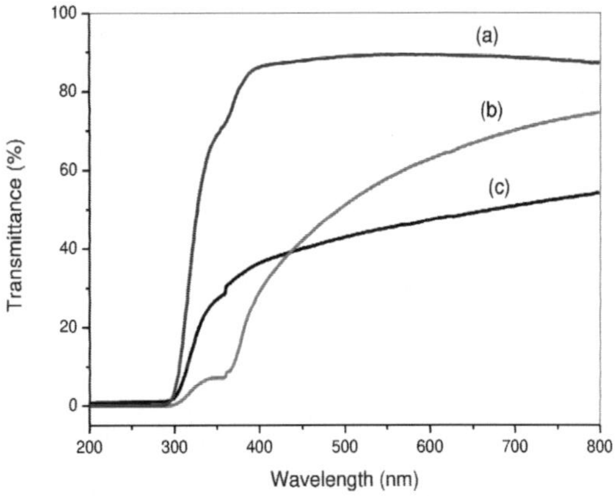

Figure 3. UV-visible transmittance spectrum of ZnO thin films deposited at 350 °C using different salts: (a) Zinc acetate-(b) Zinc nitride-(c) Zinc chloride.

The optical band gaps of films have been estimated from the plot of their absorption coefficient as a function of photon energy and using Tauc formula for direct band gap semiconductors [28]:

$$(\alpha h\upsilon)^2 = B\,(E_g - h\upsilon).$$

Were α is a absorption coefficient, B is a constant, h is Planck constant, E_g is the energy band gap and v is incident photon frequency.

Films optical band gap are found equal to 3.12 and 3.20 eV for films prepared with zinc nitrate and zinc acetate respectively .While the optical gap of film deposited with zinc chloride is close to its value for ZnO bulk material which is 3.27 eV because of the good crystallinity of this film. The films thickness, refractive index and optical band Eg deduced from optical transmittance spectra were reported in Table 1.

Table 1: thickness, band gap, refractive index, degradation rate k and conversion rate of ZnO thin films deposited with different precursors.

ZnO precursors	d (nm)	Eg (eV)	n	Rate k (10^{-4})	Conversion rate (%)
Nitrate	209.936	3.12	1.88	18	36
Chloride	202.463	3.25	1.93	6.58	17
Acetate	134.864	3.20	1.69	8.56	20

3.3. Electrical properties:

Figure 4 shows the variation of the electrical conductivity, of ZnO thin films deposited by various precursors, measured in the dark and at room temperature. As can be seen the conductivity of films obtained with zinc nitrate have the slightly larger conductivity value $7.23\times10^{-5}(\Omega cm)^{-1}$. This is due to the fact that film prepared with this precursor is dense and continuous in contrary to others ones.

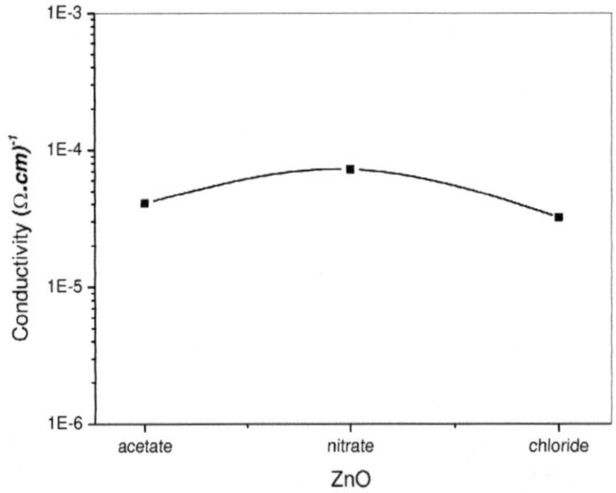

Figure 4. Variation of the conductivity for the three precursors.

3.4. Photocatalytic activity:

We have studied the variation of absorbance spectra of polluted solution after different UV light exposure times. The wavelength region is ranged from 450 to 600 nm; this broad absorption band is a characteristic of the used red dye. The absorption peak intensity is used as signature of the dye degradation. From the variation of this absorption peak with exposure time , we concluded that the

IOP Conf. Series: Materials Science and Engineering **108** (2016) 012049 doi:10.1088/1757-899X/108/1/012049

photo degradation is more significant when solution is in contact with film prepared using zinc nitrate then the ones prepared with zinc acetate and zinc chloride films.

To have more insight on the photo degradation kinetics and the influence of ZnO precursors, we have monitored the variation of the intensity of the absorption located at 542 nm. We have reported in figure 5 the variation of the ratio C/C_0, where C is the peak intensity at time t and C_0 is the peak intensity before light exposure. As can be seen, the photo degradation is more important when using film obtained with zinc nitrate salt, the ratio is reduced with increasing exposure time; it reaches 0.64 after 240 minutes of irradiation. However in the case of zinc acetate and zinc chloride films salt source, the ratio is slightly reduced up to 0.78 and 0.83 after 240 minutes, respectively. As can be seen in figure 6, the photocatalytic decomposition of red pollutant, in contact with the surface of ZnO thin films of the three precursors, follow a pseudo first-order kinetic law, it can be expressed as [31]:

$$-\text{Ln}\,(C/C_0) = kt$$

Where C and C_0 are the reactant concentration at time t and t = 0, respectively and k rate constant (reaction rate constant) [29].

In table I we have reported the calculated value of the rate k for the all ZnO thin films. Zinc nitrate film is characterized by a larger rate k then the other films.

The conversion rate defined as $\tau = (C_0\text{-}C\,(t)/C_0)\ \text{x}100$, is an interesting quantity that can yield information about the pollutant degradation, it represents the relative quantity of removed pollutant from the solution. In figure 6 we have reported the variation of the conversion rate obtained with all ZnO thin films. In the case of zinc nitrate film, 36 % of pollutant is removed after 240 minutes of exposure time, while only less than 22% and 17% are removed when using zinc acetate and zinc chloride films, respectively. The discrepancy in the photocatalysis activity of all thin films of ZnO may found explication in the difference between their microstructure. As deduced from SEM images, zinc nitrate and zinc chloride films are rough and have a porous structure while zinc acetate film is smoother and has a continuous structure. Therefore, the reactive specific surface of the first film is larger than in the others films, thereafter, more reactive surfaces are available for the water than in the other ones (acetate and chloride). The same conclusion has been outlined by Li et al [32] in ZnO/CuO structure, they reported that a material with an open and porous surface exhibit higher degradation efficiency. So, the reactive specific surface is one of the important parameters to controller the photocatalytic activity of ZnO thin films.

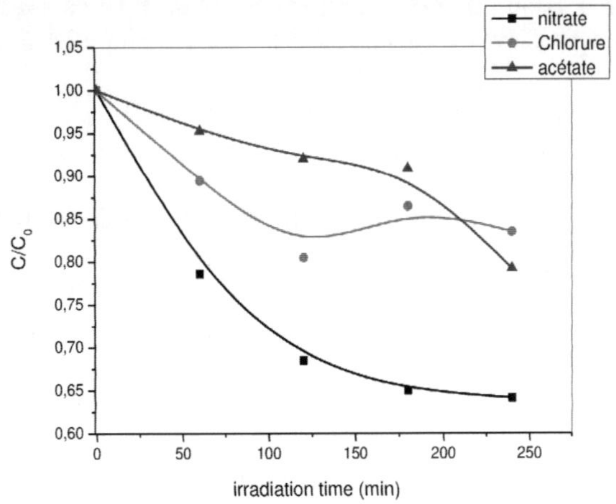

Figure 6. Evolution temporal of the conversion rate of dye red pollutant during its degradation in contact with ZnO thin films deposited by various salts.

4. Conclusion:

In order to investigate the influence of the precursor's salt nature, we have studied the ZnO thin films deposited with spray pyrolysis by different precursors. The obtained ZnO thin films are polycrystalline when using zinc nitrate and zinc chloride, however using zinc acetate salt leads to an amorphous film microstructure. From this investigation we inferred that the starting salt source nature and morphology of the Zn salt play an important role on the photocatalytic activity. Thin and porous films offer more reactive sites due the large surface to volume ratio and larger charge transfer, thereafter their better photocatalytic activity. According to the obtained films SEM images and to morphologies, we concluded that zinc nitrate source is the best precursor for the preparation of ZnO thin film suitable to the red dye photo degradation.

References:

[1] J. Zhou, N.S. Xu, Z.L. Wang, Adv. Mater. 18 (2006) 2432.
[2] C.Riggio, V.Raffa and A. Cuschieri, Micro & Nano Letters 5(2010)35
[3] M.A. Behnajady, N. Modirshahla and R. Hamzavi, J. hazard.Mater.20 (2006)133.
[4] T.J. Brunner , P Wick , P Manser , P Spohn , RN Grass , L.K.Limbach , A.Bruinink W.J. Stark ,Environ Sci Technol. 15(2006)4374.
[5] Y. Lu, Y. Lin, D. Wang, L. Wang, T. Xie and T. Jiang, J. Phys. D: Appl. Phys. 44 (2011) 315502.
[6] R. Yokoyama, S. Suzuki, K. Shirai, T. Yamauchi, N. Tsubokawa and M. Tsuchimochi, Eur. Polym. J. 42 (2006) 3221.
[7] J. M. Szarko, J. K. Song, B C. W. Lackledge, I. Swart, S. R. Leone, S. Li, Y.Zhao, Chem. Phys. Lett., 404(2005) 171.
[8] T. Ootsuka, Z. Liu, M. Osamura, Y. Fukuzawa, R. Kuroda, Y. Suzuki, N. Otogwa, T. Mise, S. Wang, Y. Hoshino, Y, Nahayama and H. Tanoue, Thin Solid Films, 476(2005)30.
[9] B. Baruwati, D.K. Kumar and S.V. Manorama, Sens. Actuators B 119 (2006) 676.
[10] G.-C. Yi, C. Wang, W.I. Park, Semicond. Sci. Technol. 20 (2005) S22.
[11] Z.S. Wang,C.H. Huang,Y.Y. Huang, Y.J. Hou, P.H. Xie, B.-W. Zhang and H. M. Cheng, Chem. Mater. 13 (2001) 678.
[12] T. Iwasaki, M. Satoh, T. Masuda, T. Fujita and J. Mater. Sci. 35 (2000)4025.
[13] X. Yu, J. Ma, F. Ji, Y. Wang, X. Zhang and H. Ma,Thin Solid Films,483(2005) 296.
[14] J. Ma, F. Ji, D. Zhang, H. Ma and S. Li, Thin Solid Films, 357(1999) 98.
[15] D. A. Lamb and S. J. C. Irvine, Journal Cryst. Growth 273(2004)111.
[16] J. Zou, S. Zhou, C. Xia, Y. Hang, J. Xu, S. Gu, and R. Zhan, J. Cryst. Growth 280(2005)185.
[17] P. Nunes, E. Fortunato, P. Tonello, F. Braz Fernandez, P. Vilarinho and R. Martins, Vacuum, 64(2002)281.
[18] A. Mosbah, M.S.Aida, J. of Alloys and Compds, 515(2012)149.
[19] F.Ynineb, A. Hafdallah, M.S. Aida, N. Attaf, J.Bougdira, H.Rinnert and S.Rahman, Mater. Sci. in Sem. Process. 16 (2013)2021.
[20] M. Messaoudi, M.S. Aida, N. Attaf, T. Bezzi, J. Bougdirab and G. Medjahdi. Mater. Sci in Sem. Process. Volume 17, (2014) 38.
[21] A. Hafdallah, F. Yanineb, M.S. Aida and N. Attaf, J. of Alloys and Compnds, 509 (2011) 7267.
[22] N. Kislov, J. Lahiri, H. Verma, D. Yogi Goswami, E. Stefanakos, and M. Batzill, Langmuir 25 (2009) 3310.
[23] S. Baruah, M. Jaisai, R. Imani1, M. M. Nazhad and J. Dutta, Sci. Technol. Adv. Mater 11 (2010).
[24] H. Zhu, R. Jiang, Y. Fu, Y. Guan , J. Yao, L. X. Guangming Zeng, Desalination 286 (2012) 41.
[25] M. Qamar and M. Muneer, Desalination 249 (2009) 535.
[26] R. Ullah and J. Dutta, Journal of Hazardous Materials 156 (2008) 194.
[27] N. Lehraki, M.S. Aida,, S. Abed, N. Attaf, A. Attaf and M. Poulain Jornal of Current Applied Physics 12(2012)1283-1287.
[28] J.J. Tauc, Amorphous and Liquid Semiconductors, Plenum, London, 1974.
[29] B. X. Li and Y. F.Wang, Super lattice Microstruct. 47(2010) 615.
[30] Q. Xiang, G. Meng, Y. Zhang, J. Xu, P. Xu, Q. Pan and W.Yu, Sens. Actuator B-Chem.143. (2010) 635.

IOP Conf. Series: Materials Science and Engineering **108** (2016) 012050 doi:10.1088/1757-899X/108/1/012050

Attractive mechanical properties of a lightweight highly sensitive bi layer thermistor: polycarbonate/organic molecular conductor

E Laukhina[1,4], V Lebedev[2], C Rovira[1,2], V Laukhin[1,2,3], and J Veciana[1,2]
[1]CIBER de Bioingeniería, Biomateriales y Nanomedicina (CIBER-BBN), Madrid, Spain
[2]Institut de Ciencia de Materials de Barcelona (ICMAB-CSIC), Campus UAB, Bellaterra, 08193, Spain
[3]Institució Catalana de Recerca i Estudis Avançats (ICREA), Barcelona, Spain

E-mail: laukhina@icmab.es

Abstract. The paper covers some of the basic mechanical characteristics of a recently developed bi layer thermistor: polycarbonate/(001) oriented layer of organic molecular conductor α'-(BEDT-TTF)$_2$I$_x$Br$_{3-x}$, were BEDT-TTF=bis(ethylenedithio)tetrathiafulvalen. The nano and macro mechanical properties have been studied in order to use this flexible, low cost thermistor in sensing applications by proper way. The nano-mechanical properties of the temperature sensitive semiconducting layer of α'-(BEDT-TTF)$_2$I$_x$Br$_{3-x}$ were tested using nanoindentation method. The value of Young's modulus in direction being perpendicular to the layer plan was found as 9.0 ±1.4 GPa. The macro mechanical properties of the thermistor were studied using a 5848 MicroTester. The tensile tests showed that basic mechanical characteristics of the thermistor are close to those of polycarbonate films. This indicates a good mechanical strength of the developed sensor. Therefore, the thermistor can be used in technologies that need to be instrumented with highly robustness lightweight low cost temperature sensors. The paper also reports synthetic details on fabricating temperature sensing e-textile. As the temperature control is becoming more and more important in biomedical technologies like healthcare monitoring, this work strongly contributes on the ongoing research on engineering sensitive conducting materials for biomedical applications.

1. Introduction

Temperature is one of the crucial parameters to be measured in numerous modern technologies. Healthcare, environment and food control, biomedical applications call for low cost, lightweight flexible thermistors for precise and accurate temperature measurements [1-4]. For example, temperature is the most important factor in controlling rate at which food go bad: cod will remain edible for up to 15 days at 0°C, whereas it may be unfit to eat after only six days at 5°C. [5] There is also a particular interest in integrating lightweight conductive sensing materials in human wearable interfaces such as fabrics, since wearable electronics could offer personalized healthcare, security and comfort [1-4]. These fabrics are able to sense and react to environmental conditions.

[4] To whom any correspondence should be addressed

Recently, we have suggested a novel approach to engineering lightweight flexible sensors that consists in covering polymeric films with organic molecular conductors (BEDT-TTF)$_2$X, were BEDT-TTF=bis(ethylenedithio)tetrathiafulvalen (figure 1) and X is trihalide ion [6-9]. These conductors are charge transfer salts with a 1/2 filled conducting band; they have very deformable layered crystal structures with strong electron–phonon coupling; due to this they demonstrate unique electronic properties that may be exploited in numerous sensing applications [10-12].

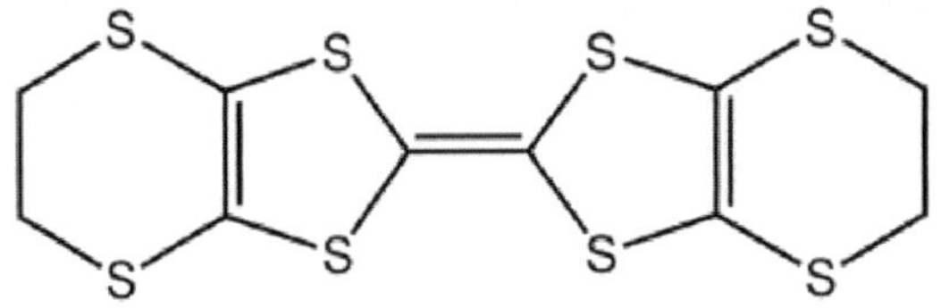

Figure 1. Skeletal formula of BEDT-TTF

In this context, we have developed a bi layer (BL) thermistor: polycarbonate/polycrystalline (001) oriented α'-(BEDT-TTF)$_2$I$_x$Br$_{3-x}$ molecular conductor (hereafter it will be as polycarbonate/α'-(BEDT-TTF)$_2$I$_x$Br$_{3-x}$ [13, 14] that can be successfully used as a highly temperature sensing component in e-textiles [15, 16]. Such textiles are capable of controlling very small temperature changes with accuracy of 0.005 degree. This is significantly better than that reported for the commonly used thermistors: the accuracy of a Pt-1000 detector is 0.01 degree [17]. Here it is necessary to recall that some of nonelectronic properties of polycarbonate/α'-(BEDT-TTF)$_2$I$_x$Br$_{3-x}$ are also equally important for its applications in sensing electronics. In this context the study of the basic mechanical characteristics of both components of the developed bi layer thermistor are very important since the similarity of the elastic properties of layered components is one of the key requirements for engineering highly robust electronic layered systems. The above conclusion motivated us to study nano and micro mechanical properties of the developed BL thermistor: polycarbonate/α'-(BEDT-TTF)$_2$I$_x$Br$_{3-x}$.

Here we report the Young's moduli of the thin (001) oriented layer of α'- (BEDT-TTF)$_2$I$_x$Br$_{3-x}$ molecular conductor measured using a nanoindentation method, as well as basic mechanical characteristics [Young's Modulus (E), Yield Strain (ε_{yield}), and Tensile Yield Strength (TYS)] of the developed BL thermistor; these characteristics were determined from stress-strain dependences measured using a 5848 MicroTester. The paper also presents a new approach to the preparation of a temperature sensing e-textile in which the polycarbonate/α'-(BEDT-TTF)$_2$I$_x$Br$_{3-x}$ thermistor was embedded as an active component.

2. Experimental

2.1. Preparation and characterization of the flexible lightweight BL thermistor: polycarbonate/α'-(BEDT-TTF)$_2$I$_x$Br$_{3-x}$

In line with a reported synthetic procedure [13, 18] the BL thermistor was fabricated as follows: first, a 25 μm thick polycarbonate (PC) film that contains a 2 wt. % of BEDT-TTF was prepared. To do this, the film was cast on a glass support at 130 °C from a 1,2-dichlorobenzene solution of polycarbonate and BEDT-TTF. Second, to cover the film with the (001) oriented layer of α'-(BEDT-TTF)$_2$I$_x$Br$_{3-x}$, we exposed the film surface to the vapors of a dichloromethane solution of IBr. The surface of the film easily swells under this treatment that facilitates the migration of BEDT-TTF molecules from the bulk of the film to the swollen film surface where they are oxidized by IBr. This redox process induces the rapid nucleation of the (BEDT-TTF)$_2$I$_x$Br$_{3-x}$ conductor with a consequent formation of the conductive polycrystalline covering layer [18]. The resulting surface-modified film was fully characterized using Scanning Electron Microscopy (SEM), Atomic Force Microscopy (AFM) and X-ray diffraction techniques; its R(T) dependence in the range of the human body temperatures was also studied.

SEM topographic images (figure 2) shows that the covering layer of the polycarbonate/α'-(BEDT-TTF)$_2$I$_x$Br$_{3-x}$ thermister consists of submicro plate-like oriented crystallites whose crystal planes are parallel to the film plan.

Figure 2. SEM image of the conductive covering layer based on (BEDT-TTF)$_2$I$_x$Br$_{3-x}$

The powder X ray diffraction data indicate the presence of only *(00l)* reflections of the α'-(BEDT-TTF)$_2$I$_x$Br$_{3-x}$ conductor suggesting that the covering layer is dominantly formed from *(001)*-oriented α'-(BEDT-TTF)$_2$I$_x$Br$_{3-x}$ crystallites that is consistent with reported data [13, 18].

Direct current resistance measurements reveal that in the human body temperature range the change of the electrical sheet resistance of BL film is 250-300 Ω per degree. Therefore, the BL film, as it was reported in [15], is capable of controlling very small temperature changes.

2.2. Nanomechanical testing

Mechanical measurements were performed by means of the Force Spectroscopy mode, which consists on using the AFM probe as a nanoindenter. The force (F) value necessary to provoke a plastic deformation was experimentally assessed and topographically detected as a hole in the sample surface (permanent deformation). Then, F value for the mechanical measurements was set to be 70% of this plastic-onset F value to ensure that all reported mechanical measurements are fully inside the elastic deformation region. In all cases F value of 250 nN was chosen. In order to obtain results as representative as possible, one force curve was performed in each sample location so as to ensure that previous mechanical tests did not change the local mechanical response of the sample and a minimum of 200 individual experiments were performed for the sample. In order to extract the E value from the extension force curve, Hertz model considering a spherical indenter of radius R was applied [19, 20].

2.3. Macromechanical testing of the BL thermistor: polycarbonate/α'-(BEDT-TTF)$_2$I$_x$Br$_{3-x}$

Mechanical properties of the thermistor were studied using a 5848 MicroTester, equipped with a 1 kg load cell (Instron). Film strips in dimension of ca. 28x2 mm^2 and free from physical imperfections, were held between two clamps positioned at a distance of ca. 18 mm. During measurement, the strips were pulled by the top clamp with velocity being equal to 2.0 μm/s. The force and elongation were monitored up to the film samples were broken. Measurements were run in two replicates of the BL thermistor: polycarbonate/α'-(BEDT-TTF)$_2$I$_x$Br$_{3-x}$.

3. Results and discussion

Taking into account that mechanical properties of both components of the developed BL thermistor: polycarbonate/α'-(BEDT-TTF)$_2$I$_x$Br$_{3-x}$ are very important for its applications in sensing

IOP Conf. Series: Materials Science and Engineering **108** (2016) 012050 doi:10.1088/1757-899X/108/1/012050

electronics we studied them using two different approaches. We use a nanoindentation method to determine Young's moduli of the (001) oriented layer of α'-$(BEDT\text{-}TTF)_2I_xBr_{3-x}$ because it is known as the most realistic mechanical testing technique to determine the elastic modulus for thin layers. [20-23]

Another approach - tensile testing – was used to determine the basic mechanical characteristics of the BL thermistor. The 25 μm thick sample of the BL thermistor of was prepared as it was early reported [18]. The sample was morphologically and structurally characterized as it is described in the Experimental part (paragraph 2.1); X-ray powder diffraction data and resistance temperature dependence of the sample fully consistent with those reported for the BL film: polycarbonate/α'-$(BEDT\text{-}TTF)_2I_xBr_{3-x}$ [13, 18]. For comparison a 25 μm thick polycarbonate film was additionally prepared using the same casting parameters. This uncovered polycarbonate film was studied as a reference sample.

2.1. Nano mechanical properties of the (001) oriented layer of α'-$(BEDT\text{-}TTF)_2I_xBr_{3-x}$

The mechanical properties of the conductive sensitive α'-$(BEDT\text{-}TTF)_2I_xBr_{3-x}$-based layer were studied using the nanoindentation method. At the arrangement of the BL-based thermistor under AFM testing (figure 3), the AFM probe – nanoindenter - was applied perpendicular to thermistor's conductive sensing layer.

As figure 4 shows, the X-ray powder pattern of thermistor's sensing layer demostrates the *(00l)* reflections of the α'-$(BEDT\text{-}TTF)_2I_xBr_{3-x}$ crystal phase; this is in agreement with previously reported data.[18] At this orientation, the molecular conducting *ab* planes of the α'-crystallites are parallel to the plane of the BL-based thermistor. Therefore, the nanoindentation study permitted us to measure Young's moduli in a direction being perpendicular to the crystallographic *ab* plan ($E_{\perp ab}$) of the α'-$(BEDT\text{-}TTF)_2I_xBr_{3-x}$ crystallites. For definiteness sake, nanoindentation was also done on a polycarbonate film that was prepared using a drop casting technique, parameters of which - solvent, concentration, temperature, support - were the same as in the case of the BL film polycarbonate/α'-$(BEDT\text{-}TTF)_2I_xBr_{3-x}$ casting.

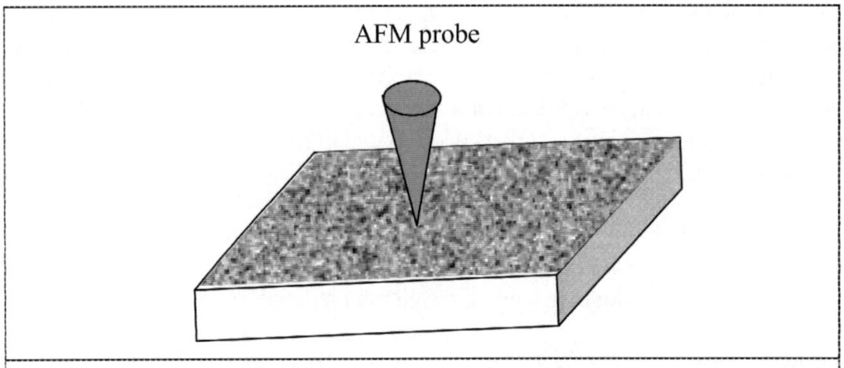

Figure 3. Schematic representation of BL-based thermistor for mechanical testing using the nanoindentation method

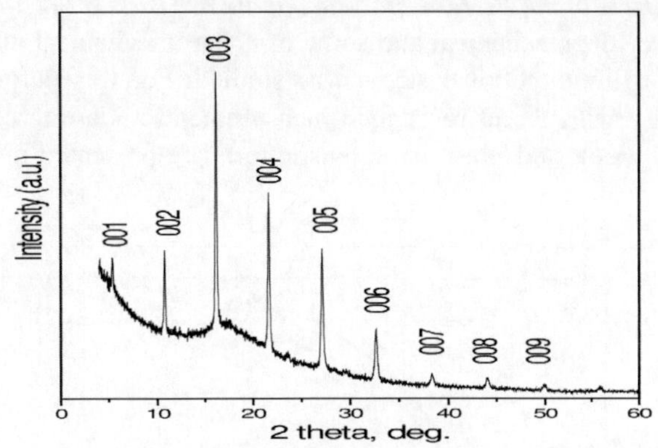

Figure 4. X-ray diffraction pattern of the conductive sensing layer of the BL thermistor: polycarbonate/α'-$(BEDT\text{-}TTF)_2I_xBr_{3-x}$

The values of Young's modulus were extracted from the force-displacement curves measured during nanoindentation; AFM probe contact points are shown in figure 5. From dispersion of Young's moduli and its Gaussian fit (figure 6) the value of $E_{\perp ab}$ of α'-$(BEDT\text{-}TTF)_2I_xBr_{3-x}$ crystallites was estimated to be 9.0±1.4 GPa.

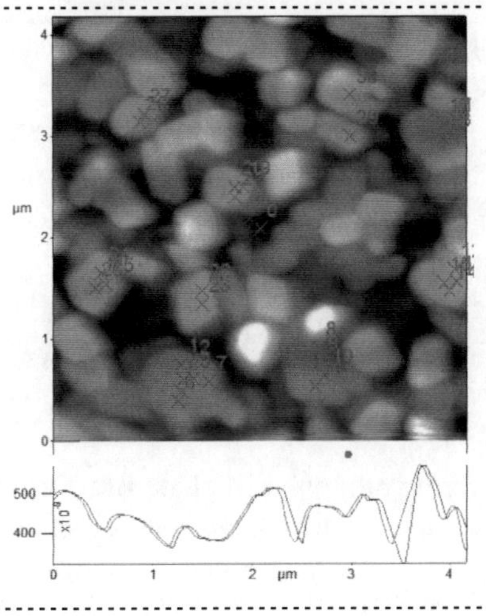

Figure 5. AFM image of the area of the sensitive conductive layer of the (001) oriented α'-$(BEDT\text{-}TTF)_2I_xBr_{3-x}$; red crosses are the AFM probe contact points during nano-mechanical testing

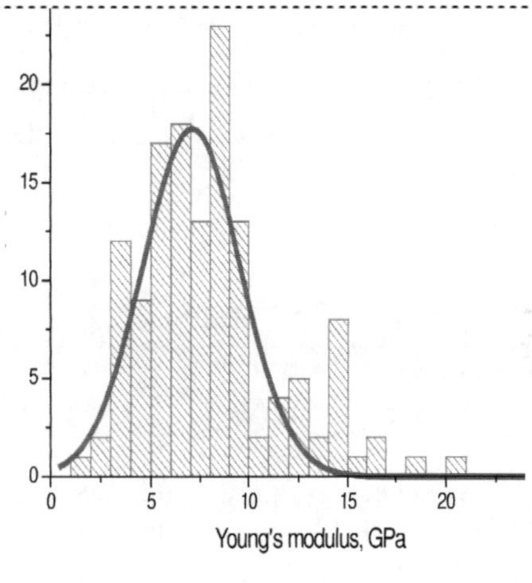

Figure 6. Dispersion of Young´s moduli and its Gaussian fit (green line) measured for the sensitive conductive layer of the BL thermistor: polycarbonate/α'-$(BEDT\text{-}TTF)_2I_xBr_{3-x}$

The value of $E_{\perp}$ of the polycarbonate film sample was found as 0.8±0.2 GPa. Although the value of $E_{\perp ab}$ of α'-$(BEDT\text{-}TTF)_2I_xBr_{3-x}$ crystallites is an order of magnitude higher than the elastic modulus of polycarbonate film, it is an order of magnitude less than that reported for the thin metallic films, for example, a gold film (78 GPa) deposited over a POLY2 layer [23]. Therefore, the robustness of the BL thermistor: polycarbonate/α'-$(BEDT\text{-}TTF)_2I_xBr_{3-x}$ significantly surpasses the same mechanical characteristics of plastics metalized with gold and other conventional metals that commonly used in flexible electronics.

3.2.. Mechanical properties of the BL film: polycarbonate/α'-(BEDT-TTF)$_2$I$_x$Br$_{3-x}$

To provide an overview of the mechanical and some of electromechanical properties of the developed thermistor, its response to unidirectional stress was studied. For this purpose tensile testing in the elastic range was coupled with direct resistance measurements. For reference, the prepared 25 μm thick polycarbonate film was tested. Results on tensile tests are presented in figure 7 and summarized in Table 1.

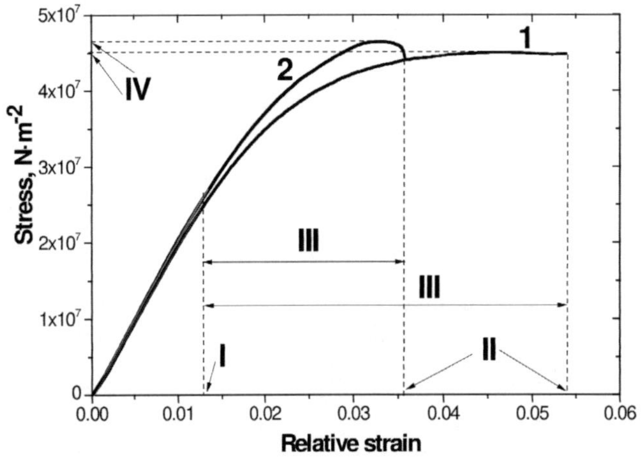

Figure 7. Full stress-strain curves for the BL polycarbonate/α'-(BEDT-TTF)$_2$I$_x$Br$_{3-x}$ thermistor (curve 1) and polycarbonate film reference (curve 2); I-Yield point, II-fracture point, III- plastic deformations, IV- ultimate tensile stress (UTS)

Table 1. Tensile and electromechanical data for the BL thermistor and polycarbonate film.

sample	E (GPa)	ε_{Yield} (%)	Yield Stress (N/m^2 x10^7)	Yield Resistance (%)	UTS (N/m^2 x10^7)
BL thermistor	1.99±0.2	1.18	2.3	1.18	4.50
Polycarbonate film	2.05±0.2	1.24	2.5	-	4.65

As data presented in figure 7 and table 1 show, the covering of a polycarbonate film with the (001) oriented α'-(BEDT-TTF)$_2$I$_x$Br$_{3-x}$ layer slightly decreases the basic mechanical characteristics of polycarbonate film. So, the values of E, Elastic Limit (ε_{Yield}) and Yield Stress decrease by ≅3 %, ≅5 %, and ≅8 %, respectively. The ultimate tensile stress (UTS) value also decreases by 3%. From the stress-strain curves presented in figure 7, only one difference stands out: the significantly wide plastic region for the developed BL thermistor. Thus, the thermistor plastic behavior is observed in the range of relative strain from 1.18 to 5,4% while the reference polycarbonate film demonstrates the plastic region between 1.24 and 3.6 % of relative strain. Therefore, the BL thermistor undergoes larger plastic deformations without cracking compared with the uncovered polycarbonate film that was prepared under similar casting parameters. In spite of the above mentioned differences, the measured basic mechanical characteristics of the BL thermistor are very close to those of the reference polycarbonate film. This imparts a good robustness to BL film: polycarbonate/α'-(BEDT-TTF)$_2$I$_x$Br$_{3-x}$ and makes possible the integration of the developed thermistor in different devices as an active component.

As figure 8 shows, the electrical resistance of the thermistor, which was simultaneously measured at tensile testing, deviates from linear proportionality at the elastic limit point of the polycarbonate/α'-(BEDT-TTF)$_2$I$_x$Br$_{3-x}$ bilayer. Here it should be noted that such coincidence was not observed for the recently reported highly piezo-resistive BL film: polycarbonate/β$_H$-(BEDT-TTF)$_2$I$_3$, the resistance of which deviates from linear proportionality versus strain at ε=0.65%, while Yield point of this BL film was observed at ε_{Yield}=1.2% [8, 24]. By analogy with the elastic limit point (Yield point), we have suggested to term the point at which the resistance of BL films: polycarbonate/(BEDT-TTF-based molecular conductor deviates from linearity versus strain as the resistance proportional limit (R_{yield}). [24] As it was shown [8, 24], the nonlinearity above R_{yield} is associated with small numerous nanocracks formed perpendicularly to an elongation direction. Relying on such data, one may conclude that the point of the resistance proportional limit of the BL thermistor corresponds to the elastic limit point of the (001) oriented α'-$(BEDT\text{-}TTF)_2I_xBr_{3-x}$ layer. At the end of this paragraph, an important conclusion have to be made: the (001) oriented α'-$(BEDT\text{-}TTF)_2I_xBr_{3-x}$ layer, formed on the surface of a polycarbonate film significantly surpass similar layers of the α-, β_H- $(BEDT\text{-}TTF)_2I_3$ organic molecular metals in elasticity. The significant enhancement of the resistance proportional limit of BL thermistor makes it very usable and expands the scope of sensing devices to which the developed thermistor might be adopted.

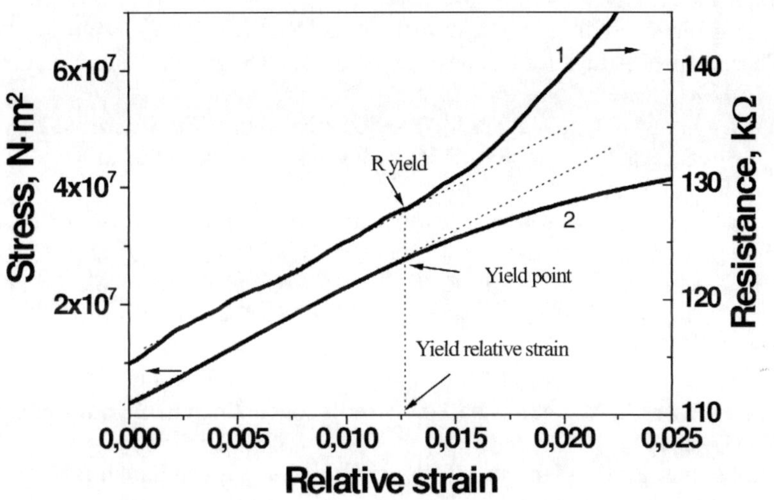

Figure 8. Resistance-strain (plot 1) and stress–strain (plot 2) curves for the BL thermistor polycarbonate/α'-$(BEDT\text{-}TTF)_2I_xBr_{3-x}$; the data were simultaneously collected at room temperature.

4. Conclusions

For the first time the anisotropy of elastic properties of the α'-$(BEDT\text{-}TTF)_2I_xBr_{3-x}$ molecular conductor was studied: the measured values of Young's moduli in the direction perpendicular to its molecular conducting *ab* plan (≅9 GPa) and along the plan (≅2 GPa) indicate that the crystal structure of this molecular conductor is a highly anisotropic mechanical system.

The measured values of Young's moduli reveal that the thin conductive α'-$(BEDT\text{-}TTF)_2I_xBr_{3-x}$-based layer is significantly softer than thin layers of conventional metals deposited on plastic substrates; this permits one to fabricate durable conductive sensing multilayer structures for their applications in sensing technologies.

It was shown that the (001) oriented α'-$(BEDT\text{-}TTF)_2I_xBr_{3-x}$ layer, formed on the surface of a polycarbonate film, significantly surpass similar layers of the α-, β_H- $(BEDT\text{-}TTF)_2I_3$ organic molecular metals in both elasticity and plasticity. The significant enhancement of the resistance proportional limit of the BL thermistor: polycarbonate/α'-$(BEDT\text{-}TTF)_2I_xBr_{3-x}$ makes it very usable and expands the scope of sensing devices to which the developed thermistor might be adopted.

5. Acknowledgements

The authors acknowledge the financial support from Instituto de Salud Carlos III, through "Acciones CIBER." The Networking Research Center on Bioengineering, Biomaterials and Nanomedicine (CIBER-BBN), an initiative funded by theVINational R&D&I Plan 2008–2011, Iniciativa Ingenio 2010, Consolider Program, CIBER Actions and financed by the Instituto de Salud Carlos IIIwith assistance from the European Regional Development Fund. The authors also appreciate the financial support through the projects: BE-WELL (CTQ2013–40480-R) granted by DGI (Spain), and GenCat (2014-SGR-17) financed by DGR (Catalunya), the European Commission's Seventh Framework Programme for Research under contracts FP7-OCEAN-2013-614155.

6. References

[1] Axisa F, et al. 2005 Flexible Technologies and Smart Clothing for Citizen Medicine, Home Healthcare, and Disease Prevention *IEEE Transactions on Information Technology in Biomedicine* **9** 325-336

[2] Lorussi F, Rocchia W, Scilingo E P, Tognetti A, and Rossi D De Wearable, 2004 Redundant Fabric-Based Sensor Arrays for Reconstruction of Body Segment Posture, *IEEE Sensors Journal* **4** 807-818

[3] Lumelsky V J, Shur M S, and Wagner S 2001 Sensitive skin *IEEE Sensors* J. **1** 41–51.

[4] Ajmera N, Priya S Dash and Chet Ram Meena *Smart Textile* presented at www.fibre2fashion.com

[5] Torry Research Station *Temperature measurement in the fish industry ,FAO Document Repository* presented at www.fao.org/wairdocs/tan/x5992e/X5992e01.htm

[6] Laukhina, E., Pfattner, R., Ferreras, L. R., Galli, S., Mas-Torrent, M., Masciocchi, N., Laukhin, V., Rovira, C., Veciana, J. 2010 Ultrasensitive Piezoresistive All-Organic Flexible Thin Films, *Adv. Mater.* **22** 977-981

[7] Ferreras L, Pfattner R, Mas-Torrent M, Laukhina E, Lopez L, Laukhin V, Rovira C, Veciana J 2011 Highly piezoresistive textiles based on a soft conducting charge transfer salt *J. Mater. Chem.* **21** 637- 641

[8] Laukhina E, Pfattner R, Mas-Torrent M, Rovira C, Veciana J, and Laukhin V 2011 Film-Based Sensors with Piezoresistive Molecular Conductors as Active Components: Strain Damage and Thermal Regeneration *Sensors and Transducers* **10** 1-12

[9] Laukhina E, Lebedev V, Laukhin V, Oncins G, Pfattner R, Rovira C. and Veciana J 2011 Lightweight Biocompatible Physical Sensors: Polymeric Films "Self-Metallized" with Organic Molecular Conductors *Solid-State Sensors, Actuators and Microsystems Conference (Transducers) 16th International* Beijing 1978 - 1981

[10] Shibaeva R. P., Yagubskii E. B. 2004 Molecular Conductors and Superconductors Based on Trihalides of BEDT-TTF and Some of Its Analogues *Chem. Rev* **104** 5347- 5378

[11] Saito G and Yoshida Y 2007 Development of Conductive Organic Molecular Assemblies: Organic Metals, Superconductors, and Exotic Functional Materials *Bulletin of the Chemical Society of Japan* **80** 1-137

[12] Williams J.M. et al. 1992 *Organic Superconductors (Including Fullerenes): Synthesis, Structure, Properties, and Theory* USA (Prentice Hall, Englewood Cliffs, NJ)

[13] Laukhina E, Tkacheva V, Khasanov S, Zorina L, Gomez-Segura J, Perezdel Pino A, Veciana J, Laukhin V, Rovira C 2006 Linked Crystallites in the Conducting Topmost Layer of Polymer Bilayer Films Controlled by Temperature: From Micro- to Nanocrystallites *ChemPhysChem* **7** 920–923

[14] R. Pfattner et al. 2015 A Highly Sensitive Pyroresistive All-Organic Infrared Bolometer, balometer *Advanced Electronic Materials* **1** Issue **8** Article first published online: 25 JUN

2015 DOI: 10.1002/aelm.201500090

[15] Laukhina E, Laukhin V, Lebedev V, Rovira, C, Veciana, J 2014 Conductive Fabric Responding to Extremely Small Temperature Changes, *28th European Conference on Solid-State Transducers. Procedia Engineering* **87** 144–147

[16] Somov A, Lebedev V, Baranov A, Laukhina E, Laukhin V, Passerone R, Rovira C, Veciana J 2014 Wireless sensor node with ultrasensitive film sensors for emergency applications. *The 28th European Conf. on Solid-State Transducers (Eurosensors'14) Proc. Engineering* **87** 520-523

[17] *PT-100 Series Platinum RTDs*: http://www.lakeshore.com/Documents/LSTC_Platinum_l.pdf

[18] Laukhina E, et al. 2001 New flexible low-density metallic materials containing the (BEDT-TTF)$_2$(I$_x$Br$_{1-x}$)$_3$ molecular metals as active components *Phys. Chem. B* **105** 11089-11097

[19] Hertz H, Reine J 1881 *Angew. Math.* **92** 156

[20] Cappella B Dietler G1999 Force-distance curves by atomic force microscopy *Surface Science Reports* **34** 1-104

[21] Roa J J, Oncins G, Diaz J, Capdevila X G, Sanz F, and Segarra M 2011 Study of the Friction, Adhesion and Mechanical Properties of Single Crystals, Ceramics and Ceramic Coatings by Afm *Journal of the European Ceramic Society* **31** 429-449

[22] Roa J J, Oncins G, Diaz J, Capdevila X G, Sanz F, and Segarra M 2011 Calculation of Young's Modulus Value by Means of Afm *Recent Patents on Nanotechnology* **5** 27-36

[23] Pamula V K, Jog A and Fair R B 2001 Mechanical Property Measurement of Thin-film Gold Using Thermally Actuated Bimetallic Cantilever Beams *Nanotech.* **1** 410 – 413

[24] (a) Lebedev V, Laukhina E, Laukhin V, Rovira C and Veciana J 2014 Tuning the Electronic Properties of Piezoresistive Bilayer Films Based on α-(BEDT-TTF)$_2$I$_3$ *EurJIC* **24** 3927-3932; (b)Lebedev V 2014 PhD Thesis: *Towards plastic electronic sensing devices with TTF-based molecular conductors as active components* Universitat Autònoma de Barcelona Departament de Química at http://hdl.handle.net/10803/2851382011